T0136836

Asset Analytics

Performance and Safety Management

Series Editors

Ajit Kumar Verma, Western Norway University of Applied Sciences, Haugesund, Rogaland Fylke, Norway

P. K. Kapur, Centre for Interdisciplinary Research, Amity University, Noida, India

Uday Kumar, Division of Operation and Maintenance Engineering, Luleå University of Technology, Luleå, Sweden

The main aim of this book series is to provide a floor for researchers, industries, asset managers, government policy makers and infrastructure operators to cooperate and collaborate among themselves to improve the performance and safety of the assets with maximum return on assets and improved utilization for the benefit of society and the environment.

Assets can be defined as any resource that will create value to the business. Assets include physical (railway, road, buildings, industrial etc.), human, and intangible assets (software, data etc.). The scope of the book series will be but not limited to:

- Optimization, modelling and analysis of assets
- Application of RAMS to the system of systems
- Interdisciplinary and multidisciplinary research to deal with sustainability issues
- Application of advanced analytics for improvement of systems
- Application of computational intelligence, IT and software systems for decisions
- Interdisciplinary approach to performance management
- Integrated approach to system efficiency and effectiveness
- Life cycle management of the assets
- Integrated risk, hazard, vulnerability analysis and assurance management
- Adaptability of the systems to the usage and environment
- Integration of data-information-knowledge for decision support
- Production rate enhancement with best practices
- Optimization of renewable and non-renewable energy resources

More information about this series at http://www.springer.com/series/15776

P. K. Kapur · Gurinder Singh · Saurabh Panwar
Editors

Advances in Interdisciplinary Research in Engineering and Business Management

 Springer

Editors
P. K. Kapur
Amity Center for Interdisciplinary Research
Amity University
Noida, Uttar Pradesh, India

Gurinder Singh
Group VC Amity International Business
School
Amity University
Noida, Uttar Pradesh, India

Saurabh Panwar
Department of Operational Research
University of Delhi
New Delhi, Delhi, India

ISSN 2522-5162   ISSN 2522-5170 (electronic)
Asset Analytics
ISBN 978-981-16-0039-5   ISBN 978-981-16-0037-1 (eBook)
https://doi.org/10.1007/978-981-16-0037-1

This Springer imprint is published by the registered company Springer Nature Singapore Pte Ltd.
The registered company address is: 152 Beach Road, #21-01/04 Gateway East, Singapore 189721,
Singapore

Preface

Advances in Interdisciplinary Research in Engineering and Business Management is a book on interdisciplinarity with its role in information technology, industrial engineering, software engineering, and business management. When solving complex industrial and business queries, field-specific research will not be adequate. A more robust and factual approach as an interdisciplinary approach is required to solve such problems strategically. In this book, a collection of recent studies in the field of engineering and management are presented to help readers gain a better understanding and build skills to solve industrial and scientific problems.

This book offers some advances in interdisciplinary research and serves as a reference work for academicians and business analysts in the field of engineering and management. The purpose of this book is to help academia and professionals to develop effective tools to address business issues linked with different areas of study through practical solutions. This book focuses on rendering interdisciplinary approach to decision-makers and managers by providing research articles on Software Quality and Reliability, Digital Marketing, Cloud Computing, Power Distribution System, Performance Management, Inventory Management, Organizational Behavior, Process Optimization, Data Mining, Total Quality Management, Time Series Analysis, Multi-criteria Decision Making, Machine Learning, Evolutionary Algorithm, Dynamic Programming, Software Maintainability, and System Modeling.

We believe this book will yield essential new insights and benefits to policy-makers, managers, researchers, lecturers, market analysts, industrial experts, and all those who are looking for pragmatic solutions to industrial problems with an interdisciplinary approach.

Noida, India
Noida, India
New Delhi, India

P. K. Kapur
Gurinder Singh
Saurabh Panwar

Acknowledgment

This book is developed with extraordinary help and cooperation from a large number of people whose invariable assistance provided immense support and encouragement. We would like to acknowledge all the researchers and scholars who are involved in this book to make it a success. Our earnest gratitude goes to all the authors who have contributed their time and knowledge in this book. We would also like to express our enormous debt of appreciation to the reviewers and experts for their insightful comments and invaluable suggestions that assisted in improving the quality, coherence, and presentation of the research work.

We would like to express our gratitude to Prof. Uday Kumar, Luleå University of Technology, Sweden, and Prof. A. K. Verma, Western Norway University of Applied Sciences, Norway, for their constant support during the compilation of the book. The book editors also wish to acknowledge the Springer team for their sincere efforts and professional support in the successful publication of the book. Our special appreciation goes to Mr. Vivek Kumar, Ph.D. scholar, Department of Operational Research, University of Delhi, Delhi, India, for his ineffable assistance in completion of the book. Finally, we want to thank our family and friends for their unparalleled support.

Lastly, we apologize for any omissions.

P. K. Kapur
Gurinder Singh
Saurabh Panwar

Contents

About the Editors

P. K. Kapur is Director of Amity Center for Interdisciplinary Research, Amity University, Noida. He is Former Dean of the Faculty of Mathematical Sciences, and Former Head of the Department of Operational Research, University of Delhi. His vast research experience in the areas of Software Reliability, Optimization, Innovation Diffusion Modeling in Marketing, and Multi-Criteria Decision Making (MCDM), Big Data Projects Adoption, and other areas of management, is illustrated through his work with nearly 400 research paper publications in top international and national journals/proceedings of repute, and supervision of over 44 Ph.D. & 25 M.Phil scholars. Presently, he is also a member of the University Research Council, Amity University, Noida.

He has authored and edited more than 10 books/conference volumes, many of which have been published by Springer. He is also editor-in-chief of the *International Journal of Systems Assurance Engineering and Management (IJSAEM)* published by Springer. He is Founder President of the Society for Reliability Engineering, Quality and Operations Management (SREQOM) since 2000 and former President of the Operational Research Society of India (ORSI). He has executed various research projects from UGC, DRDO in the areas of Software Reliability and Innovation Diffusion Modeling; has delivered keynote addresses at various prestigious international conferences; and has been invited to deliver lectures in various universities in Sweden, Denmark, South Africa, Russia, Serbia, Nepal, Dubai, to name some.

Gurinder Singh is Group Vice Chancellor, Amity Universities, Director General, Amity Group of Institutions, India, and Vice Chairman, Global Foundation for Learning Excellence, and has an experience of more than 27 years in institutional building, teaching, research and industry. He holds a doctorate and a postgraduate degree from Jamia Millia & Indian Institute of Foreign Trade, Delhi. He is well acknowledged author and researcher who has published more than 100 research papers, 15 books and is an Editor of 10 prestigious research journals. He has given key note lectures at various forums which includes speaking at Harvard Business School, NYU, University of Leeds, University of Berkeley, National University of Singapore, NTU, MIT, and many more.

Saurabh Panwar is a Ph.D. scholar in the Department of Operational Research, University of Delhi, Delhi, India. He holds an M.Phil and Postgraduate degree in Operational Research from the University of Delhi. His research interests include mathematical modeling, new product development, technological forecasting, software reliability, and optimization. He is a lifetime member of the Society for Reliability Engineering, Quality, and Operations Management (SREQOM) since 2015. He is also an Associate Editor of the *International Journal of Systems Assurance Engineering and Management* (IJSAEM), Springer. He has been judged as a Young Promising Researcher of international repute by the President of SREQOM in 2018. He has authored several papers presented to conferences and published quality research papers in international scientific journals of high repute such as *Annals of Operations Research, Information Technology & Management, Journal of Modelling in Management, International Journal of Reliability, Quality and Safety Engineering*, and *International Journal of Product Development*, to name a few. He has also contributed as a co-author in book chapters published by Springer.

A Study of Barriers Faced by Consumers in Using UPI-Based Apps

Shalini Gautam, Kokil Jain, and Vibha Singh

1 Introduction

Over the last 10 years India has made steady but slow growth toward E-payments. Till now many methods are invented in E-payments to digitalize the current banking system—unified payment interface (UPI) is one of the latest one launched.

Reserve Bank of India (RBI) has taken efficient steps to sponsor digital payments presence in India and has started the National Payments Corporation of India (NPCI) as a coordinating association to progress small-scale digital payment presence in India. In August 2016, NPCI launched the unified payment interface (UPI), a next-generation and peer group smart phone payment system that allows real-time bank payments. UPI takes advantage of the high telephone density in India to make the mobile phone the main means of payment for consumers as well as merchants and to revolutionize digital payments domestically. The UPI ensures that the transaction of the two different parties creates no human intervention. In such a case, the payment is done through mobile in such a way that the bank accounts of two parties are interconnected through the transaction that has taken place. The UPI system is something which has paved the way for the manual system of recording the transaction. The major idea behind inventing this system by the government was for the smooth and secured flow of payment. With the growing population and emerging difference, the need for the protection is a major concern. The UPI system ensures that the data interchange uses the virtual payment address (an authorized unique IP

S. Gautam (✉) · K. Jain · V. Singh
Amity International Business School, AUUP, Noida, India
e-mail: shalinigautam2412@gmail.com

K. Jain
e-mail: kjain@amity.edu

V. Singh
e-mail: vsingh20@amity.edu

© The Author(s), under exclusive license to Springer Nature Singapore Pte Ltd. 2021
P. K. Kapur et al. (eds.), *Advances in Interdisciplinary Research
in Engineering and Business Management*, Asset Analytics,
https://doi.org/10.1007/978-981-16-0037-1_1

address required by the bank). The UPI system is directly connected with the Aadhar card of an individual as a proof for one's identity.

The UPI system works as an online digital platform that is available 24×7, even during public holidays. The use of UPI system has eradicated the traditional method of payment system. The services provided by the UPI can wisely make the work done within a fraction of seconds with a click on the mobile screen. It is something which the businessmen and the other group of society were looking from past year for effective and efficient coordination with the involved parties. The idea of UPI has made a global presence for a flow of payment interacting widely across the territories. The services also include receiving money, sending money, balance inquiry and change MPIN. UPI is a type of electronic data interchange whereby the business party transfers their document over the internet. For an individual to be a member of UPI system, first the UPI apps such as BHIM, SBI UPI and Google TEZ need to be downloaded. Further, an account should be made in the downloaded app and thus one can work through online portal. The bank is somewhat directly related while the payment takes place as the payment from the party is deposited in the bank account of an individual.

The various important features of UPI are as follows.

1.1 Account Security

Clients can make a one of a kind money-related personality that would be connected to their financial balance at the bank's backend. For making exchanges, a person would need to share just this character and not the financial balance points of interest. Utilizing the one of a kind personality, NPCI would guide the exchange to the connected ledger and the bank would apply the important exchange at its end. UPI rejects the need to disclose completely sensitive personal and financial details, such as account details, personal information and sensitive PINs. In contrast, the accounts receivable providers create a virtual address in the process, which results in unique identifiers. These operations do not need to be permanent; they can be made for an exact period or a single transaction. As a result, a customer may have multiple unique identifications and arrange the payment or collection appropriately. Setting an upper payment limit on specific guidelines, such as certain controls, and restrictions on merchants or outlets where a fixed identifier can be used, vertical payment instructions make the whole process very useful for consumers.

1.2 One Identity, More Accounts

With UPI, the bank can now become a single entity for a consumer, completely eliminating the need to juggle different banking applications with different IDs and passwords. The promise of interoperability is probably the biggest attraction for consumers when adopting the process. Financial institutions that become the first to evolve in the UPI space by offering interoperable platforms are like consumer loyalty and benefit from commissions in other banking transactions on their platform.

1.3 Automated Collection Process

UPI will go about as a supplement to the current Electronic Clearance Service (ECS)/National Automated Clearing House (NACH) charge process. UPI could mechanize the accumulation procedure for the biller and assets can be asked for utilizing the interface. Merchants breathe a sigh of consolation as they will no longer be under the administrative scanner since the incidents of fraudulent transactions may be significantly reduced, in addition to the ones of E-commerce players who have cash-on-delivery options for payments to shipping carts acquired more securely with UPI banks and other financial institutions are all too eager to be the primary to deliver on customer demands.

1.4 Instant Transaction

Earlier, one may have had a couple of exchanges on sites, particularly shopping destinations, where exchange was either fizzled or was declined, because of numerous layer associations. These are associations that exist between the shopping site, installment portal and the bank. UPI tries to make the associations more straightforward and speedier with the goal that the procedure turns out to be more proficient and customers can make moment installments.

There are many players who have started offering the UPI services on their platform. The prominent among them are BHIM, Google Tej, PhonePe, SBI Payapp and so on. According to RBI, UPI-based apps in India have seen a rise of 40% in the transaction volume month-on-month. But a closer look at the numbers show that most of the transactions which are happening on the UPI platforms are peer-to-peer or the deal seekers. The adoption by merchants has still not happened at large scale. Although the UPI-based apps are gaining traction among the Indian consumers, there are still certain barriers faced by the consumers in using these UPI apps. The exhaustive literature review was done to identify the barriers that the consumers would face while accepting any new technology. The barriers identified to study the acceptance of the UPI-based apps were usage barrier, value barrier, risk barrier, tradition barrier and image barrier. The purpose of the present study is to understand the various barriers faced by the consumers in the intention for using these UPI-based apps. An attempt has been made to measure the association of these various barriers with the behavioral intention of the individuals for using the UPI-based apps.

2 Review of Literature

The key drivers of UPI-based apps are simplicity, innovation, adoption, security and cost [1]. The simplicity of the operations of UPI-based apps is expected to lead more people to start using UPI-based apps for all kind of payments, thereby reducing cash transactions. The innovations to transfer money exist both on the payer and the payee side. There is no need to know the IFSC code of the account holder or the credit card number of the person to which money needs to be transferred. The two-factor authentication of UPI enables to transfer money in a secure way without any transaction cost. The highly secured features, high innovation and virtually nil cost have led to more and more people adopting these UPI apps. The UPI services are until now being allowed to be provided by the banks only. As a part of the psyche of the Indian consumers, people generally trust banks more than the private service providers offering the mobile wallets [2]. This gives a great opportunity to banks to leverage upon consumers' transactions and optimize the trust bestowed upon them by creating a secured and smooth payment ecosystem. The system is beneficial for the merchants also as it simplifies the collection process for them. They can send the reminders online to their customers to make the payments. The customers also need not worry about the security of their data being stored at the merchants' system.

The indigenously developed payment systems like BHIM and Rupay card have cornered over 60% of digital transactions. These are giving tough competition to the global giants such as Mastercard and Visa. As a matter of fact, these companies are losing the market share to these startups. In November 2018, UPI had 500 mn transactions, which was more than Amex's global transactions [3]. There are many big firms like Amazon.com, Ola, Uber, Big Bazar, PayTM which have integrated UPI into their apps to facilitate payments from the customers. In India, there is a dominance of cash but with the rising mobile and data penetration, this dominance can be reduced and UPI apps would play a significant role in reducing this dominance. The UPI transactions are expected to grow ten times in the next 12 months. As the UPI is expected to play an important role in the Indian financial sector in the future, it is important to know if the consumers are facing any barriers in using these apps.

The concept of innovation resistance was explained in detail by [4]. They gave two major reasons why a consumer resists an innovation. First, the innovation may disrupt the established routines of a consumer and may create a high degree of change in his day-to-day existence. Second, a consumer prior belief structure might come in conflict with an innovation. The resistance to innovation is dependent upon three parameters. First, it depends upon the characteristics of the consumers themselves. As defined by [5], the adopters of the new innovations are classified as innovators, early adopters, early majority, late majority and laggards. The innovators would have the least resistance toward the new innovation as compared to the laggards, who would have the maximum resistance. Second, innovation resistance varies in degrees. The consumers would resist the innovation if it is contrary to the group norms, or societal and family values [6]. The innovation would also be resisted if the consumer thinks it to be risky to adopt [4]. In the studies related to the adoption of innovations related with money matters, security plays a significant role. The security system in mobile banking impacts the behavioral intention of adopting it [7]. The security system is the important motivating factor for the Chinese customers to adopt mobile banking [8]. As per the study done by [9], the security system is inversely related to the internet banking adoption. In [10] the authors identified three types of risk: (a) aversive physical, social or economic; (b) performance uncertainty; and (c) perceived side-effects associated with innovation. The major components of the consumer resistance were identified by [11] as rejection, postponement and opposition. The resistance occurs across all the product categories whether it is continuous or discontinuous [4]. The resistance increases with the increase in the discontinuity of innovation. The two broad barriers toward resistance to innovation were identified by [4] in their study. These were functional barriers and psychological barriers. The functional barriers can further be subdivided into usage, value and risk barrier. The psychological barrier can be subdivided into tradition and image barrier.

3   Hypothesis Development

Drawing from the literature review, the five barriers were identified for further study and accordingly hypotheses were developed. The five hypotheses developed are as follows:

(1)   Usage barrier and behavior intention: Usage barrier is an important functional barrier defined by [4]. Any innovation that is not compatible with the existing usage patterns of an individual would face resistance. This was further corroborated by the study done by [6], in which it was suggested that the innovation's incompatibility with the existing workflows creates a consumer acceptance barrier. In [12] the reasons why Finnish banking customers resist the usage of internet banking were found out. The major reasons were that the consumers found the internet banking to be difficult and inconvenient to use. They also found it difficult to memorize their PIN codes for the usage of the internet banking. The consumers also resist the innovation even if the medium via what it is offered is incompatible with their existing usage pattern [13]. Thus, the following hypothesis is proposed:

H1: There is a positive relationship between usage barrier and behavioral intention of an individual while using UPI-based apps.

(2)   Value barrier and behavior intention: The consumer would accept the innovation only if it offers a strong performance to price value in comparison with the substitute of the product [4]. If consumers believe that the cost of learning the innovation is far more than the benefits that the innovation offers, they would resist the innovation [14]. At the same time, innovations that threaten the core values of the user are also significant subject for resistance [15]. The compatibility was one of the major factors which defined the adoption of innovation in the study done by [5]. The compatibility is defined as the degree to which

an innovation is perceived as consistent with the existing values and habits of the consumer. The second hypothesis proposed thus is as follows:

H2: There is a positive relationship between value barrier and behavioral intention of an individual while using UPI-based apps.

(3) Risk barrier and behavior intention: The risk of the innovation stems from its uncertainty [4]. The consumer would try to postpone his adoption until and unless he learns more about the risk. The risk could be physical in which the innovation could harm the person or the property. It could be economic, especially in case of innovation related with money matters. The fear of making mistakes while doing banking transactions online make many of the consumers resist this technology when it was first introduced [16]. The consumers are also worried about the connection breaks that pose a major security threat, especially in case of mobile banking [12]. The confidentiality of the customer data is also found to be a dominant issue in case of internet banking [17]. The risk could also be functional where the consumer is not comfortable using the innovation until and unless it is fully tested [10]. Thus, the hypothesis proposed is as follows:

H3: There is a positive relationship between risk barrier and behavioral intention of an individual while using UPI-based apps.

(4) Tradition barrier and behavior intention: When a consumer is required to deviate from the proven traditions and culture, he would resist the innovation [10]. The resistance increases with the increase in deviation. An innovation may cause the customer to change its routines. But if these routines are important to a consumer, resistance would be high [12]. The consumers may resist the electronic banking because of the way in which they are accustomed to paying bills [18]. Many of the consumers would like to interact with the bank employee personally instead of dealing with the impersonal technologies [19]. Thus, the hypothesis proposed is as follows:

H4: There is a positive relationship between tradition barrier and behavioral intention of an individual while using UPI-based apps.

(5) Image barrier and behavior intention: The image barrier is the perceptual problem which arises out of the stereotyped thinking of the consumer [10]. The stereotype can arise because of the industry or product class to which the innovation belongs or the country in which the innovation is manufactured. Image barrier in electronic banking emerges from the 'hard-to-use' image of the computers and internet channel [18]. The consumers use image as an extrinsic cue to base their decision to adopt the innovation or not. These extrinsic product cues are important for consumers to assess new products [20]. Some studies also suggest that the negative media coverage can induce negative image perceptions of innovations [21]. Thus, the hypothesis proposed is as follows:

H5: There is a positive relationship between image barrier and behavioral intention of an individual while using UPI-based apps.

Based on the above hypothesis, a research model as shown in Fig. 1 is proposed.

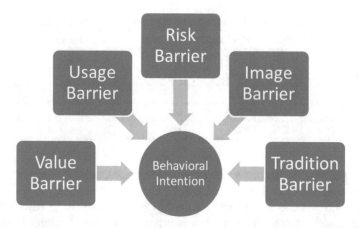

Fig. 1 The conceptual model

Table 1 Reliability measures of scale items

Name of the scale	Items	Cronbach's Alpha
Usage barrier	5	0.911
Value barrier	3	0.856
Risk barrier	5	0.914
Image barrier	3	0.860
Tradition barrier	1	–
Behavior intention	3	0.920

4 Methodology

A structured questionnaire using existing scales was used to test the above-mentioned hypothesis. The sample was collected using convenience and snow-ball sampling method and the data was collected between August and November 2018. A total of 300 questionnaires were collected and 252 were found to be fit for further analysis. The step-wise multiple regression was done between the identified barriers and the behavior intention to find out relationship between them. Table 1 summarizes the reliabilities of the scales used in the study.

5 Data Analysis and Findings

The sample distribution shows that most of the respondents (74%) are males and are in the age group of 18–25 years (67%). The results of the step-wise regression depicts that usage barrier, tradition barrier and risk barrier have the significant relationship with the behavior intention for using UPI apps. The relationship between the other

Table 2 Step-wise linear regression results on behavior intention of using UPI apps

Factors	Beta	Significant	Hypothesis	Hypothesis accepted/rejected
Usage barrier	0.304	0.000	H1	Accepted
Value barrier	0.095	0.287	H2	Rejected
Risk barrier	0.231	0.005	H3	Accepted
Tradition barrier	0.461	0.000	H4	Accepted
Image barrier	0.094	0.421	H5	Rejected

two barriers, viz., image and value barrier, is not significant. The summary of the regression analysis is depicted in Table 2.

There is a positive significant relationship between the usage barrier and the behavior intention to use UPI apps. This is as per the existing literature. The positive relationship implies that if an individual is not comfortable using the various UPI apps and it would not be convenient for him to use the apps, he would not intend to use that app for his day-to-day functions.

The second hypothesis is rejected as there is no significant relationship between value barrier and intention to use various UPI apps. The value barrier basically arises when cost of adopting the new technology is higher than the technology which the customer is using presently. In this case, the cost of the new technology, i.e., using UPI apps is negligible as an individual can download the app free of cost. Because of the economy of the new technology, value barrier is not impacting the behavior intention of an individual to use UPI apps.

The third hypothesis is accepted and there found to be a significant relationship between risk barrier and intention to use UPI apps. The result supports the past research. As the money is involved in using these apps, customers are wary of the risks involved. There is a concern toward the confidentiality of the data and the ability to do the transactions in a proper way so that there would be no loss of money.

The fourth hypothesis is accepted. There is a significant relationship between the tradition barrier and the intention of an individual in using UPI apps. This supports the existing literature. The UPI apps would effectively change the way in which an individual is doing the money transactions. If an individual is comfortable doing the transactions which are not as per his past way of doing the same, then only he would be intending to use the various UPI apps.

There is no significant relationship between image barrier and intention to use UPI apps and therefore, the fifth hypothesis is rejected. The image barrier usually arises because of the negative perception about a product in the minds of the consumer. It can be because of the media coverage or the experience of an individual. In case of UPI apps, there is quite positivity around UPI apps. The media is also encouraging the use of cashless transactions. Therefore, the image barrier is not impacting the intention of an individual to use the UPI apps.

6 Conclusions

The UPI apps are gaining the prominence in the Indian economy. The number of money transactions which are being done on these platforms are increasing month on month. But it's still a long way to go. There are many players in this segment who are vying for the customer attention. They all want their respective apps to be downloaded by the consumers. Generally, the individual would download only one or two apps and that would be sufficient enough for doing transactions.

Therefore, it is imperative for the individual players to understand the barrier which a consumer might face. If the barriers could be removed, then the propensity of an individual to use the app would increase. The present study is an attempt to understand these barriers. It throws light on various barriers, which, if removed, can increase the intention of an individual to use the UPI app.

The most important barriers which hamper the intention to use these apps are usage barrier, risk barrier and tradition barrier. If the particular app has consumer-friendly features and is considered to be easy to use, then the intention of using that app increases. Second, if the customer feels that using a particular UPI app would not lead to any monetary loss for him and there is no risk of him losing his personal and financial data, then again the intention of using the app increases. Last but not the least, if an individual is ready to leave the traditional way of doing the transactions and is ready to accept the usage of UPI apps, then he is more likely to use these apps.

Thus, the challenges for the various UPI players are manifold. They will have to develop the apps which are easy and convenient for the individuals to use. They should not feel intimidated with the various technical terms associated with these apps. They will also have to assure the consumers that the data which is being used in these apps is absolutely secure and there is no risk in the financial transactions which an individual is doing over these apps. They will have to make the transition of the consumers from the traditional way of doing the transactions to these new age apps as seamless as possible.

7 Limitations

The research has its limitations as it is confined to a particular area, i.e., Delhi NCR in India. The sample size is also limited. The results can be more pronounced if the sample size would have been increased. The future research can study these barriers in context of various socio-demographic factors like gender, age and education. These factors would give the detailed insights about the behavior intentions of the consumer. It is also possible to do a comparison study between various online modes like mobile banking, internet banking and these UPI apps, and study if the results replicate in all the modes.

References

1. Sharma, S. (2016). Unified payments interface: The recent Indian financial innovation demystified. *Apeejay Journal of Management & Technology, 11*(2), 17–27.
2. Thomas, R., & Chatterjee, A. (2017). Unified payment interface (UPI): A catalyst tool supporting digitalization- utility, prospects and issues. *International Journal of Innovative Research and Advanced Studies, 4*(2), 192–195.
3. Sharma, S. (2018). UPI transactions exceeds AMEX global figure: The State of the Indian startup ecosystem summit
4. Ram, S., & Sheth, J. N. (1989). Consumer resistance to innovations: The marketing problem and its solutions. *The Journal of Consumer Marketing., 6*(2), 5–14.
5. Rogers, E. M. (1983). *Diffusion of innovation* (3rd ed.). New York NY: Free Press.
6. Herbig, P. A., & Day, R. L. (1992). Customer acceptance: The key to successful introductions of innovations. *Marketing Intelligence & Planning, 10*(1), 4–15.
7. Boonsiritomachai, W., & Pitchayadejanant, K. (2017). Determinants affecting mobile banking adoption by generation Y based on the unified theory of acceptance and use of technology model modified by the technology acceptance model concept. *Kasetsart Journal of Social Sciences,* 1–10
8. Laforet, S., & Li, X. (2005). Consumers' attitude towards online and mobile banking in China. *International Journal of Bank Marketing, 23*(5), 362–380.
9. Lee, M. C. (2009). Factors influencing the adoption of internet banking: An integration of TAM & TPB with perceived risk and perceived benefit. *Electronic Commerce Research & Applications, 8*(3), 130–141.
10. Sheth, J. N. (1981). Psychology of innovation resistance: The less developed case (LDC) in diffusion research. *Research in Marketing, 4*(3), 273–282.
11. Kleijnen, M., Lee, N., & Wetzels, M. (2009). An Exploration of consumer resistance to innovation and its antecedents. *Journal of Economic Psychology, 30,* 344–357.
12. Kuisama, T., Laukkanen, T., & Hiluten, M. (2007). Mapping the reasons for resistance to internet banking: A means-end approach. *International Journal of Information Management, 27,* 75–85.
13. Szmigin, I., & Foxall, G. (1998). Three forms of innovation resistance: The case of retail payment methods. *Technovation, 18*(6/7), 459–468.
14. Dunphy, S., & Herbig, P. A. (1995). Acceptance of innovations: The customer is the key. *Journal of High Technology Management Research, 6*(2), 193–209.
15. Arnould, E., Price, L., & Zinkhan, G. (2004). *Consumers* (2nd ed.). New York: McGraw Hill.
16. Laukkanen, T., Sinkkonen, S., Kivijarvi, M., & Laukkanen, P. (2007). Innovation resistance among mature consumers. *Journal of Consumer Marketing, 24*(7), 419–427.
17. Liao, Z., & Cheung, M. T. (2002). Internet based e-banking and consumer attitudes: An empirical study. *Information & Management, 39,* 283–295.
18. Fain, D., & Roberts, M. L. (1997). Technology V. Consumer behavior: The battle for the financial services customer. *Journal of Direct Marketing, 11*(1), 44–54
19. Heinonen, K. (2004). Time and location as customer perceived value drivers. Economi och Samhalle 124, Swedish School of Economics and Business Administration, Helsinki
20. Bearden, W. O., & Shimp, T. A. (1982). The use of extrinsic cues to facilitate product adoption. *Journal of Marketing Research, 19*(2), 229–239.
21. Fortin, D. R., & Renton, M. S. (2003). Consumer acceptance of genetically modified foods in New Zealand. *British Food Journal., 105*(1/2), 42–58.

Disparity in Perception of Male and Females About Employees Welfare Programs in the Outsourcing Industry

Sonal Pathak

1 Introduction

Outsourcing industry has always been counted as one of the most stressed industry due to its 24 × 7 working hours and stressful deadlines. Managing stress among employees is a challenge for management to face. The organizations that fail to recognize the need to manage stress effectively have been at the brink of closure. Every organization and individual are suffering from some amount of stress despite of its invisibility. Efficiency of employees influence organizational effectiveness. Taking a glimpse into their growth and development in various spheres, they do their best but are the organizations up to the challenge in satisfying the employees is a big question. Employee welfare practices are part of management of human resources. It is a function in the organizations designed to maximize employee performance in service of an employer's strategic objectives. HR is primarily concerned with the management of people within organizations, focusing on policies and on systems. In the present corporate scenario, most of the research and corporate training programs are focused toward managerial and technical aspects only. This objective can be fulfilled by offering effective employee welfare practices. Hence, the author sheds light on effectiveness of employee welfare programs for successful stress management among employees and improve productivity. Just like any other sector, outsourcing industry is also at increasing need to undergo constant enhancement in employee's welfare policies in order to enhance harmonious environment.

However, a lot of work have been published on effective employee welfare practices, but still the gap exists between research on the role of gender in the planning of welfare practices and its impact on the performance of outsourcing sector. Unfortunately, significant research programs based on the requirement of the different stress

S. Pathak (✉)
Manav Rachna International Institute of Research and Studies, Faridabad, Haryana, India
e-mail: Pathak25@gmail.com

© The Author(s), under exclusive license to Springer Nature Singapore Pte Ltd. 2021   11
P. K. Kapur et al. (eds.), *Advances in Interdisciplinary Research in Engineering and Business Management*, Asset Analytics,
https://doi.org/10.1007/978-981-16-0037-1_2

management techniques for each gender have not been encouraged and undertaken so far. Therefore, to value a stress-free work environment, this study will be a helpful tool for the effective supervision. The study to assess the impact of this novel issue on the BPO sector is also required.

2 Literature Survey

The concept of "Employee welfare" differs and is flexible widely with times, industry, region–country, social values and customs, and is based on general social economic development of people, on the degree of industrialization and political ideologies prevailing at those moments. The term welfare refers to an act of seeking physical, mental, moral and emotional well-being of an individual. According to the traditional economic theory, labor can be defined as "a factor of production which consists of manual and mental exertion and receives some return in form of wages, salaries or professional fees". An employer secures the benefits of high efficiency and low employee absenteeism and minimum employee turnover. Thus, improving the quality of working life by providing the employee welfare facilities would go a long way in achieving the goals of the organization. Chaplin et al. [1] explored some theories regarding effective welfare practices that are presently settled and accepted; others are still being debated and researched. The companies most likely to be successful in making change in work to their advantage are the ones that no longer view change as a discrete event to be managed but as a constant opportunity to evolve business. HRM practices refer to organizational activities directed at managing the pool of human resource and ensuring that the resources are employed toward the fulfillment of organizational goals.

Gianakos [2, 3] wrote that social gender may affect the workplace stress. Due to increased number of women in the BPO industry, some researchers have started to take their research subject on females to study the field of work stress . BPO Employee—Satisfaction Survey (2010) disclose that all is not well on the BPO front. Indian BPO employees are suffering from a variety of health-related issues, right from anxiety attacks to sleep-related disorders.

Burke [4] had done a research on Taiwanese bank employees. They found that the masculine characteristics exhibited less stress level as compared to the feminine characteristics.

Bekker and Nijssen [5] reveal that the basic reason and a decisive factor for the subsistence of gender differences, concerning stress manifestation, is the subsistence of differences between how the workplace encounters the two genders, the stress stimulus and their experience to different stimulus types.

Melissa [6] has done a survey-based analysis of workers on a diverse group and abridged the old notion about female workers in the expression of emotions at work and investigated that women express anger less and happier than men in the work-place. Sudhashree et al. [7] described that everything is not in a good situation in the

BPO industry. Sloan [8] found that management role in an organization is one of the important factors which is related to stress generation among employees.

Kristiansen et al. [9] have stated that as legislation, attitudes and norms are growing over a period of time, gender differences in the place of work are always changing.

Alexandros-Stamatios et al. [10] illustrated that reduction in the environmental violence, increased job satisfaction and management of moral stress are directly dependent on the variables such as suspension and reflection, education and perfect leadership.

Nelson and Burke [11] found that differences in the responses of women and men to stress are not limited to neurological evidences only.

Taylor [12] have revealed in a research on stress responses for 30 years and recognized that women respond differently to stressful situations. Compared to males, females' physical aggression and fear-related behaviors are less intense.

Taylor et al. [13] described that responses in stress outcomes are basically dissimilar in each gender, which can be often characterized as tend and befriend among women, whereas fight or flight among men.

Fozia [14] bring into notice that males often develop stress due to their career, whereas females often encounter stress due to their interpersonal relationship. Presently, organizations are competing through implementing the unique employee welfare practices, and due to globalization, the organizations must adopt the most up-to-date employee welfare practices in order to accomplish the organizational goals.

Ryan and Frederick [15] disclose in their study that best employee welfare practices are advantageous for both employee and employer; it plays an important role in constructive growth of the organization.

The data have been collected from different sources and high-quality databases have been used to generate the conceptual framework of the paper. We are aware of the fact that employee welfare programs are the centralized subsystem of any business organization. These practices play a vital role in organizational performance, and ultimately give better results, through which any business organization can achieve good position in their functional area. Effective employee welfare practices need to inculcate the stress-free environment industry.

3 Objectives of the Research Paper

The objective is to analyze the impact of gender on employees' opinion about effectiveness of common employee welfare programs offered by the organizations.

4 Research Methodology

The universe of the study is confined to all BPOs operating in NCR, which are working in the process outsourcing of business for the third party. The survey population of the study includes only those business process units working for information technology (IT) sector and situated in the National Capital Region (NCR). The study has been conducted based on a sample of 100 respondents in the area of NCR. Descriptive research is a suitable choice when the purpose of the research is to classify frequencies, characteristics, categories and trend. The present study is also about to identify the perception of category; therefore, descriptive research design has been chosen. As population size for this study might be too large, convenience sampling has been used for the research. As the study deals with the gauging public opinion and perceptions of employees, therefore survey approach was employed for data collection with the help of a structured questionnaire. Some items were changed and cancelled to ensure the reliability and the validity of the instrument. Variables in the questionnaire was selected based on the studies published by Hayman et al. [16]. The questionnaire was distributed at workplaces and the responses of employees were collected. A total of 110 employees from the outsourcing sector responded to the survey. Since the nature of the proposed study is descriptive, therefore, the primary data has been used to reach at the results by applying suitable statistical techniques. To conduct this study, a total of 120 questionnaires were distributed among the employees working in the BPOs in the NCR. But after the completion of the survey, only 120 employees gave their responses, and in total 50 male respondents and 50 female respondents were chosen for the present study. Therefore, in total only 100 questionnaires are included in this study. The purpose of the study was explained to the employees and were assured of confidentiality on their responses. The respondent was asked to rate the questions on a three-point scale which ranges from: Agree (1), Neutral (2), and Disagree (3). As the variable's category are mutually exclusive and to deal with group differences when the dependent variable is measured at a nominal level, Chi-square test is suitable to investigate the difference in respondents' point of views. To analyze and compare the responses related to perception difference among male and female BPOs employees, the following null hypothesis was framed.

Null Hypothesis H0: There are no significant differences in the opinion about effectiveness of common employee assistant programs among male and female employees in the outsourcing industry.

Alternative Hypothesis HA: There are significant differences in the opinion about effectiveness of common employee assistant programs among male and female employees in the outsourcing industry.

Chi-square test has been employed at 5% significance level to check the acceptance or rejection of the null hypothesis. p-value method has been used to check if H0 can be rejected.

5 Data Analysis and Interpretation

In this study, analysis of data is done through interferential statistics. Data have been examined by using statistically significant tools such as Chi-square test on SPSS version 16.0 The results of crosstab have been shown through percentage analyses. The requirement of gender-based stress management programs has also been analyzed by examining positive responses for a welfare program. It has also been concluded that all existing employee assistance programs are not perceived equally effective by both males and female employees.

5.1 Crosstab and Chi-Square Test on the Males and Females' Responses for Employees Assistant Program to Check the Association Between Each Gender

(i) **Gender * Relaxation Exercises and Sports**

From the above data, it is observed that 54.2% females responded to agree in comparison to 45.8% male respondents who depicts the differences between the significance of this EAP among male and female employees (Table 1).

Interpretation: At the 5% level of significance, p-value is less than $\alpha = 0.05$, thus Chi-square fails to get significant value for this statement; therefore, the null hypothesis is rejected. Hence, it can be implied that the question "Relaxation Exercises and Sports" is not equally responded by males and female employees (Table 2).

(ii) **Gender * Mentoring Programs**

From the data, it is observed that 53.4% females responded to agree in comparison to 46.6% male respondents which depicts the differences between the significance of this EAP among male and female employees (Table 3).

Table 1 Percentage responses of male and female respondents

Gender	Agree (%)	Neutral (%)	Disagree (%)
Male	45.8	49.7	35.7
Female	54.2	50.3	64.3

Table 2 Result of Chi-square test (N = 100)

Chi-square statistics	Chi-square value	df	p-value
Values	12.974	4	0.011*

$^*p < 0.05$

Table 3 Percentage responses of male and female respondents

Gender	Agree (%)	Neutral (%)	Disagree (%)
Male	46.6	63.1	78.5
Female	53.4	36.9	21.5

Interpretation: At the 5% level of significance, p-value is less than $\alpha = 0.05$, thus chi-square fails to get significant value for this statement; therefore, the null hypothesis is rejected. Hence it can be implied that the question "Mentoring Programs" is not equally responded by male and female employees (Table 4).

(iii)   Gender * Workshop on Work–Life Balance

From the data, it is observed that 48.3% females responded to agree in comparison to 51.7% male respondents, which depicts the differences between the significance of stressor—Workshop on Work–Life Balance among male and female employees (Table 5).
   Interpretation: At the 5% level of significance, p-value is less than $\alpha = 0.05$, thus chi-square fails to get significant value for this statement; therefore, the null hypothesis is rejected. Hence, it can be implied that the question "Workshop on Work—Life Balance" is almost equally responded by male and female employees (Table 6).

(iv)   Gender * Opening of BPO's Forum

From the above data, it is observed that 41.1% females responded to agree in comparison to 58.9% male respondents which depicts the differences between the significance of stressor—Opening of BPO's Forum among males and females (Table 7).
   Interpretation: At the 5% level of significance, p-value is less than $\alpha = 0.05$, thus chi-square fails to get significant value for this statement; therefore, the null hypothesis is rejected. Hence, it can be implied that the question "Opening of BPO's Forum" is not equally responded by male and female employees (Table 8).

Table 4 Result of Chi square test (N = 100)

Chi-square statistics	Chi-square value	df	p-value
Values	1.291	4	0.000*

$^{*}p < 0.05$

Table 5 Percentage responses of male and female respondents

Gender	Agree (%)	Neutral (%)	Disagree (%)
Male	51.7	50.4	52.9
Female	48.3	49.6	47.1

Table 6 Result of chi-square test (N = 100)

Chi-square statistics	Chi-square value	df	p-value
Values	2.503	4	0.644*

*$p < 0.05$

Table 7 Percentage responses of male and female respondents

Gender	Agree (%)	Neutral (%)	Disagree (%)
Male	58.9	31.7	28.7
Female	41.1	68.3	71.3

Table 8 Result of Chi-square test (N = 100)

Chi-square statistics	Chi-square value	df	p-value
Values	1.263	4	0.000*

*$p < 0.05$

(v) Gender * Training for Building Manager/Leaders

From the above data, it is observed that 34.7% female responded to agree in comparison to 65.3% male respondents which depicts the differences between the significance of stressor—Training for Building Manager/Leaders among male and female respondents (Table 9).

Interpretation: At the 5% level of significance, p-value is less than $\alpha = 0.05$, thus Chi-square fails to get significant value for this statement; therefore, the null hypothesis is rejected. Hence it can be implied that the question "Training for Building Manager/Leaders" is not equally responded by males and female employees (Table 10).

(vi) Gender * Opening of Crèches

From the above chart, it is observed that 67.3% females responded to agree in comparison to 32.7% male respondents which depicts the differences between the significance of stressor—Opening of Crèches among male and female employees (Table 11).

Table 9 Percentage responses of male and female respondents

Gender	Agree (%)	Neutral (%)	Disagree (%)
Male	65.3	58.5	31.4
Female	34.7	41.5	68.6

Table 10 Percentage responses of male and female respondents

Gender	Agree (%)	Neutral (%)	Disagree (%)
Male	32.7	66.7	72.4
Female	67.3	33.3	27.6

Table 11 Percentage responses of male and female respondents

Gender	Agree (%)	Neutral (%)	Disagree (%)
Male	32.7	66.7	72.4
Female	67.3	33.3	27.6

Table 12 Result of Chi-square test (N = 100)

Chi-square statistics	Chi-square value		df	p-value
Values	57.735		4	0.000*

*$p < 0.05$

Interpretation: At the 5% level of significance, p-value is less than $\alpha = 0.05$, thus Chi-square fails to get significant value for this statement; therefore, the null hypothesis is rejected. Hence it can be implied that the question "Opening of Crèches" is not equally responded by male and female employees (Table 12).

(vii) Gender * Work from Home

From the above data, it is observed that 69% females responded to agree in comparison to 31% male respondents which depicts the differences between the significance of stressor—Work from Home among male and female employees (Table 13).

Interpretation: At the 5% level of significance, p-value is less than $\alpha = 0.05$, thus Chi-square fails to get significant value for this statement; therefore, the null hypothesis is rejected. Hence, it can be implied that the question "Work from Home" is not equally responded by male and female employees (Table 14).

(viii) Gender * Arrangement of Safe Transport

From the above data, it is observed that 62.3% female responded to agree in comparison to 37.7% male respondents which depicts the differences between the significance of stressor—Arrangement of Safe Transport among male and female employees (Table 15).

Interpretation: At the 5% level of significance, p-value is less than $\alpha = 0.05$, thus Chi-square fails to get significant value for this statement; therefore, the null

Table 13 Percentage responses of male and female respondents

Gender	Agree (%)	Neutral (%)	Disagree (%)
Male	31	58.6	87.8
Female	69	41.4	12.2

Table 14 Result of Chi-square test (N = 100)

Chi-square statistics	Chi-square value	df	p-value
Values	1.096	4	0.000*

*$p < 0.05$

Table 15 Percentage responses of male and female respondents

Gender	Agree (%)	Neutral (%)	Disagree (%)
Male	37.7	62	78
Female	62.3	38	22

Table 16 Result of Chi-square test (N = 100)

Chi-square statistics	Chi-square value	df	p-value
Values	1.190	4	0.000*

$^*p < 0.05$

Table 17 Percentage responses of male and female respondents

Gender	Agree (%)	Neutral (%)	Disagree (%)
Male	65.2	50	24.2
Female	34.8	50	75.8

hypothesis is rejected. Hence it can be implied that the question "Arrangement of Safe Transport" is not equally responded by male and female employees (Table 16).

(ix) Gender * Sabbatical Leaves.

From the above data, it is observed that 34.8% females responded to agree in comparison to 65.2% male respondents which depicts the differences between the significance of stressor—Sabbatical Leaves for male and female (Table 17).

Interpretation: At the 5% level of significance, p-value is less than $\alpha = 0.05$, thus Chi-square fails to get significant value for this statement; therefore, the null hypothesis is rejected. Hence it can be implied that the question "Sabbatical Leaves" is not equally responded by male and female employees (Table 18).

(x) Gender * Paid Maternity Leaves

From the above data, it is observed that 70.8% females responded to agree in comparison to 29.2% male respondents which depicts the differences between the significance of stressor—Paid Maternity Leaves among male and female employees (Table 19).

Interpretation: At the 5% level of significance, p-value is less than $\alpha = 0.05$, thus Chi-square fails to get significant value for this statement; therefore, the null hypothesis is rejected. Hence it can be implied that the question "Paid Maternity Leaves" is not equally responded by male and female employees (Table 20).

Table 18 Result of Chi-square test (N = 100)

Chi-square statistics	Chi-square value	df	p-value
Values	99.554	4	0.000*

$^*p < 0.05$

Table 19 Percentage responses of male and female respondents

Gender	Agree (%)	Neutral (%)	Disagree (%)
Male	29.2	66.7	93.1
Female	70.8	33.3	6.9

Table 20 Result of Chi-square test (N = 100)

Chi-square statistics	Chi-square value	df	p-value
Values	1.595	4	0.000*

$^*p < 0.05$

Table 21 Percentage responses of male and female respondents

Gender	Agree (%)	Neutral (%)	Disagree (%)
Male	40.5	62.5	84.3
Female	59.5	37.5	15.7

Table 22 Result of Chi-square test (N = 100)

Chi-square statistics	Chi-square value	df	p-value
Values	1.486	4	0.000*

$^*p < 0.05$

(xi)   Gender * Higher Education Programs

From the above data, it is observed that 59.5% females responded to agree in comparison to 40.5% male respondents which depicts the differences between the significance of stressor—Higher Education Programs among male and female employees (Table 21).

   Interpretation: At the 5% level of significance, p-value is less than $\alpha = 0.05$, thus Chi-square fails to get significant value for this statement; therefore, the null hypothesis is rejected. Hence, it can be implied that the question "Higher Education Programs" is not equally responded by male and female employees (Table 22).

(xii)   Gender * Health Awareness Camp/ Guidance Dietician

From the above data, it is observed that 35.7% females responded to agree in comparison to 64.5% male respondents which depicts the differences between the significance of stressor—Health Awareness Camp/ Guidance Dietician among male and female employees (Table 23).

   Interpretation: At the 5% level of significance, p-value is less than $\alpha = 0.05$, thus Chi-square fails to get significant value for this statement; therefore, the null

Table 23 Percentage responses of male and female respondents

Gender	Agree (%)	Neutral (%)	Disagree (%)
Male	64.3	47	32.9
Female	35.7	53	67.1

Table 24 Result of
Chi-square test (N = 100)

Chi-Square Statistics	Chi-square value	df	p-Value
Values	1.049	4	0.000*

*$p < 0.05$

Table 25 Percentage
responses of male and female
respondent

Gender	Agree (%)	Neutral (%)	Disagree (%)
Male	66.7	51.9	97
Female	33.3	48.1	2.4

hypothesis is rejected. Hence, it can be implied that the question "Health Awareness Camp/Guidance Dietician" is not equally responded by male and female employees (Table 24).

(xiii)   Gender * Provision of LTC

From the above data, it is observed that 33.3% female responded to agree in comparison to 66.7% male respondents which depicts the differences between the significance of stressor—Provision of LTC among male and female employees (Table 25).

Interpretation: At the 5% level of significance, p-value is less than $\alpha = 0.05$, thus Chi-square fails to get significant value for this statement; therefore, the null hypothesis is rejected. Hence it can be implied that the question "Provision of LTC" is not equally responded by male and female employees (Table 26).

(xiv)   Gender * Enforcement of Antidiscrimination Laws

From the above data, it is observed that 70.9% females responded to agree in comparison to 29.1% male respondents which depicts the differences between the significance of stressor—Enforcement of Antidiscrimination Laws among male and female employees (Table 27).

Interpretation: At the 5% level of significance, p-value is less than $\alpha = 0.05$, thus Chi-square fails to get significant value for this statement; therefore, the null hypothesis is rejected. Hence it can be implied that the question "Enforcement of

Table 26 Result of
Chi-square test (N = 100)

Chi-square statistics	Chi-square value	df	p-value
Values	53.622	4	0.000*

*$p < 0.05$

Table 27 Percentage
responses of male and female
respondents

Gender	Agree (%)	Neutral (%)	Disagree (%)
Male	29.1	60	79.4
Female	70.9	40	20.6

Table 28 Result of
Chi-square test (N = 100)

Chi-square statistics	Chi-square value	df	p-value
Values	84.918	4	0.000*

$^*p < 0.05$

Table 29 Percentage
responses of male and female
respondents

Gender	Agree (%)	Neutral (%)	Disagree (%)
Male	34.8	55.4	80.2
Female	78.9	44.6	19.8

Antidiscrimination Laws" is not equally responded by male and female employees (Table 28).

(xv)   Gender * Recreational/Cultural Programs

From the above data, it is observed that 78.9% female responded to agree in comparison to 34.8% male respondents which depicts the differences between the significance of stressor—Recreational/Cultural Programs among male and female employees (Table 29).

Interpretation: At the 5% level of significance, p-value is less than $\alpha = 0.05$, thus Chi-square fails to get significant value for this statement; therefore, the null hypothesis is rejected. Hence, it can be implied that the question Recreational/Cultural Programs is not equally responded by male and female employees (Table 30).

(xvi)   Gender * Medical Insurance

From the above data, it is observed that 39.4% female responded to agree in comparison to 60.6% male respondents which depicts the differences between the significance of stressor—Medical Insurance among male and female employees (Table 31).

Interpretation: At the 5% level of significance, p-value is less than $\alpha = 0.05$, thus Chi-square fails to get significant value for this statement; therefore, the null hypothesis is rejected. Hence it can be implied that the question "Medical Insurance" is not equally responded by male and female employees (Table 32).

Table 30 Result of
Chi-square test (N = 100)

Chi-square statistics	Chi-square value	df	p-value
Values	89.645	4	0.000*

$^*p < 0.05$

Table 31 Percentage
responses of male and female
respondents

Gender	Agree (%)	Neutral (%)	Disagree (%)
Male	60.6	67.6	12.3
Female	39.4	32.4	87.7

Table 32 Result of Chi-square test (N = 100)

Chi-square statistics	Chi-square value	df	p-value
Values	1.040	4	0.000*

$^*p < 0.05$

Table 33 Percentage responses of male and female respondents

Gender	Agree (%)	Neutral (%)	Disagree (%)
Male	64.2	30.5	49.3
Female	35.8	69.5	50.7

(xvii) Gender * Fun Games/Activities at Office

From the above data, it is observed that 35.8% female responded to agree in comparison to 64.2% male respondents which depicts the differences between the significance of stressor—Fun Games / Activities at Office among male and female employees (Table 33).

Interpretation: At the 5% level of significance, p-value is less than $\alpha = 0.05$, thus Chi-square fails to get significant value for this statement; therefore, the null hypothesis is rejected. Hence it can be implied that the question "Fun games / Activities at Office" is not equally responded by male and female employees (Table 34).

(xviii) Gender * Awareness Session on Employment Laws/Taxes

From the above data, it is observed that 56.6% females responded to agree in comparison to 46.4% male respondents which depicts the differences between the significance of stressor—Awareness Session on Employment Laws / Taxes among male and female employees (Table 35).

Interpretation: At the 5% level of significance, p-value is more than $\alpha = 0.05$, thus Chi-square test is successful to get significant value for this statement; therefore, the null hypothesis is accepted. Hence it can be implied that the question "Awareness Session on Employment Laws/Taxes" has been equally responded by male and female employees (Table 36).

Table 34 Result of Chi-square test (N = 100)

Chi-square statistics	Chi-square value	df	p-value
Values	53.343	4	0.000*

$^*p < 0.05$

Table 35 Percentage responses of male and female respondents

Gender	Agree (%)	Neutral (%)	Disagree (%)
Male	46.4	48.5	45.8
Female	56.6	51.5	54.2

Table 36 Result of
Chi-square test (N = 100)

Chi-square statistics	Chi-square value	df	p-value
Values	9.204	4	0.056*

$^{*}p > 0.05$

Table 37 Percentage
responses of male and female
respondents

Gender	Agree (%)	Neutral (%)	Disagree (%)
Male	64	67.6	15.7
Female	36	32.4	84.3

Table 38 Result of
Chi-square test (N = 100)
respondents

Chi-square statistics	Chi-square value	df	p-value
Values	64.023	4	0.000*

$^{*}p < 0.05$

(xix)   **Gender * Programs on Career Counseling**

From the above data, it is observed that 36% female responded to agree in comparison to 64% male respondents which depicts the differences between the significance of stressor—Programs on Career Counseling among male and female employees (Table 37).

Interpretation: At the 5% level of significance, p-value is less than $\alpha = 0.05$, thus Chi-square fails to get significant value for this statement; therefore, the null hypothesis is rejected. Hence it can be implied that the question "Programs of Career Counseling" is not equally responded by male and female employees (Table 38).

(xx)   **Gender * Employee's Grievance Cell**

From the above data, it is observed that 52% female responded to agree in comparison to 48% male respondents which depicts the differences between the significance of stressor—Employee's Grievance Cell among male and female employees (Table 39).

Interpretation: At the 5% level of significance, p-value is less than $\alpha = 0.05$, thus Chi-square fails to get significant value for this statement; therefore, the null hypothesis is rejected. Hence it can be implied that the question "Employee's Grievance Cell" is not equally responded by male and female employees (Table 40).

Table 39 Percentage
responses of male and female
respondents

Gender	Agree (%)	Neutral (%)	Disagree (%)
Male	48	63.7	81.8
Female	52	36.3	18.2

Table 40 Result of Chi-square test (N = 100)

Chi-square statistics	Chi-square value	df	p-value
Values	66.639	4	0.000*

*$p < 0.05$

5.2 Inferences

From the above discussed analysis of data, we can observe that Chi-square test fails to show significant value for 19 statements, with $p < 0.05$, out of 20 statements at α = 0.05 level of significance; therefore, null hypothesis is rejected.

Hence the statement that there are significant differences in the opinion about effectiveness of common employee assistant programs among male and female employees in the outsourcing industry.

Further, data analysis based on mean value of responses has shown that female employees place more emphasis on the need to following employee's welfare/stress management programs:

- Provision of Work from Home: Female (M = 4.29), male (M = 4.05)
- Opening of Crèches: Female (M = 4.38), male (M = 3.05)
- Arrangement of Safe and Comfortable Office Transport: Female (M = 4.18), male (M = 2.20)
- Enforcement of Antidiscrimination Legislation and Implementation of No-nonsense/Sexual Harassment policies: Female (M = 3.28), male (M = 2.25)
- Provision of paid maternity leaves/childcare leaves (at least 6 months to 2 years): Female (M = 4.11), male (M = 2.85)
- Cultural/Recreational Programs: Female (M = 2.28), male (M = 1.85)
- Higher Education Programs: Female (M = 2.59), male (M = 2.01)
- Awareness Camp for Health Problems/Guidance of Dietician: Female (M = 3.48), male (M = 2.15).

Today, management does take care of their employee's well-being in all respect. This may be due to good philosophy of top management or pressure by the employee's trade union, human rights association, enforcement by employment laws or provision of corporate social responsibility. BPO's firms are providing number of employee's wellness/stress management programs, but effectiveness of various EAPs is not guaranteed. Hence, the present research is supporting new interventions in these programs by suggesting the concept of gender-based EAPs. In view of the present study, male employees place more emphasis on the need to following employee's welfare/stress management programs:

- Relaxation/Excursion/Sports Activities (at least for 5–6 days): Male (M = 4.06), female (M = 3.20)
- Mentoring Programs.: Male (M = 3.88), female (M = 3.15)
- Sabbatical Leavers: Male (M = 4.15), female (M = 2.54)
- Sports Tournaments: Male (M = 4.29), female (M = 3.20)

- Medical Insurance: Male (M = 4.57), female (M = 3.97)
- Opening of BPO Forum: Male (M = 3.48), female (M = 2.85)
- Training for Building Leaders: Male (M = 4.19), female (M = 3.15)
- Provision of LTC: Male (M = 4.38), female (M = 1.97)
- Training/Awareness Session on Employment Laws/Taxes: Male (M = 4.25), female (M = 3.09).

In the view of the above observations, the study reveals that female employees place more emphasis on the need to 8 employee's welfare/stress management programs out of 20 EAPs included in the questionnaire, whereas male employees place more emphasis on the need to 9 employee's welfare/stress management programs out of 20. Only three factors were found almost common in them. These factors are: Provision of Employee Grievance Cell, Career Counselling and Workshops on Work–Life Balance.

6   Conclusion

The belief is that the common EAPs/stress management programs are beneficial for each gender. Past researches on gender differences in stress outcomes have been conducted mostly in the context of developed countries. Hence, there is a need to investigate this problem in less affluent societies, like ours. Therefore, this is an attempt to fill this gap. Increase number of female employees at all levels in BPO industry opened up an unexplored arena for research which is gender difference in the effectiveness of EAPs/stress management programs.

An attempt has been made to make some necessary recommendations for the effective implementation of gender-based stress management strategies under cost-effective employees' assistance programs for the BPO's employees. As stressors for men and women and their impact on each gender are not the same, therefore same stress management programs cannot be lucrative for each gender. Consequently, if management is spending gigantic amount on such programs, its effectiveness must be ensured in terms of providing benefits to both the genders. In today's competitive working environment, if management want to go for output-based employee welfare programs to lower down stress level of their employees, it must include gender-based techniques, otherwise success rate of such EAPs may not be ensured. Management must know the requirement of gender-based stress management techniques to get more benefits out of these well-being programs. Therefore, it is recommended that EAPs should not be generalized for both genders. A gender-based recreational program may become a result-oriented program. It would be beneficial for employees and management as well.

Based on the findings in the present study, it is recommended that the management may provide some different and some common employee welfare programs to the male and female employees to make it more cost-effective and beneficial for both the genders. Based on the present study, an attempt has been made to make some necessary recommendations on employee assistant programs for female employees:

For Female Employees

- Promotion of policies on gender sensitivity
- Special provision for women employees to provide them equality/benefits/opportunities as a compensation of the time period loss due to maternity phase (which can help women in regaining their jobs)
- Formation of (action oriented) small committee to keep a check on enforcement of antidiscrimination legislation and women's security law
- Senior leadership engagement programs to build a woman leader
- Multiple communication channels
- Training and counseling on women's rights/human rights
- Annual organization climate survey (rating boss)
- Special (gender-based) assistance programs for employee
- Sense of security/family taken care
- At least 2-year sabbatical leave for higher studies
- Reduced work hours half-day/half-work week for new mothers.

Thus women can learn to perform their responsibilities at the office along with their roles at home. BPO companies need to make attractive policies for incentives to women workers and make the industry as a whole an attractive and preferred carrier destination for them. Though Indian society has given a bit more liberty to the males, but even they are also under stress due to increased competitive environment which has forced employees to work for long hours. Due to dealing with UK and USA-based companies, technological advancement, economic and social standard and family requirement, male employees also need to perform on high metrics, and male employees also get mental and physical exhaustion due to stress. The following recommendations have been proposed in employee assistant programs for the male employees:

For Male Employees

- Provision of paid paternity leaves (at least two weeks)
- Fast track program for promoting middle managers
- Part time work
- Yoga classes, health café, stress management workshops
- Restricted work structure and flat leave structure
- Foreign trips for spouses
- More monetary benefits in the form of promotions and incentives
- Incorporate (part time) higher education programs
- Sessions on gender—egalitarian to redirect the males' perception about gender inclusivity
- Training session on management of role overloading and role ambiguity
- Cross-functional job rotation
- Social recognition and upward mobility.

In the present study, we described that there is a significant relationship between gender and the preferences of employees welfare practices. If stressors for men and women and their impact on each gender are not the same, therefore same stress management programs cannot be lucrative for each gender. Consequently, if management is spending gigantic amount on such programs, its effectiveness must be ensured in terms of providing benefits to both the genders. It was stated that in the twenty-first century, physicians are treating their patients—especially for stress depression, anxiety and post-traumatic stress disorder through gender-based medical treatments as according to medical community, biological differences do matter. Thus, if this fraternity has started recognizing the requirements to incorporate gender differences in their medical and psychological treatment, it is the right time for corporates to apply the same concepts in their planning and policy related to employee wellness programs. To create a harmonious, vivacious workplace, such new interventions should be applied in the corporate sector. Management must critically consider their employee's needs along with their machinery, methods, materials and money. This study makes a significant contribution across BPO industry and the development of a conceptual framework that explains and predicts the factors that influence the employee assistance, stress management programs and its application regarding gender-based stress management programs in the BPO sector in NCR.

7   Limitations of the Study

Due to time and cost constraints, a limited number of BPOs (working only for IT sector) situated in NCR region were selected. BPOs from various cities and other sectors are not selected for the study. A sample size of 100 respondents in NCR region may not be the true representation of Indian society. Due to budgetary constraints, the acceptance of gender-based stress management programs might not get easy acceptance by top management in some BPOs.

8   Scope of the Study

Although an attempt has been made in the paper to examine the impact of gender on welfare practices in the outsourcing industry, yet there is an ample scope for further research. This study is based on content which is available up to a specific period, but research is going on and many new results may be on the way to completion which may present a new vista in the domain of study. Therefore, there is scope for the future researchers to add on the new concepts and this study can be further utilized by future researchers in the various functional areas of management, contrasting the employees working in public versus private sector.

References

1. Chaplin, T. M., Hong, K., Bergquist, K., & Sinha, R. (2008). Gender differences in response to emotional stress: An assessment across subjective, behavioral, and physiological domains and relations to alcohol craving. *Alcoholism, Clinical and Experimental Research, 32*(7), 1242–1250.
2. Gianakos, I. (2000). Gender roles and coping with work stress. *Sex Roles, 42*(11/12), 1059–1079.
3. Gianakos, I. (2002). Predictors of coping with work stress: The influences of sex, gender role, social desirability, and locus of control. *Sex Roles, 46*(5/6), 149–158.
4. Burke, R. J. (2003). Work experiences, stress and health among managerial women: Research and practice. In *The Handbook of Work and Health Psychology* (pp. 259–278).
5. Bekker, M. H. J., & Nijssen, A. (2001). Stress prevention training: Sex differences in types of stressors, coping, and training effects. *Stress and Health, 17,* 207–218.
6. Melissa, M. S. (2012). Controlling anger and happiness at work: An examination of gender differences. *Gender, Work and Organization 19*(4).
7. Sudhashree, V. P., Rohith, K., & Shrinivas, K. (2004). Issues and Concerns of Health among call center employees. *Indian Journal of Occupational and Environment Medicine., 9*(3), 120–125.
8. Sloan, M. (2010). Controlling anger and happiness at work: An examination of gender differences. *Gender, Work and Organization.* Retrieved October 11, 2010, from https://online.library.wiley.com.
9. Kristiansen, L., Hellzé, O., & Asplund, K. (2006). Swedish assistant nurses' experiences of job satisfaction when caring for persons suffering from dementia and behavioral disturbances; An interview study. *International Journal of Qualitative Studies on Health and Well-Being, 1*(4), 245–256.
10. Alexandros-Stamatios, G. A., Matilyn, J. D., & Cary L. C. (2003). Occupational stress, job satisfaction, and health state in male and female junior hospital doctors in Greece. *Journal of Managerial Psychology, 18*(6), 592–621.
11. Nelson, D. L., & Burke, R. J. (2002). A framework for examining gender, work stress and health. *Gender, Work Stress, and Health Washington, 48,* 3–18.
12. Taylor, S. (2011). Gender difference in stress management: 'Fight or Flight' versus 'Tend and Befriend' stress responses. Article Retrieved from https://www.health.wikinut.com.
13. Taylor, S. E., Klein, L. C., Lewis, B. P., Gruenewald, T. L., Gurung, R. A. R., & Updegraff, J. A. (2000). Biobehavioral responses to stress in females: Tend-and-befriend, not fight-or-flight. *Psychological Review, 107,* 441–429.
14. Fozia, K. (2013). Gender differences in coping strategies and Life Satisfaction among Cardiac Patients. *Interdisciplinary Journal of Contemporary Research in Business, 5*(5), 537–552.
15. Ryan & Frederick. (1997). The effects of gender role on perceived job stress. *The Journal of Human Resource and Adult Learning, 6*(2), 74–79.
16. Hayman, G. J., Blaschke, T., Marceau, D. J., & Bouchard, A. (2003). A comparison of three image-object methods for the multiscale analysis of landscape structure. *ISPRS Journal of Photogrammetry and Remote Sensing, 57*(5–6), 327–345.

Augmented Reality: An Upcoming Digital Marketing Tool in India

Himanshu Matta and Ruchika Gupta

1 Introduction

The term augmented reality (AR) is defined as combination of technologies that enable real-time mixing of computer-generated content with live video display. "AR is the combination of virtual objects and real world and the users are given a chance to interact with these objects in real time". "AR provides users with sub-immersive experience by allowing interactions to occur between the actual and virtual worlds". First AR system was created by Ivan Sutherland. Since then many researches have been done to explore AR.

AR is being used as a digital marketing tool by various brands like Walmart, PepsiCo., Facebook and so on, to market their brands and products. AR-based advertising is attracting customers more efficiently than traditional advertising methods. There are various advantages of augmented reality in marketing:

Real-time interaction: Augmented reality builds an interactive experience, making customer engage more rapidly. Customer experiences naturalness and uniqueness and wants to experience it again.

Instant feedback: AR provides instant and real-time feedback of customers about their product and services. Feedback is an important information for digital marketer.

Customer Personalization: AR provides personalized experience to the consumers, take for example LensKart application which allows users to try different styles of spectacles using their AR-based Android or Apple mobile or PC application.

H. Matta (✉) · R. Gupta
Amity University, Noida, Uttar Pradesh, India
e-mail: mattahimanshu@gmail.com

R. Gupta
e-mail: rgupta@gn.amity.edu

© The Author(s), under exclusive license to Springer Nature Singapore Pte Ltd. 2021   31
P. K. Kapur et al. (eds.), *Advances in Interdisciplinary Research in Engineering and Business Management*, Asset Analytics,
https://doi.org/10.1007/978-981-16-0037-1_3

Fig. 1 Benefits of AR in marketing

Attracting through AR Games: Through the use of AR in markets, customers in various games can indirectly promote their product and services too. Through games customers experience the brand and earn their way toward coupons, points and rewards that are highly customized.

ROI tracking: AR enables brands to track ROI by providing a real-time information of how products are performing in the market [1] (www.kestone.in, 2017) (Fig. 1).

1.1 Rationale of the Study

Since AR in marketing is uprising and successfully attracting consumers worldwide, it becomes important for Indian consumers as well as marketing firms to be aware about this upcoming revolutionary technology. MNCs like Coca Cola, Walmart, Hasbro, Nintendo have already applied augmented reality in promoting their brands and products and experienced the high potential of AR in marketing through growth in their market share. Through this paper awareness regarding uses of AR technology in marketing as well as future scope of AR will be spread for its uses in Indian market scenario.

2 Literature Review

There are various researches on AR in marketing, like Elham Baratali [2] in his paper has explored the uses of augmented reality in digital marketing and advertisement. He concluded that the augmented reality is an interactive way to attract customers. The research paper's aim was to spread awareness about the uses of augmented reality in advertisement and explore the future scope of this technology. Ooi Jin [3] in his paper has explored the augmented reality in marketing and stated the advantages of AR in attracting consumer attention and in creating brand associations. He also explored the evolution of AR and VR technologies and their uses in marketing. Desti

Kannaiah [4] in his paper has studied the attitudes of Chennai consumers toward AR-based e-commerce applications. This empirical research concluded the positive attitude of consumers in Chennai toward e-commerce websites using AR to let them virtually try some products like caps and so on. In addition, Javornik [5] in his paper has explored the different uses of augmented reality in marketing and further scope of AR technology which is still untapped in marketing application.

Mehdi Mekni [6] in his paper has explored the applications of AR, challenges of AR technology and their future scope. He also stated the problems that augmented reality technology is facing at present, especially difficulties in developing AR applications. He suggested that augmented reality technology has vast potential to grow in future. Yuan [7] explored the potential of augmented reality and virtual reality technology and their boom and increasing popularity and usage in today's world. He stated that AR and VR technologies have potential to change the future world. Kim [8] in his paper reviewed papers published in IEEE in the last three years on augmented reality and virtual reality. The paper described the trends of how user studies have been incorporated into AR and VR as well as future scope of AR and VR technologies.

Many researchers have suggested prototype applications based on AR like Zhang [9] in his paper described the development of augmented reality-based technology for direct marketing system which uses eyeball tracking cameras to give 3D images. This technology will simulate the sales agent presence with customers through webcams. While Wang [10] in his paper has explored the use of augmented reality in women hair styling try out to provide convenience to women customers as well as to generate more sales for the salon. He suggested that AR-based women hair styling try out will attract more customers toward salons who uses such technology. Lai [11] in his paper has explored the new application of augmented reality for educational book publishing, which aims to bring interactive learning experience to life. Using AR technology readers can see the live 3D models of images of the books in their smartphone or iPads. According to them this technology will revolutionize the learning experience for kids. Pantile [12] in his paper described the multimedia solutions designed and developed by ETT S.p.A. and explained the potential of the use of technology in museum and tourist attractions. Emphasis is placed on augmented reality, virtual reality, and immersive 3D projection projects to attract more visitors to museums. Moreover, Hakim [13] in his paper describes the importance of augmented reality in marketing and context awareness. Context-aware branding and customer preference marketing will present users to view marketing content at their own convenience and driven by utility. Valjus [14] in his paper presented an Adobe Flash-based AR video streaming application and its application in web marketing. The application enables augmenting the content of a webcam view by adding video content to it. This paper explored the usage of this application in marketing and attracting customers. Lin [15] in his study explored the use of mobile devices in AR as well as presented a concept of introducing AR in toy industry so as to make toys more attractive and interesting. In this study "a human-computer interactive interface between physical toys and virtual objects was designed with the aim of incorporating augmented reality". On the other hand, Günes [16] in their paper presented a marker less 3D AR application for virtual accessory try-on applications around human arm. According to them this technology based on Kinect and AR can be used in multiple areas of marketing like trying wristwatches virtually.

Gap analysis: There are many papers on augmented reality uses in marketing but rarely any explored the Indian scenario.

3   Research Methodology

3.1   Type of Research

This research is a descriptive research as we describe the present uses and future scope of augmented reality technology in marketing in Indian context.

3.2   Objectives of Study

- To spread awareness about uses of augmented reality technology in marketing in India.
- To explore the challenges faced by AR in marketing in India.
- To explore the future scope of augmented reality in marketing in Indian context.
- To propose an AR-based application for Indian market.

3.3   Data Collection Tool

This study is done using secondary data. Secondary data are collected from websites, journals and so on, like www.kestone.in, *Journal of Marketing and Consumer Research*.

4   Findings and Analysis

AR technology is widely used nowadays either through filters like Facebook, Snapchat and other social media platforms or games like Pokemon Go as well as to smartphone companies adding AR features to phone cameras to improve image quality and adding AR-based beatification effects in selfie. Last five years of commercial application of such technologies have helped in building the trend of developing utilities which are providing screen-enhanced information. AR-based innovations are pacing up like Google Glass, Microsoft Holo Lens.

4.1 Present Scenario in India

In present scenario in India, the most basic commercial application of AR so far has been for events and activation programs. Brands like Coca Cola, Tata Tiago, Renault KWID in India are using AR-based personalized experiences in on-ground mall spaces and public event spaces for mass. Mahindra back in 2012 used AR in auto Expo, while launching XUV 500, to let people experience a virtual cheetah which was a visual treat and gained an impressive traction.

Some examples of AR in marketing in India are:

- *Lenskart*: It is one of the pioneers of AR-based application in India, which enables users to try spectacles and sunglasses virtually.
- *Dulux paints*: Dulux paints have launched their application based on AR which allow users to try different colors and designs for wall painting virtually.
- *Mahindra*: Mahindra back in 2012 used AR in auto Expo, while launching XUV 500, to let people experience a virtual cheetah, which was a visual treat and gained an impressive traction.
- *Caratlane*: Caratlane.com, a brand dealing in jewelry, has launched AR-based application working on AR and facial recognition which allows users to try jewelries virtually.
- *Makaan*: This application lets users look for houses by simply moving their phone around, users can point the phone at a tower to see how many units are available for rent in the building.
- *South India Bank*: South India Bank has launched an application named SIB Mirror based on AR technology which let users to locate ATMs and their branches.
- *Tinkle*: Tinkle, an Indian comic book application, lets users to interact with comic book images by just scanning pages of comics.
- *Snapchat*: Snapchat is providing facial stickers based on AR.
- *Instagram*: Instagram is providing facial stickers and masks based on AR.
- *Facebook*: Facebook like Instagram is also providing facial stickers based on AR.

Indian tech-start-ups are joining the AR bandwagon with creative solutions to offer. Start-ups like Shopsense are helping apparel stores where customers can zero down their choice by virtually trying out on screen consisting in-store inventory.

4.2 Challenges of AR Marketing in India

- Digital illiteracy among Indians especially among rural people. In India, across over 650,000 villages and 250,000 panchayats are represented by 3 million panchayat members. About 40% population is living below the poverty line, so digital literacy is almost non-existent among more than 90% of India's population. (https://defindia.org/national-digital-literacy-mission/, 2019).

- AR-compatible smartphones are still expensive in India; average starting price of smartphones in India is about 10,000 rupees.
- Internet data packs are still not much affordable in India.
- AR applications still not perfected, it's in progressing stage (Ratnottar, 2015).
- Applications supporting are difficult to develop as well as limited in number; iOS and Android Play store has limited AR applications as compared to conventional applications available for users.

4.3 Future Scope of AR in India

According to MarketsandMarkets report titled "Augmented Reality and Virtual Reality Market—Global Forecast to 2022", AR market is internationally projected to reach $117.4 billion by 2022 with growth rate of 76%. "The increasing demand for AR and VR technology-based products in various verticals such as consumer, aerospace, defense, commercial and medical is responsible for the market growth," said the MarketsandMarkets' report. The Indian AR and VR market is projected to grow at a CAGR of over 55% during 2016–2021. Use of AR in gaming is also on the rise in the country as per the report.

AR opens up unexplored field for marketeers as it provides new space for three-dimensional thinking and idea execution. AR applications have huge scope in future in various areas like a prototype is been developed to stimulate personal selling in remote areas through the use of AR working through just simple webcams [17] (MiCompass, n.d.).

5 Proposed AR Applications for Indian Market

5.1 Proposed AR Application for Gyms/Fitness Centers

An AR application based on fitness programs preloaded by just placing the camera on a fitness gym equipment should show how to correctly use that equipment through AR technology. Moreover, marketers can extend this application in other areas of fitness, like Zumba, Yoga and so on, where just by placing camera on the yoga mat, the application will show the techniques to do some personalized suggested yoga postures based on the preloaded user's age, BMI, fitness level and so on. Moreover, a virtual personal trainer can be added to this application to stimulate personal training in gym. Big Indian gym brands like THE GYM, Gold Gym India, Body mechanics and so on can implement this AR application so as to provide personal training facilities to gym customers virtually using AR, especially to those customers who cannot afford personal training. Through the use of this AR-based applications users can correctly use gym equipment's and yoga postures without human help just through virtual personal trainer help.

5.2 Proposed AR Application for Indian Metro Railways

An AR-based application can be designed for Indian Metro railways which will provide AR-based real-time Metro route maps as well as important information like locations of fire exits, lifts, washrooms and so on just by using smartphone camera. This application will improve passenger's travel experience as well as will provide convince and generate more sales for Metro railways.

6 Conclusions

Augmented reality (AR) technology is evolving around the world with varying uses, most prominent being in digital marketing and advertising. AR technology in marketing is on the rise in Indian market with many companies like LensKart, Dulux paints using it to attract customers by promoting AR-based mobile applications and are successfully gaining attention from Indian consumers. AR in marketing is on the rise despite facing challenges in Indian market and has ample scope in future too. Researchers can suggest companies to educate Indian consumers about using AR applications and websites so as to strengthen the pace of AR adoption in marketing and promotion.

Acknowledgements The authors would like to give special thanks to Amity University faculties for their help and support in writing this paper.

References

1. (2017, January). Retrieved from www.kestone.in; https://www.kestone.in/5WaysAugmentedReality-Jan2017.aspx.
2. Elham Baratali, M. H. (2016, April). Effective of augmented reality (ar) in marketing communication; a case study on brand interactive advertising. *International Journal of Management and Applied Science, 2*(4).
3. Ooi Jin, R. Y. (2015). The review of the effectivity of the augmented reality experiential marketing tool in customer engagement. *Global Journal of Management and Business Research: E Marketing, 15*(8), 1–18.
4. Desti Kannaiah, D. R. (2015). The Impact of Augmented Reality on E-commerce. *Journal of Marketing and Consumer Research, 8.*
5. Javornik, A. (2014). [Poster] classifications of augmented reality uses in marketing. *IEEE International Symposium on Mixed and Augmented Reality—Media, Art, Social Science,* 67–68.
6. Mehdi Mekni, A. (2014). Augmented reality: Applications, challenges and future trends. *Applied Computational Science.*
7. Yuan, Y. (2017, January). Changing the world with virtual\augmented reality technologies. *IEEE Consumer Electronics Magazine, 6,* 40–41.
8. Kim, S. J. (2012). A user study trends in augmented reality and virtual reality research: A qualitative study with the past three years of the ISMAR and IEEE VR conference papers. In *International Symposium on Ubiquitous Virtual Reality* (pp. 1–5).

9. Zhang, X., et al. (2000). E-commerce direct marketing using augmented reality. In *IEEE International Conference on Multimedia and Expo., 1* (pp. 88–91).
10. W. Wang, Y. C. (2011). Integrating augmented reality into female hairstyle try-on experience. In *IEEE Seventh International Conference on Natural Computation* (pp. 2125–2127).
11. Lai, A. S. Y., Wong, C. Y. (2015). Applying augmented reality technology to book publication business. In *IEEE 12th International Conference on e-Business Engineering* (pp. 281–286).
12. Pantile, D., Frasca, F. (2016). New technologies and tools for immersive and engaging visitor experiences in museums: The evolution of the visit-actor in next-generation storytelling, through augmented and virtual reality, and immersive 3D projections. In *12th International Conference on Signal-Image Technology & Internet-Based Systems (SITIS)* (pp. 463–467).
13. Hakim, A. H. (2016). Context aware augmentational marketing. In *IEEE SAI Computing Conference (SAI)* (pp. 1227–1231).
14. Valjus, V., Järvinen, S. (2012). Web-based augmented reality video streaming for marketing. In *IEEE International Conference on Multimedia and Expo Workshops* (pp. 331–336).
15. Lin, C., Pa, P. (2013). Mobile application of interactive remote toys with augmented reality. In *Asia-Pacific Signal and Information Processing Association Annual Summit and Conference* (pp. 1–6).
16. Günes, O. S. (2015). Augmented reality tool for markerless virtual try-on around human arm. In *IEEE International Symposium on Mixed and Augmented Reality—Media, Art, Social Science, Humanities and Design* (pp. 59–60).
17. MiCompass, T. (n.d.). Retrieved from www.medium.com; https://medium.com/@marketing orbit/future-of-marketing-with-technologies-like-augmented-reality-13a8582e5ae2.

Dual Warehouse Inventory Management of Deteriorating Items Under Inflationary Condition

Amrina Kausar, Ahmad Hasan, Prerna Gautam, and Chandra K. Jaggi

1 Introduction

The upshot of prices affects the livelihood, minimizes the abilities of the businesses to respond to varying environments and every other sphere of the economy. Many countries across the globe experience a high annual inflation rate. The demand for many items gets affected by inflation. When inflation occurs, the value of money goes down which erodes the future worth of saving. Generally, these expenditures are on peripherals and expensive items that provide rise to the demand for these items. So, the effect of inflation and the time value of money cannot be ignored. In [4] the authors presented the EOQ model with inflation under different types of pricing policies. Reference [3] discussed an article on inflation and time discounting with linear time-dependent rates and shortages. Reference [28] developed the model for deteriorating items under the finite replenishment rate taking into account the time value of money. Reference [5] studied an inventory model under inflation with permissible delay in payment for deteriorating items. Thereafter, [10] studied optimal order policy for deteriorating items with inflation-induced demand. Thereafter, many

A. Kausar
Department of Management Studies, Shaheed Sukhdev College of Business Studies, University of Delhi, Delhi-110089, India
e-mail: amrinakausar@sscbsdu.ac.in

A. Hasan · P. Gautam · C. K. Jaggi (✉)
Department of Operational Research, Faculty of Mathematical Sciences, University of Delhi, Delhi-110007, India
e-mail: ckjaggi@yahoo.com

A. Hasan
e-mail: ahmadhasan2161@gmail.com

P. Gautam
e-mail: prerna3080@gmail.com

P. K. Kapur et al. (eds.), *Advances in Interdisciplinary Research in Engineering and Business Management*, Asset Analytics,
https://doi.org/10.1007/978-981-16-0037-1_4

researchers contributed to this direction by incorporating different factors, namely, [12, 13] and [15, 18, 21–23, 25, 27]. Recently, [24] developed an inventory model for deteriorating items with time and price-dependent demand under inflationary conditions.

Conventionally, it is believed that all the products have an infinite shelf life and remains the same in terms of their usability. But this assumption may not always be true for many items, for example, blood, radioactive product, vegetables, and chemical, and so on may not remain the same after a while due to deterioration. The effect of deterioration should not be ignored while computing EOQ else it will lead to disingenuous results. Reference [7] was the first to develop an inventory model for exponentially declining inventory. Ten years later, [6] extended [7] model by introducing a variable rate of deterioration. Thereafter, many researchers have contributed to this area. The authors of [8] developed "an inventory model for deteriorating items". Reference [19] proposed an inventory model under credit financing for deteriorating items with imperfect quality. Reference [20] extended their model by incorporating price sensitivity in demand. Recently, [16, 17] proposed inventory models for deteriorating items with exponential increasing and decreasing demand rates under credit policy.

Furthermore, in all the above-mentioned articles, it is assumed that the system's storage capacity is unlimited. However, due to various reasons, many times the retailer orders more than his storing capacity. In this situation, the retailer may rent a warehouse to accumulate the exceeding units. Managing inventory under two warehousing is important as it directly affects the total cost. Many researchers have extensively studied the problems of EOQ under two storage facilities. The phenomenon of "two-warehouse" was first introduced by [9] who assumed that an inventory carrying cost in the rented-warehouse is greater than that of own-warehouse and transportation cost from the rented-warehouse to own-warehouse is negligible. Reference [11] delivered a two-warehouse inventory model for deteriorating items with a linear trend in demand and shortages under inflationary conditions. Thereafter, [1] studied a two-warehouse deterministic inventory model with a linear trend in time-dependent demand over a finite time horizon for deteriorating items. Reference [14] presented ordering policy in a two-warehouse system for deteriorating items in inflationary conditions. After that, [2] investigated an inventory problem in an interval environment under inflation via particle swarm optimization under the two-warehouse facility. Recently, [29] studied "the impact of inflation in a two-warehouse inventory model for deteriorating items with time-varying demand and shortages". Reference [26] developed an inventory model for deteriorating items with time-dependent demand and partial backlogging under inflationary conditions. Furthermore, all the above-mentioned literature assumed a finite planning horizon, which may not hold in a pragmatic situation.

Considering the above-discussed realistic assumptions, the present paper develops an inventory replenishment model in a two-warehousing scenario. The deterioration of products is considered with inflation over a finite planning horizon. The DCF approach is implemented. It has been presumed that the demand rate increases because of inflation. Further, the demand is first met by the inventory kept in a rented-warehouse and then from the own-warehouse; backorders are permissible with complete backlogging. The numerical analysis along with sensitivity analysis is given to exhibit the model characteristics.

2 Assumptions and Notations

Assumptions

1. The demand rate $D(t) = D_0 e^{\alpha t}$, which is exponentially increasing function, where D_0 is the initial demand rate and $0 \leq \alpha \leq 1$ is the inflation rate.
2. The OW has a fixed capacity of W unit, whereas the RW has unlimited capacity.
3. The replenishment rate is finite and lead time is negligible.
4. Backorders are permissible with full backlogging.
5. A DCF approach is used to consider the various costs at various times, $r (r > \alpha)$ is the discount rate.
6. The planning horizon is finite.

Notations

$I_{i0}(t)$	The inventory level of own-warehouse at any time t_i, where $i = 1, 2 \ldots, n$
$I_{iR}(t)$	The inventory level of rented-warehouse at any time t_i, where $i = 1, 2 \ldots, n$
$I_{is}(t)$	Backlogged level at any time t_i, where $i = 1, 2 \ldots, n$
Q_i	Lot size, where $i = 1, 2 \ldots, n$
S_i	Maximum inventory level per cycle, where $i = 1, 2 \ldots, n$
W	The storage capacity of own-warehouse
θ_1	The deterioration rate for rented-warehouse (RW)
θ_2	The deterioration rate for own-warehouse (OW), where $\theta_1 < \theta_2$
t_{iR}	*The time point when inventory level exhausts in rented-warehouse, where $i = 1, 2 \ldots, n$*
t_{ii}	*The time point when inventory level exhausts in own-warehouse, where $i = 1, 2 \ldots, n$*
L	Length of the finite planning horizon
T	Cycle length
n	Number of replenishments over the planning horizon, $n = \frac{L}{T}$
r	The discount rate (%)
α	Inflation rate (/unit/unit time), where $\alpha < r$
r-α	The net discount rate (/unit /unit time)

3 Mathematical Formulation

The continuous compounding of the inflation has been assumed, thus the cost of ordering, purchase, storage and backordering at any time t (Bozacott, (1975)) are:

$$A(t) = A_0 e^{\alpha t}$$

$$C(t) = C_0 e^{\alpha t}$$

$$H_{RW}(t) = C_{RW} e^{\alpha t}$$

$$H_{OW}(t) = C_{OW} e^{\alpha t}$$

$$C_S(t) = C_S e^{\alpha t} \tag{1}$$

The planning horizon (L) has been divided into n cycle of length T $(i.e. T = L/n)$. Let us assume that the ith cycle, that is, $t_{i-1} \le t \le t_i$, where $t_0 = 0, t_n = L, t_i - t_{i-1} = T$ and $t_i = iT (i = 1, 2, 3, \ldots, n)$. Lot size Q_i enters in the system at the starting of the ith cycle. Shortages are met from Q_i first. After meeting the shortages of the previous cycle, (S_i) units remained as inventory. Out of remaining units (S_i), W units are stored in the own-warehouse (OW) and $(S_i - W)$ units are kept in a rented-warehouse (RW). Here the items are instantaneously deterioration in nature. Thus, in the period (t_{i0}, t_{iR}) inventory level of the rented-warehouse (RW) decreases due to the demand as well as deterioration, and the inventory level at own-warehouse decreases only due to deterioration. At t_{iR} rented-warehouse's inventory level reaches zero. The demand is now been satisfied by the inventory of own-warehouse. The inventory level in own-warehouse now declines because of the demand as well as deterioration till time t_{i1}. At time point t_{i1} both warehouses are empty and shortages start building up till the end of the cycle. $t_{i1} = t_i - kT = (i - k)\frac{L}{n}, (i = 1, 2, 3, 4 \ldots, n), (0 \le k \le 1)$, where kT is the fraction of the cycle having shortages and $t_{iR} = t_{i-1} + bT = (i - 1)T + bT = (i - 1 + b)\frac{L}{n}, (i = 1, 2, \ldots, n), (0 \le b \le 1)$, where bT is the fraction of the cycle length. The performance of the model over the period (t_{i-1}, t_i) is presented in Fig. 1.

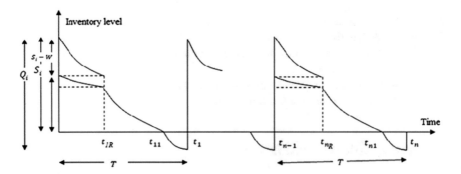

Fig. 1 Graphical representation of inventory system

For the period (t_{i-1}, t_i), the differential equations representing inventory level in RW and OW are as follows:

$$\frac{dI_{iR}(t)}{dt} = -\theta_1 I_{iR}(t) - D_0 e^{\alpha t} \quad t_{i-1} \le t \le t_{iR} \quad i = 1, 2, 3 \ldots, n \qquad (2)$$

$$\frac{dI_{i0}(t)}{dt} = -\theta_2 I_{i0}(t) \quad t_{i-1} \le t \le t_{iR} \quad i = 1, 2, 3 \ldots, n \qquad (3)$$

$$\frac{dI_{i0}(t)}{dt} = -\theta_2 I_{i0}(t) - D_0 e^{\alpha t} \quad t_{iR} \le t \le t_{i1} \quad i = 1, 2, 3 \ldots, n \qquad (4)$$

$$\frac{dI_{is}(t)}{dt} = D(t) = -D_0 e^{\alpha t} \quad t_{i1} \le t \le t_i \quad i = 1, 2, 3 \ldots .n \qquad (5)$$

The solution for the above differential equations along with the boundary condition $I_{iR}(t_{iR}) = 0$, $I_{i0}(t_{i-1}) = W$, $I_{i0}(t_{i1}) = 0$ and $I_{is}(t_{1i}) = 0$, respectively, are.

$$I_{iR}(t) = \frac{D_0}{\alpha + \theta_1} e^{-\theta_1 t} \left[e^{(\alpha - \theta_1)t_{iR}} - e^{(\alpha - \theta_1)t} \right] \quad t_{i-1} \le t \le t_{iR} \qquad (6)$$

$$I_{i0}(t) = W e^{\theta_2(t_{i-1}-t)} \quad t_{i-1} \le t \le t_{iR} \qquad (7)$$

$$I_{i0}(t) = \frac{D_0}{\alpha + \theta_2} e^{-\theta_2 t} \left[e^{(\alpha - \theta_1)t_{i1}} - e^{(\alpha - \theta_1)t} \right] \quad t_{iR} \le t \le t_{i1} \qquad (8)$$

$$I_{is}(t) = \frac{-D_0}{\alpha} \left[e^{\alpha t} - e^{\alpha t_{i1}} \right] \quad t_{i1} \le t \le t_i \qquad (9)$$

At time point $t = t_{i-1}$, $I_{iR}(t_{i-1}) = (S_i - W)$; then from Eq. (6), we get

$$\frac{D_0}{\alpha + \theta_1} e^{-\theta_1 t_{i-1}} \left[e^{(\alpha - \theta_1)t_{iR}} - e^{(\alpha - \theta_1)t_{i-1}} \right] = (S_i - W)$$

$$S_i = W + \frac{D_0}{\alpha + \theta_1} e^{-\theta_1 t_{i-1}} \left[e^{(\alpha - \theta_1)t_{iR}} - e^{(\alpha - \theta_1)t_{i-1}} \right] \quad i = 1, 2, 3, 4 .., n \qquad (10)$$

Since $I_{is}(t_i) = -B_i$, from Eq. (9) we get

$$\frac{-D_0}{\alpha} \left[e^{\alpha t_i} - e^{\alpha t_{i1}} \right] = -B_i$$

$$B_i = \frac{D_0}{\alpha} \left[e^{\alpha t} - e^{\alpha t_{i1}} \right] \quad i = 1, 2, 3, 4 \ldots, n \qquad (11)$$

Further, order quantity for ith cycle is

$$Q_i = S_i + B_i$$

$$Q_i = W + \frac{D_0}{\alpha + \theta_1} e^{-\theta_1 t_{i-1}} \left[e^{(\alpha - \theta_1) t_{iR}} - e^{(\alpha - \theta_1) t_{i-1}} \right] + \frac{D_0}{\alpha} \left[e^{\alpha t} - e^{\alpha t_{i1}} \right]$$

$$i = 1, 2, 3, 4, \ldots, n \tag{12}$$

As the model considers backlogging, various costs are calculated as.

1. Ordering cost for ith cycle is

$$A_i = A(t_{i-1})e^{-rt_{i-1}} = A_0 e^{(\alpha - r)t_{i-1}} \quad i = 1, 2, 3 \ldots, n \tag{13}$$

2. Purchasing cost for i th cycle is

$$P_i = Q_i C(t_{i-1})e^{-rt_{i-1}} = Q_i C_0 e^{(\alpha - r)t_{i-1}} \quad i = 1, 2, 3 \ldots, n \tag{14}$$

3. Inventory storage cost of RW for ith cycle is

$$H_{iRW} = C_{RW}(t_{i-1})e^{-rt_{i-1}} \int_{t_{i-1}}^{t_{iR}} I_{iR}(t)e^{-rt}dt$$

$$= \frac{D_0}{\alpha + \theta_1} C_{RW}(t_{i-1})$$

$$e^{-rt_{i-1}} \left[\frac{e^{(\alpha + \theta_1) t_{i-1}}}{\alpha + \theta_1} \left(e^{-(\alpha + \theta_1) t_{i-1}} - e^{-(\alpha + \theta_1) t_{iR}} \right) - \frac{1}{\alpha - r} \left(e^{-(\alpha - r) t_{iR}} - e^{-(\alpha - r) t_{i-1}} \right) \right] \tag{15}$$

4. Inventory storage cost of OW for ith cycle is

$$H_{iow} = C_{ow}(t_{i-1})e^{-rt_{i-1}} \left[\int_{t_{i-1}}^{t_{iR}} I_{i0}(t)e^{-rt}dt + \int_{t_{iR}}^{t_{i1}} I_{i0}(t)e^{-rt}dt \right]$$

$$= C_{ow}(t_{i-1})e^{-rt_{i-1}} \left[\frac{W e^{\theta_2 t_{i-1}}}{\alpha + \theta_2} \left(e^{-(\theta_2 - r)t_{i-1}} - e^{-(\theta_2 - r)t_{iR}} \right) \right.$$

$$+ \frac{D_0}{\alpha + \theta_2} \left\{ \frac{e^{(\alpha + \theta_2)t_{i1}}}{\alpha + \theta_2} \left(e^{-(\alpha + \theta_2)t_{iR}} - e^{-(\alpha + \theta_2)t_{i1}} \right) \right.$$

$$\left. \left. - \frac{1}{\alpha - r} \left(e^{(\alpha - r)t_{i1}} - e^{(\alpha - r)t_{iR}} \right) \right\} \right] \tag{16}$$

5. Shortage cost for ith cycle is

$$\pi_i = \frac{C_s D_0}{\alpha} \left[\frac{e^{(\alpha - r)t_i} - e^{(\alpha - r)t_{i1}}}{\alpha - r} + \frac{e^{\alpha t_{i1}}}{r} \left(e^{-rt_i} - e^{-rt_{i1}} \right) \right] e^{(\alpha - r)t_{i-1}} \tag{17}$$

Therefore, the total cost $TC_L(k, n, b)$ for ith cycle is

$$TC_i = A_i + P_i + H_{iRW} + H_{iOW} + \pi_i \quad i = 1, 2 \ldots, n \tag{18}$$

The total cost function of the system during the time horizon L is

$$TC_L(k, n, b) = \sum_{i=1}^{n} TC_i = \sum_{i=1}^{n}(A_i + P_i + H_{iRW} + H_{iOW} + \pi_i) \tag{19}$$

$$
\begin{aligned}
TC_L(k, n, b) = {} & \frac{A_0 e^{(\alpha-r)L}}{1 - e^{\frac{(\alpha-r)L}{n}}} \\
& + \left[\frac{C_0 D_0}{\alpha+\theta_2} \left(e^{(\alpha-\theta_2)(1-k)\frac{L}{n}} - e^{(\alpha-\theta_2)\frac{bL}{n}} \right) \right. \\
& + \frac{C_0 D_0}{\alpha+\theta_1} \left(e^{(\alpha-\theta_2)(1-k)\frac{L}{n}} - 1 \right) + \frac{C_0 D_0}{\alpha} \left(e^{\alpha\frac{L}{n}} - e^{\alpha(1-k)\frac{L}{n}} \right) \\
& + \frac{C_{RW} D_0}{\alpha+\theta_1} \left\{ \frac{e^{(\alpha-\theta_1)\frac{bL}{n}}}{\theta_1+r} \left(1 - e^{-(\theta_1+r)\frac{bL}{n}} \right) - \frac{1}{\alpha-r} \left(e^{(\alpha-r)\frac{bL}{n}} - 1 \right) \right\} \\
& + \frac{C_{ow} W}{r+\theta_2} \left(1 - e^{-(\theta_2+r)\frac{bL}{n}} \right) \\
& + \frac{C_{ow} W}{r+\theta_2} \left\{ \frac{e^{(\alpha-\theta_2)(1-k)\frac{L}{n}}}{\theta_2+r} \left(e^{-(\theta_2+r)\frac{bL}{n}} - e^{-(\theta_2-r)(1-k)\frac{L}{n}} \right) \right. \\
& \left. - \frac{1}{\alpha-r} \left(e^{(\alpha-r)(1-k)\frac{L}{n}} - e^{(\alpha-r)\frac{bL}{n}} \right) \right\} \\
& + \frac{C_s D_0}{\alpha(\alpha-r)} \left(e^{(\alpha-r)\frac{bL}{n}} - e^{(\alpha-r)(1-k)\frac{L}{n}} \right) \\
& \left. + \frac{C_s D_0}{\alpha r} \left(e^{(\alpha-k\alpha-r)\frac{bL}{n}} - e^{(\alpha-r)(1-k)\frac{L}{n}} \right) \right] \left\{ \frac{(1-e^{(2\alpha-r)L})}{(1-e^{(2\alpha-r)\frac{L}{n}})} \right\}
\end{aligned} \tag{21}
$$

4 Optimality

Necessary conditions of the total cost function to be optimal are as follows:

$$\frac{\partial TC_L(k, n, b)}{\partial k} = 0, \quad \frac{\partial TC_L(k, n, b)}{\partial k} = 0, \quad \frac{\partial TC_L(k, n, b)}{\partial n} = 0 \quad \frac{\partial TC_L(k, n, b)}{\partial b} = 0$$

The sufficient conditions for minimizing the total cost function using the Hessian matrix H, which is the matrix of second-order partial derivatives are as follows:

$$
H = \begin{bmatrix}
\frac{\partial^2 TC_L(k,n,b)}{\partial k^2} & \frac{\partial^2 TC_L(k,n,b)}{\partial k \partial n} & \frac{\partial^2 TC_L(k,n,b)}{\partial k \partial b} \\
\frac{\partial^2 TC_L(k,n,b)}{\partial k \partial n} & \frac{\partial^2 TC_L(k,n,b)}{\partial n^2} & \frac{\partial^2 TC_L(k,n,b)}{\partial n \partial b} \\
\frac{\partial^2 TC_L(k,n,b)}{\partial k \partial b} & \frac{\partial^2 TC_L(k,n,b)}{\partial b \partial n} & \frac{\partial^2 TC_L(k,n,b)}{\partial b^2}
\end{bmatrix}
$$

where

$$D_1 = \frac{\partial^2 TC_L(k, n)}{\partial k^2} > 0, \quad D_2 = \det \begin{bmatrix} \frac{\partial^2 TC_L(k,n,b)}{\partial k^2} & \frac{\partial^2 TC_L(k,n,b)}{\partial n \partial k} \\ \frac{\partial^2 TC_L(k,n,b)}{\partial n \partial k} & \frac{\partial^2 TC_L(k,n,b)}{\partial n^2} \end{bmatrix} > 0$$

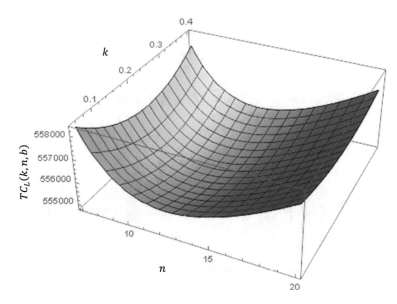

Fig. 2 Total cost function with respect to k, TC_L (k, n, b) and n

$$D_3 = \det \ H = \begin{bmatrix} \frac{\partial^2 TC_L(k,n,b)}{\partial k^2} & \frac{\partial^2 TC_L(k,n,b)}{\partial k \partial n} & \frac{\partial^2 TC_L(k,n,b)}{\partial k \partial b} \\ \frac{\partial^2 TC_L(k,n,b)}{\partial k \partial n} & \frac{\partial^2 TC_L(k,n,b)}{\partial n^2} & \frac{\partial^2 TC_L(k,n,b)}{\partial n \partial b} \\ \frac{\partial^2 TC_L(k,n,b)}{\partial k \partial b} & \frac{\partial^2 TC_L(k,n,b)}{\partial b \partial n} & \frac{\partial^2 TC_L(k,n,b)}{\partial b^2} \end{bmatrix} > 0$$

The values of k, n and b (say k^*, b^* and n^*) obtained from the necessary condition will be optimal, provided they meet the sufficient conditions. Due to the nonlinear nature of cost function, it is very difficult to prove the optimality of the total cost function mathematically. Thus, the convexity of the cost function is established through graph using **Mathematica 11** as in **Figs. 2 and 3.** The expressions for first-order partial derivatives are given in **Appendix** (Table 1).

5 Numerical Example

Using the above data, we get the optimal values of n^*, k^*, b^* and T^* as 12, 0.2701, 0.1782 and 61 days with optimal total cost as \$555,295.14. Order lot size Q_i^* for $i = 1, 2, 3, \ldots, 12$ have been shown in Table 2.

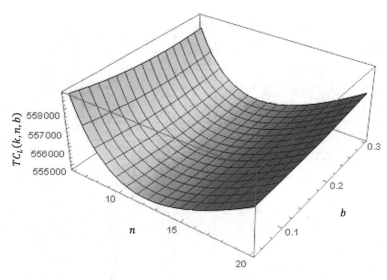

Fig. 3 Total cost function with respect to n, TC_L (k, n, b) and b

Table 1 Numerical data

Parameters	A_0	D_0	r	C_0	C_{OW}	C_{RW}
Value	$700	$1100 unit/year$	$0.12/year$	$250	$25	$30
Parameters	θ_1	θ_2	L	α	W	C_S
Value	$0.05/year$	0.07	$2 year$	0.05	70 units	$100

Table 2 .

No. of cycles (i)	Q_1	Q_2	Q_3	Q_4	Q_5	Q_6	Q_7	Q_8	Q_9	Q_{10}	Q_{11}	Q_{12}	Total
Q_i	153	189	194	202	212	224	239	256	277	300	325	355	2926

6 Sensitivity Analysis

This section investigates the impact of variation in the parameter values θ_1, θ_2, r and α on the optimal policies. The results are as shown in Tables 3, 4 and 5.

Observations from Table 3

- With an increase in the value of θ_2, optimal lot size (Q^*) and cycle length (T^*) get reduced while the total cost $(TC_L(k^*, n^*, b^*))$ increases.
- As the increase of θ_2, the value of T^*, Q^* and $TC_L(k^*, n^*, b^*)$ rises.
- With the increase in the inflation rate (α), both the value of T^* and Q^* rise and the total cost also rises.

Table 3 Sensitivity analysis with basic parameters

Parameter	Change in parameter	n	T	Q	$TC_L(k^*, n^*, b^*)$
	0	12	62	3028	555,191.24
θ_1	0.05	12	61	2927	555,295.14
	0.1	12	61	2875	555,350.76
	0.15	12	60	2843	555,385.45
	0	10	71	2503	553,309.6
θ_2	0.07	12	61	2927	555,295.14
	0.1	12	59	3082	555,936.68
	0.15	13	56	3343	556,820.54
	0.01	13	56	2583	515,374.82
α	0.05	12	61	2927	555,295.14
	0.1	10	72	3682	611,123.51
	0.15	8	96	5681	674,138.26
r	0.09	11	68	3339	570,665.69
	0.12	12	61	2927	555,295.14
	0.15	13	56	2630	540,372.87
	0.18	14	52	2404	525,913.36

Table 4 Effect of α and θ_1 on the optimal cost function

$\alpha \downarrow$	$\theta_1 \rightarrow$	**0**	**0.1**	**0.2**	**0.3**
0.01	n^*	12	13	13	13
	T^*	57	56	55	55
	Q^*	2647	2549	2514	2495
	$TC_L(k^*, n^*, b^*)$	515,306	515,411.88	515,451	515,471
0.05	n^*	11	12	12	12
	T^*	63	61	60	59
	Q^*	3028	2875	2821	2793
	$TC_L(k^*, n^*, b^*)$	555,191.243	555,350.76	555,409.2	555,439
0.1	n^*	10	10	11	11
	T^*	75	71	69	68
	Q^*	3898	3575	3467	3414
	$TC_L(k^*, n^*, b^*)$	610,934.48	611,222.76	611,325.6	611,378.5
0.15	n^*	7	8	8	8
	T^*	103	90	88	86
	Q^*	6518	5323	4994	4840
	$TC_L(k^*, n^*, b^*)$	673,718.55	674,347.9	674,558.7	674,664.8

Table 5 Effect of α and θ_2 on the total optimal cost

$\alpha \downarrow$	$\theta_2 \rightarrow$	0	0.1	0.2	0.3
0.01	n^*	12	13	14	14
	T^*	63	54	51	55
	Q^*	2215	2722	3211	2354
	$TC_L(k^*, n^*, b^*)$	513,714.7	515,922.1	517,303.7	518,465.6
0.05	n^*	10	12	13	14
	T^*	71	59	55	51
	Q^*	2503	3082	3683	4083
	$TC_L(k^*, n^*, b^*)$	553,309.2	555,936.7	557,518.9	559,432.2
0.1	n^*	8	10	12	13
	T^*	90	68	63	66
	Q^*	3186	3852	4710	5662
	$TC_L(k^*, n^*, b^*)$	608,474.1	611,939.4	613,848.1	615,743.4
0.15	n^*	6	8	9	10
	T^*	105	87	79	71
	Q^*	5915	5725	7302	8212
	$TC_L(k^*, n^*, b^*)$	669,785.9	675,283.3	677,657.3	678,754.5

- With the increase in the discount rate (r), the result of T^* and Q^* are reduced and the total cost is also declined.

Observation from Table 4

- When the deterioration rate of RW fixed and the value of α rises, the number of cycles n^* declines but $TC_L(k^*, n^*, b^*)$ and Q^* rise extensively.
- When the deterioration rate of RW rises and the value of α fixed, the number of cycles n^* increases but $TC_L(k^*, n^*, b^*)$ and Q^* declines.

Observation from Table 5

- For a given value of θ_2 when the α rises, the number of cycles n^* reduces but T^*, Q^* and $TC_L(k^*, n^*, b^*)$ increase significantly.
- For a given value of α, rise in the value of θ_2 results in a rise in the number of cycles n^*; however, T^*, Q^* and $TC_L(k^*, n^*, b^*)$ get reduced.

7 Conclusion

The present paper develops a deteriorating inventory replenishment model with inflationary conditions in a two-warehouse facility using a discounted cash flow approach. The finite planning horizon is considered. The aim of this article lies in minimizing the total cost and order quantity (Q). The optimality is represented graphically. Sensitivity analysis shows that the efficiency of products declines over time under the effect of deterioration in both warehouses, which in turn suggests smaller cycle length. Also, the effect of inflation on demand suggests a bigger cycle length. The model suggests important decision-making insights for the inventory managers who are dealing with deteriorating items. The model can be extended by considering an investment in preservation technology to deal with deterioration effectively. It would be a worthwhile contribution if the environmental aspect could be incorporated. Further, the behavior of the model could be studied under different trade-credit environment.

Appendix

The expressions obtained for the necessary conditions for minimizing the total cost are given as:

$$\frac{\partial TC_L}{\partial k} = \left[+\frac{C_0 D_0}{\alpha+\theta} \left(\frac{-\dfrac{C_0 Le^{\frac{(1-k)L(\alpha+\theta_2)}{n}}}{n} + \dfrac{C_0 D_0 Le^{\frac{\alpha(1-k)L}{n}}}{n}}{-\dfrac{Le^{\left(\frac{(1-k)L(\alpha+\theta_2)}{n} - \frac{(1-k)L(\theta_2+r)}{n}\right)}}{n} + \dfrac{Le^{\frac{(1-k)L(\alpha-r)}{n}}}{n}}{L(\alpha+\theta_2)e^{\frac{(1-k)L(\alpha+\theta_2)}{n}} \left(e^{-\frac{Lz(\theta+r)}{n}} - e^{-\frac{(1-k)L(\theta_2+r)}{n}}\right)}{n(\theta_2+r)} \right) + \frac{C_s D_0 L}{\alpha n} e^{\frac{(1-k)L(\alpha-r)}{n}} + \frac{C_s D_0}{\alpha r} \left(\frac{L(\alpha-r)}{n} e^{\frac{(1-k)L(\alpha-r)}{n}} - \frac{\alpha L}{n} e^{\frac{L(\alpha-\alpha k-r)}{n}} \right) \right] = 0$$

$$\frac{\partial TC_L(n,k,b)}{\partial n}$$

$$= -\frac{A_0 e^{\frac{L(\alpha-r)}{n}}\left(1 - e^{L(\alpha-r)}\right)L(\alpha-r)}{\left(1 - e^{\frac{L(2\alpha-r)}{n}}\right)n^2}$$

$$-\frac{\left(1 - e^{L(2\alpha-r)}\right)L(2\alpha-r)}{\left(1 - e^{\frac{L(2\alpha-r)}{n}}\right)n^2}e^{\frac{L(2\alpha-r)}{n}}\left(\begin{array}{l} C_{OW}\left(\frac{\left(1 - e^{\frac{Lb(r+\theta_1)}{n}}\right)W}{r+\theta_1} + \frac{D_0}{\alpha+\theta_1}\left(\frac{e^{\frac{(1-k)L(\alpha+\theta_1)}{n}}}{r+\theta_1}\left(-e^{\frac{-(1-k)L(r+\theta_1)}{n}} + e^{\frac{-Lb(r+\theta_1)}{n}}\right)\right) \right. \\ \left. \qquad - \frac{1}{\alpha-r}\left(e^{\frac{(1-k)L(\alpha-r)}{n}} - e^{\frac{Lb(\alpha-r)}{n}}\right)\right) \\ + \frac{C_0 D_0}{\alpha}\left(e^{\frac{L\alpha}{n}} - e^{\frac{(1-k)L\alpha}{n}}\right) + \frac{C_S D_0}{r\alpha}\left(-e^{\frac{(1-k)L(\alpha-r)}{n}} + e^{\frac{L(-r-k\alpha+\alpha)}{n}}\right) + \frac{C_0 D_0}{\theta_2+\alpha}\left(-1 + e^{\frac{L z(\theta_2+\alpha)}{n}}\right) \\ + \frac{C_{OW} D_0}{\alpha(\alpha-r)}\left(e^{\frac{L(\alpha-r)}{n}} - e^{\frac{(1-k)L(\alpha-r)}{n}}\right) + \frac{C_{RW} D_0 e^{\frac{Lb(\theta_2+\alpha)}{n}}}{(r+\theta_2)(\theta_2+\alpha)}\left(1 - e^{\frac{L(r+\theta_2)b}{n}}\right) \\ + \frac{C_0 D_0}{\alpha+\theta_1}\left(e^{\frac{(1-k)L(\alpha+\theta_1)}{n}} - e^{\frac{Lb(\alpha+\theta_1)}{n}}\right) - \frac{C_{RW} D_0}{(\alpha-r)(\theta_2+\alpha)}\left(-1 + e^{\frac{Lb(\alpha-r)}{n}}\right) \end{array}\right)$$

$$+\left(1 - e^{L(2\alpha-r)}\right)\left(\begin{array}{l} C_{OW}\left\{\frac{D_0}{\alpha+\theta_1}\left[\frac{e^{\frac{(1-k)L(\alpha+\theta_1)}{n}}}{r+\theta_1}\left(\frac{e^{\frac{Lb(r+\theta_1)}{n}}Lb(r+\theta_1)}{n^2} - \frac{1}{\alpha-r}\left(\frac{e^{\frac{Lb(\alpha-r)}{n}}Lb(\alpha-r)}{n^2}\right) - \frac{e^{\frac{Lb(r+\theta_1)}{n}}WLb}{n^2}\right)\right. \right. \\ \left.\left. \qquad - \frac{e^{\frac{(1-k)L(r+\theta_1)}{n}}(1-k)L(r+\theta_1)}{n^2}\right) - \frac{e^{\frac{(1-k)L(\alpha-r)}{n}}(1-k)L(\alpha-r)}{n^2}\right)\right. \\ \left. \qquad - \frac{e^{\frac{(1-k)L(\alpha+\theta_1)}{n}}(1-k)L(\alpha+\theta_1)}{n^2(r+\theta_1)}\left(-e^{\frac{(1-k)L(r+\theta_1)}{n}} + e^{\frac{Lb(r+\theta_1)}{n}}\right)\right\} \\ + \frac{C_0 D_0}{\alpha}\left(\frac{e^{\frac{(1-k)L\alpha}{n}}(1-k)L\alpha}{n^2} - \frac{e^{\frac{L\alpha}{n}}L\alpha}{n^2}\right) + \frac{C_S D_0}{r\alpha}\left(\frac{(1-k)L(\alpha-r)}{n^2}e^{\frac{(1-k)L(\alpha-r)}{n}} - \frac{L(-r-k\alpha+\alpha)}{n^2}e^{\frac{L(-r-k\alpha+\alpha)}{n}}\right) \\ + \frac{C_S D_0}{\alpha(\alpha-r)}\left(\frac{(1-k)L(\alpha-r)}{n^2}e^{\frac{(1-k)L(\alpha-r)}{n}} - \frac{L(\alpha-r)}{n^2}e^{\frac{L(\alpha-r)}{n}}\right) + \frac{C_{OW} D_0 Lb}{n^2(\theta_2+\alpha)}e^{\frac{Lb(\alpha-r)}{n}} - \frac{C_{OW} D_0 Lb}{n^2(\theta_2+\alpha)}e^{\frac{Lb(\theta_2+\alpha)}{n}}e^{\frac{L(r+\theta_2)b}{n}} \\ + \frac{C_0 D_0}{\alpha+\theta_1}\left(\frac{Lb(\alpha+\theta_1)}{n^2}e^{\frac{Lb(\alpha+\theta_1)}{n}} - \frac{(1-k)L(\alpha+\theta_1)}{n^2}e^{\frac{(1-k)L(\alpha+\theta_1)}{n}}\right) - \frac{C_0 D_0 Lb}{n^2}e^{\frac{Lb(\theta_2+\alpha)}{n}} - \frac{C_{OW} D_0 Lb e^{\frac{Lb(\theta_2+\alpha)}{n}}}{n^2(r+\theta_2)}\left(1 - e^{\frac{L(r+\theta_2)b}{n}}\right) \end{array}\right)$$

$$= 0$$

$$\frac{\partial TC_L(n,k,b)}{\partial b}$$

$$= \left(\begin{array}{l} \frac{C_{RW} D_0 L\left(1 - e^{\frac{Lb(r+\theta_2)}{n}}\right)e^{\frac{Lb(\alpha+\theta_2)}{n}}}{n(r+\theta_2)} + \frac{C_{RW} D_0 L e^{\frac{Lb(\alpha+\theta_2)}{n}}}{n(\alpha+\theta_2)}\frac{Lb(r+\theta_2)}{n} - \frac{C_{RW} D_0 L e^{\frac{Lb(\alpha-r)}{n}}}{n(\alpha+\theta_2)} + \frac{C_0 D_0 L e^{\frac{Lb(\alpha+\theta_2)}{n}}}{n} \\ - \frac{C_0 D_0 L e^{\frac{Lb(\alpha+\theta_1)}{n}}}{n} + C_{OW}\left(\frac{WLe^{\frac{Lb(\theta_1+r)}{n}}}{n} + \frac{D_0\left(\frac{Le^{\frac{Lb(\alpha-r)}{n}}}{n} - \frac{Le^{\frac{(1-k)L(\alpha+\theta_1)}{n}}\frac{Lb(\theta_1+r)}{n}}{n}\right)}{\alpha+\theta_1}\right) \end{array}\right)$$

$$= 0$$

References

1. Bhunia, A., Shaikh, A., Maiti, A., & Maiti, M. (2013). A two-warehouse deterministic inventory model for deteriorating items with a linear trend in time dependent demand over finite time horizon by elitist real-coded genetic algorithm. *International Journal of Industrial Engineering Computations, 4*(2), 241–258.

2. Bhunia, A. K., & Shaikh, A. A. (2016). Investigation of two-warehouse inventory problems in interval environment under inflation via particle swarm optimization. *Mathematical and Computer Modelling of Dynamical Systems, 22*(2), 160–179.

3. Bose, S., Goswami, A., & Choudhuri, K. S. (1995). An EOQ model for deteriorating items with linear time-dependent demand rate and shortages under inflation and time discounting. *Journal of the Operational Research Society, 46,* 771–782.

4. Buzacott, J. A. (1975). Economic order quantity with inflation. *Operations Research Qtr., 26*(3), 553–558.

5. Chang, C.-T. (2004). An EOQ model with deteriorating items under inflation when supplier credits linked to order quantity. *International Journal of Production Economics, 88,* 307–316.

6. Covert, R. P., & Phillip, G. C. (1973). An EOQ model with weibull distribution deterioration. *AIIE Transactions, 5,* 323–326.

7. Ghare, P. M., & Schrader, S. K. (1963). A model for exponentially decaying inventory. *Journal of Industrial Engineering, 14*(5), 238–243.

8. Goyal, S. K., & Giri, B. C. (2001). Recent trends in modelling of deteriorating inventory. *European Journal of Operational Research, 134*(1), 1–16.

9. Hartley, V. R. (1976). *Operations Research—A Managerial Emphasis* (pp. 315–317). Santa Monica, California: Good Year.

10. Jaggi, C. K., Aggarwal, K. K., & Goel, S. K. (2006). Optimal order quantity for deteriorating items with inflation induced demand. *International Journal of Production Economics, 103*(2), 707–714.

11. Jaggi, C. K., & Khanna, A. (2010). Supply chain model for deteriorating items with stock-dependent consumption rate and shortages under inflation and permissible delay in payment. *International Journal of Mathematics in Operational Research, 2*(4), 491–514.

12. Jaggi, C. K., & Verma, P. (2010). Two-warehouse inventory model for deteriorating items with linear trend in demand and shortages under inflationary conditions. *International Journal of Procurement Management, 3*(1), 54–71.

13. Jaggi, C. K., Goel, S. K., & Mittal, M. (2011). Pricing and replenishment policies for imperfect quality deteriorating items under inflation and permissible delay in payments. *International Journal of Strategic Decision Sciences (IJSDS), 2*(2), 20–35.

14. Jaggi, C. K., Pareek, S., Khanna, A., Sharma, R., (2014). Ordering policy in a two- warehouse environment for deteriorating items under inflationary conditions. In analytical approches to strategic decision-making: interdisciplinary considerations (pp. 320–338). IGI Global.

15. Jaggi, C., Khanna, A., & Nidhi, N. (2016). Effects of inflation and time value of money on an inventory system with deteriorating items and partially backlogged shortages. *International Journal of Industrial Engineering Computations, 7*(2), 267–282.

16. Jaggi, C. K., Gautam, P., & Khanna, A. (2018). Credit policies for deteriorating imperfect quality items with exponentially increasing demand and partial backlogging. In *Handbook of Research on Promoting Business Process Improvement Through Inventory Control Techniques* (pp. 90–106). IGI Global.

17. Jaggi, C. K., Gautam, P., & Khanna, A. (2018). Inventory decisions for imperfect quality deteriorating items with exponential declining demand under trade credit and partially backlogged shortages. In *Quality, IT and Business Operations* (pp. 213–229). Springer, Singapore.

18. Jain, D., & Aggarwal, K. K. (2012). The effect of inflation-induced demand and trade credit on ordering policy of exponentially deteriorating and imperfect quality items. *International Transactions in Operational Research, 19*(6), 863–889.

19. Khanna, A., Mittal, M., Gautam, P., & Jaggi, C. (2016). Credit financing for deteriorating imperfect quality items with allowable shortages. *Decision Science Letters, 5*(1), 45–60.

20. Khanna, A., Gautam, P., & Jaggi, C. K. (2017). Inventory modeling for deteriorating imperfect quality items with selling price dependent demand and shortage backordering under credit financing. *International Journal of Mathematical, Engineering and Management Sciences, 2*(2), 110–124.

21. Kumar, S., & Rajput, U. S. (2016). A probabilistic inventory model for deteriorating items with ramp type demand rate under inflation. *American Journal of Operational Research, 6*(1), 16–31.

22. Mittal, M., Khanna, A., & Jaggi, C. K. (2017). Retailer's ordering policy for deteriorating imperfect quality items when demand and price are time-dependent under inflationary conditions and permissible delay in payments. *International Journal of Procurement Management, 10*(4), 461–494.
23. Rastogi, M., & Singh, S. R. (2018). A production inventory model for deteriorating products with selling price dependent consumption rate and shortages under inflationary environment. *International Journal of Procurement Management, 11*(1), 36–52.
24. Saha, S., & Sen, N. (2019). An inventory model for deteriorating items with time and price dependent demand and shortages under the effect of inflation. *International Journal of Mathematics in Operational Research, 14*(3), 377–388.
25. Singh, S., Dube, R., & Singh, S. R. (2011). Production model with Selling Price dependent demand and Partial Backlogging under inflation. *International Journal of Mathematical Modelling & Computations, 1*(1), 1–7.
26. Singh, S., Sharma, S., & Pundir, S. R. (2018). Two-Warehouse Inventory Model for Deteriorating Items with Time-Dependent Demand and Partial Backlogging Under Inflation. *International Journal of Mathematical Modelling and Computations, 8*(2), 73–88.
27. Thangam, A., & Uthayakumar, R. (2010). An inventory model for deteriorating items with inflation induced demand and exponential partial backorders—A discounted cash flow approach. *International Journal of Management Science and Engineering Management, 5*(3), 170–174.
28. Wee, H. M., & Law, S. T. (1999). Economic production lot size for deteriorating items taking account of time value of money. *Computers and Operations Research, 26*, 545–558.
29. Yadav, A. S., Tyagi, B., Sharma, S., & Swami, A. (2017). Effect of inflation on a two-warehouse inventory model for deteriorating items with time varying demand and shortages. *International Journal of Procurement Management, 10*(6), 761–775.

Appraising Impact of Re-Classification on Growth of Mutual Fund Industry in India

Mandakini Garg and Shobhit

1 Introduction

In the era of financial intermediation, investors are more inclined toward financial products rather than physical assets. In support to this, the announcement of demonetization in 2016 pushes its growth into height of success. Contemporary financial instruments such as shares, debentures, bonds, and mutual funds are available in financial market as compared to traditional products, like insurance, post office savings, bank deposits and so on. The most efficient and feasible investment avenue in today's era is mutual funds. It is a pool of funds collected by many investors to invest in financial market. This instrument is very useful in mitigating the risk factor of investors. It is most demanded by investors in the financial market as it possesses the quality of debt and/or equity. Some of these instruments even provide tax benefit under Sect. 80C of Income Tax Act 1961. The biggest motivating factor for investors to invest in mutual funds is that investors can exit and enter whenever they wish to.

As per the above chart, it is clearly visible about the positive inclination of investors toward mutual funds. Mutual funds possess the quality of all other financial instruments. Demonetization has given a kick to growth of mutual fund industry as RERA (Real Estate Regulation and Development Act 2016) and Benami Transaction Prohibition Amendment Act 2016 were imposed by Government of India.

With the help of above diagram, the clear difference can be figured out. From 2014 to 2018, the percentage contribution of mutual funds has risen significantly in every sector, that is, equity sector (8.5–18.4%), debt market (9.7–12.1%) and

M. Garg (✉)
Pranveer Institute of Technology, Kanpur, India
e-mail: Mandakini_garg@yahoo.co.in

Shobhit
Amity University, Lucknow, India
e-mail: drshobhitgoel@gmail.com

© The Author(s), under exclusive license to Springer Nature Singapore Pte Ltd. 2021
P. K. Kapur et al. (eds.), *Advances in Interdisciplinary Research in Engineering and Business Management*, Asset Analytics,
https://doi.org/10.1007/978-981-16-0037-1_5

in corporate debt market as per the RBI bulletin, 2017. This regulation of SEBI (Securities and Exchange Board of India) promotes a boost to the growth of mutual funds. In addition to this, SEBI with AMFI (Association of Mutual Fund in India) are very prominent in formulating new regulations on timely basis, even it injects more favorable conditions for investors and companies both. Few of the regulatory regulations are listed are as follows:

In 2018, re-classification and rationalization of open-ended schemes is carried out by SEBI in lieu to bring uniformity and standardize the attributes of mutual fund schemes. This regulatory regulation reduces the complexity of schemes. Now, open-ended schemes are categorized into six heads: equity, debt, hybrid, solution oriented, index funds, and exchange traded funds. SEBI, after re-classification stated that inclination of investors toward open-ended schemes has shifted in upward directions. The segregation has been carried out based on the below characteristics.

The parameters mentioned above highlights the classification approach of SEBI.

It can be newer concept in Indian context. But this concept was already in existence in global markets. Many countries have already adopted it after recession (financial crisis) in 2008. This approach gives investors little confidence to enter again in capital market through mutual funds. Europe was the first nation to execute this concept with the help of EFAMA (European Fund and Asset Management Association). After adopting this procedure in India, mutual fund industry has shown a dramatic increase in investment corpus.

It is a new concept in economy, so too much research is not conducted in India. Still, the researchers are in process to analyze the impact of re-classification and rationalization of open-ended mutual fund schemes. There is a huge gap between the approachability of re-classification and investor's inclination toward it. Still, loopholes are prevalent in the industry regarding its importance and promotion among the existing and new investors. The basic problem faced by companies is that they are not able to convince the investors regarding the benefits, and investors are not keen to know the basic attribute of these regulatory changes.

This research is concerned about the growth pattern of mutual fund industry after re-classification of open-ended schemes. The basic objective of pursuing this research is to showcase the significance of this change in mutual fund industry. This change directly and indirectly impacted the growth of mutual fund industry, so it becomes very challenging to judge about the various implication of this change in mutual fund industry. As a researcher, the focus area will be in two perspectives. First will be in view of industry through which the growth of assets under management (AUM) will be judged before and after application of this regulatory change. So, as a researcher laid focus on whether open-ended schemes are also moving in the same direction or opposite direction as compared to the growth of mutual fund industry. Second area of research will throw some light on investor's perceptive toward this desired change. As re-classification brings uniformity and clarity in mutual fund schemes, investors are more inclined toward mutual funds or not. Investor's attitude will be judged by their investment pattern after this regulatory act.

2   Research Methodology

This study will be descriptive and exploratory research. As re-classification is a recent phenomenon in mutual fund industry, its impact is not noticeable in terms of scheme-specific (open-ended schemes) and AUM of mutual fund industry.

The research objectives are as follows:

To study the growth pattern of asset under management before and after re-classification of open-ended schemes.

To analyze the perception of investors toward the re-classification of open-ended schemes.

As a researcher, its impact will be analyzed with the help of primary and secondary data. Primary data will be assembled through questionnaire. Respondents will be examined about their investment pattern after embarking of this regulatory change. Respondents will be new as well as existing investors. Investors will be questioned based on their expectation and understanding of the desired change in mutual fund industry. The pattern of growth of AUM will be analyzed with the available data on AMFI, SEBI and RBI (Reserve Bank of India). Data for the last four years was taken for analyzing the trend, that is, two years prior and post amendments. Sample unit will be respondents (investors) at various geographical locations to understand their investment pattern.

The hypotheses for the research are as follows:

Hypothesis 1

H0: There is no impact of re-classification on AUM of various AMC.

H1: There is impact of re-classification on AUM of various AMC.

Hypothesis 2

H0: There is no association between the investments pattern and re-classification of open-ended schemes.

H1: There is association between the investment pattern and re-classification of open-ended schemes.

To validate the data as per the hypothesis framed collected software like SPSS (Statistical Package for the Social Sciences).

The limitation of the study is time constraint. As time was limited, so sampling area was confined only to three places. As this concept was executed before two years, so lack of qualitative and quantitative data was available for further research.

3   Analysis and Interpretation

Analysis of collected data is carried out in two aspects:

In terms of asset under management of asset management companies:

This data was collected with the help of AMFI website. The data was collected to highlight the significance of growth pattern of top 10 assets management companies (AMC) in terms of assets under management (AUM). There is a continuous growth in AUM for many years because of the inclination of investors in mutual funds.

If we look toward the trend of growth of AUM, there is significant growth every year irrespective of the new amendments. As investors are more inclined toward investment in mutual funds because of many other factors due to which any changes hardly play any direct impact on its investments.

In terms of investors: Investors are inclined toward mutual funds as they have many advantages that are enjoyed by investors. Benefits like tax advantage, returns, regular income and safety of funds as compared to equity market and so on. Few investors are interested in re-classification of mutual funds schemes. Investors are interested in open-ended funds as they give investors more freedom to exit and enter from the funds. Investors are interested on doing the investment in open-ended schemes because of the flexibility feature. According to investors, re-classifications have given some clarity for open-ended schemes but it does not initiate any new investments.

As per the result, Hypothesis 1 is rejected as there is no impact of re-classification of AUM of various AMC. As per the result, Hypothesis 2 is rejected as there is no association between investment pattern and re-classification of open-ended schemes.

4   Conclusions

This research will be helpful both for investors and mutual fund companies for longer span of time. As the concept of re-classification is already a successful phenomenon in European nation since decades, now Indian mutual fund industry wishes to recipro-cate the same success rate in our country. In India, investors are most unpredictable in terms of choosing the investment alternatives, so it is very challenging for the companies to attract the investors toward their financial products. So, this type of regulatory change will boost the motivation of investors to invest in mutual funds. As this change is investor-friendly and all the companies are bound to apply it, investors cannot be manipulated. AMC will be adhering to showcase all the attributes of every open-ended scheme. For AMC, they will be getting new opportunities to develop the new schemes according to the norms of SEBI and AMFI. As they will be able to find the desire of investors for open-ended schemes, they can alter some characteristics of the schemes. AUM has been increased tremendously but re-classification was not the only one factor. Tax advantage under Sect. 80C of Income Tax Act has boosted its growth. Still, investors are only concerned about their returns, flexibility, conve-nience, approachability and so on. Various awareness programs should be organized for the investors by SEBI (Securities and Exchange board of India), AMFI (Asso-ciation of Mutual Funds of India) and various AMCs to highlight the advantages of re-classification to the investors. As it is a new concept, there is a huge scope for the researchers to judge about its impact on the investment pattern of investors.

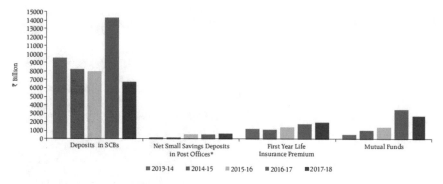

Fig. 1 *Source* RBI, IRDAI and SEBI, 2017

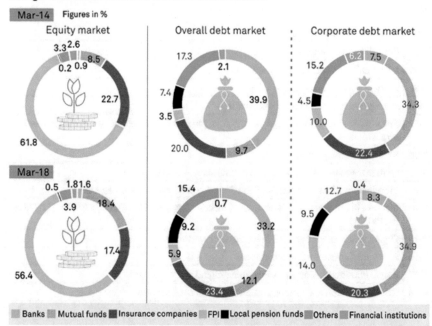

Fig. 2 Source: National Securities Depository Limited (NSDL), RBI, 2018

Fig. 3   Source: CRISIL
research, 2018

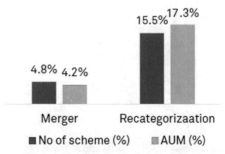

Mergers and recategorisation post the rule

Month-end AUM considered for the month of March 2018
Only open-ended schemes considered

Table 1   Source: primary source

Year	Regulatory regulations
1964	Introduction of UTI (Unit Trust of India)
1987	Entry of public sector undertakings
1993	Emergence of private sector undertakings
1997	Entry for foreign players solely or through, joint ventures, collaboration, mergers and acquisition
2009	Removal of entry load
2012	Cash investment was permitted in mutual fund industry
2013	Introduction of direct plans, uniform dividend policy for debt holders etc
2014	Increase in amount under Sect. 80C of Income Tax Act 1961, that is, from Rs.100,000 to Rs.150,000
2015	Digital platform initiated for investors for investment
2016	Rules are flexible for foreign players
2017	Rs. 50,000 investment is allowed under digital wallet in a year
2018	Re-classification and rationalization of mutual fund schemes

Table 2   Source: SEBI, CRISIL Research2018

Classification criteria	Equity	Debt	Hybrid	Solution oriented	Others
Market capitalization	Yes				
Investment strategy	Yes		Yes		
Specific objective	Yes	Yes			
Macaulay duration		Yes			
Credit quality		Yes			
Asset allocation			Yes		
Specific investors				Yes	
Index/ETF					Yes
Fund of funds					Yes

Table 3   Source: Primary data

	AUM (in Lakhs)			
	2018	2017	2016	2015
HDFC MUTUAL FUND	129,139,041	111,357,956.3	86,609,267.41	69,122,476.38
ICICI PRUDIENTIAL FUND	125,068,970	113,974,254.1	88,079,186.78	66,851,693.04
SBI MF	103,641,213.2	78,142,155.17	55,151,263.88	38,136,165.29
ADITYA BIRLA SUN LIFE MUTUAL FUND	99,321,492.65	92,012,684.7	69,497,318.43	53,312,953.41
RELIANCE MUTUAL FUND	95,772,386.32	94,580,968.7	76,053,697.68	61,687,623.68
UTI MUTUAL FUND	63,640,953.8	60,431,293.53	50,570,122.48	40,924,588.09
KOTAK MAHINDRA MUTUAL FUND	55,224,596.24	45,712,892.65	3,085,698.15	21,895,229.48
FRANKLIN TEMPLETON MUTUAL FUND	44,726,049.56	39,221,562.91	30,230,675.88	29,318,252.57
AXIS MUTUAL FUND	33,836,854.45	28,360,613.71	19,531,590.74	13,265,043.41
DSP BLACK ROCK	34,246,805.43	32,420,925.39	21,786,679.25	15,502,218.9

References

1. Ibef. (2019, January). Retrieved July 2019, from, www.ibef.org: https://www.ibef.org/dow nload/financial-services-jan-2019.pdf.
2. Company, U. M. (2018). Categorization and Rationalization of UTI Mutual Fund Schemes and Merger of UTI Mutual Fund Schemes- Hybrid Schemes.
3. Shri Anand Prakash, S. S. (2018, October 89–103). Recent Developments in India's Mutual Fund Industry. RBI Bulletin.

4. Vidyadharan, J. (2018, August). Digital evolution: Can technology be the propelling factor for the industry? Retrieved July 2019, from, www.crisil.com: https://www.google.com/search?q=crisil-amfi-mutual-fund-report-digital-evolution%2520&oq=crisil-amfi-mutual-fund-report-digital-evolution%2520&aqs=chrome..69i57.268j0j7&sourceid=chrome&ie=UTF-8.
5. (2018, March). Retrieved June 2019, from, www.utimf.com: https://www.utimf.com/.../2017.../54-17-18-categorisation-and-rationalisation-of-hybrid.
6. (2018, April). Retrieved July 2019, from, www.tatamutualfund.com: www.tatamutualfund.com/.../tmf-schemes-categorization-summary-april-2018.pdf?
7. India, S. A. (2017). Categorization and Rationalization of Mutual Fund Schemes.
8. Jiju Vidyadharan, P. G. (2017, June). Retrieved July 2019, from, www.crisil.com: https://www.crisil.com/content/dam/crisil/.../Quantum-leap-beckons-June05-2017.pdf.
9. PWC. (2017). Categorisation of mutual fund schemes.
10. Chandra, A. P. (2017, November). Retrieved June 2019, from, www.pwc.in: https://www.pwc.in/assets/pdfs/financial.../categorisation-of-mutual-fund-schemes.pdf.
11. GUJJAR, R. (2017, October). Retrieved July 2019, from, www.amfiindia.com: https://www.amfiindia.com/Themes/Theme1/downloads/1507291273374.pdf.

Lean Implementation Within Construction SMEs: A Review

U. C. Jha

1 Introduction

Lean philosophy has become very vital for the survival of organizations. The construction industry performs poorly in India, faced with several issues, including contract management, through lengthy and complicated payment processes, and payments overdue for project execution. Large numbers of small and medium-sized businesses dominate the construction sector (SMEs) in India and are characterized by high attrition rates employment [7]. It is also estimated that SMEs will contribute about 70 percent to India's GDP and approximately 92 percent of businesses in India [3]. Therefore, the contribution of this sector to the economy of India cannot be overemphasized.

Any method that seeks to make an enterprise Lean always reduces waste and maximizes value in the company, thus enhancing core resources and establishing a corporate culture devoted to the recognition and continuous growth promoting customer satisfaction [9].

2 Research Methodology

The paper was developed on the basis of a study of already published empirical and theoretical studies. Past research on Lean construction and SMEs was obtained primarily from research databases, including Research Gate, IGLC Science Direct (Elsevier), Emerald Insight, Taylor and Francis, Google Scholar and other internet

U. C. Jha (✉)
Lovely Professional University, Phagwara, Punjab, India
e-mail: udai.22511@lpu.co.in

© The Author(s), under exclusive license to Springer Nature Singapore Pte Ltd. 2021
P. K. Kapur et al. (eds.), *Advances in Interdisciplinary Research in Engineering and Business Management*, Asset Analytics,
https://doi.org/10.1007/978-981-16-0037-1_6

sources. The original search phrases used were "Lean construction tools," "construction SMEs," "Lean and construction SMEs. The initial descriptors were used to search the databases. A total of 75 articles were reviewed for the research. This paper is based on four hypotheses, which are related to the construction SMEs capacity to implement Lean as efficiently as large enterprises [8]. The hypotheses are: Does the size of the firm matter in Lean adoption and implementation?; Can SMEs apply the Lean package partially in order to reap its benefits?; Can SMEs still benefit from Lean considering their organizational structure, culture, financial and human resource capacities? and Is the organizational culture of Indian construction SMEs supportive of Lean construction? The paper sought to answer these questions and continues to suggest Lean tools that can be implemented by the construction SMEs in India.

2.1 Need for Lean Implementation in Construction SMEs

There is a need for the construction SMEs to be more aware of Lean and be able to inspire their employees, clients and partners in order to attain greater joint performance [27]. This, the authors believe, can be achieved through carrying out Lean construction principles in the Indian construction SMEs [14]. Studies conducted within the Indian construction industry suggest a low level of familiarity and application of Lean construction among practitioners within the industry [6]).

Rose et al. [35] pointed out that there is currently no standard measure for Lean implementation that SMEs can adopt, as different researchers take varying perspectives, but they do show that SMEs should go for the least costly practices, such as 5S and visual management (VM) [10]. SMEs have challenges with their processes or production, thus making the implementation of Lean in construction worth exploring.

2.2 Lean Applicability Within Construction SMEs

This paper is based on four hypotheses, which are related to the construction SMEs capacity to implement Lean as efficiently as large enterprises.

2.3 Does the Size of the Firm Matter in Lean Adoption and Implementation?

There is still a lot of interest as to if the applicability of Lean does vary between large enterprises (LE) and SMEs (e.g. [35]). The size of a firm continues to be discussed whether it is a critical factor in Lean implementation. Many authors have indicated that Lean is best suited to large enterprises than SMEs (e.g. [2, 34, 38, 45]). The researchers argue that there is difficulty for SMEs to cope with Lean due to their lack of required resources and capabilities. However, other researchers disagree to say that size does not affect the capacity of an organization to implement Lean, and that SMEs can such structures be applied as well as large organizations. This is in line with Rose et al. [35] who suggested that SMEs can implement Lean but should go in for the least costly practices [22]. It is the preliminary proposition of the authors that the size of the firm does not really matter. Small firms by virtue of their size can implement some Lean tools to their own benefit but need to go in for tools that requires less investment [28].

2.4 Can SMEs Apply the Lean Package Partially in Order to Reap Its Benefits?

Although some researchers have suggested that Lean practices should be implemented as a full package, Golicic and Medland argued otherwise [24]. The researchers believed that Lean can be applied partially. The application of some Lean tools will lead to gradual performance improvement of SMEs, which can then lead to more advanced practices [35].

2.5 Can SMEs Still Benefit from Lean Considering Their Organizational Structure, Culture, Financial and Human Resource Capacities?

SMEs can implement some of the tools such as the last planner and 5S, to enhance their performance and productivity. The SMEs can do this without having to invest huge sums of monies [25]. As the SMEs grow, they can implement other tools which require substantial investments such as building information management (BIM).

Lee and Oakes [23] showed that Lean is applicable to SMEs as they have the ability to build the required culture in terms of leadership and workforce involvement much more easily than large companies, since they have relatively little functional differentiation, which makes the management very close to workers. Antony and Kumar shared the same view, suggesting that the organizational culture in SMEs is full of energy and that they are willing to "learn and change" rather than "control"; the SME culture is friendlier, since relationships between workers and top management are loose and informal [30].

Is the organizational culture within Indian construction SMEs supportive of Lean construction?

Evidence from literature suggests that an organization cannot succeed in Lean unless it has a healthy culture. In the UK, only 10 percent of firms succeed in their

Lean implementation efforts [29]. The reason behind the low success rate is culture and management [41]. Culture is a vital factor for a successful Lean implementations. Dahlgaard and Dahlgaard-Park argued that appropriate culture cannot be compromised if a company wants to adopt Lean successfully. According to [5], small firms in India have a dominant hierarchy culture focused on the internal structures [33]. The relationship between contractors and suppliers within the industry is also short-term and only based on the needs of current projects. Lean construction requires a culture of employee empowerment, teamwork and enhanced relationship with employees and suppliers [44]. Companies that have successfully implemented Lean have argued that it would not have been possible without a sustained employee and supplier engagement and support at all levels of the organization. There will be a need for culture change in the Indian construction industry to accommodate Lean initiatives [37].

3 Lean Practices

Table 1 identifies Lean practices applicable to construction, with a brief description of the tools and references based on the literature. The classification was mostly based on analyzing the papers published in IGLC proceedings, although other studies published elsewhere were also consulted.

Table 1. Identified Lean practices applicable to construction, description of the tools and references

Based on the Lean construction tools identified within the literature in Table 1, the authors classified these tools into three groups, namely

- Tools that require less monetary investment to implement.
- Tools that can be fully implemented by construction SMEs.
- Tools that can be partially implemented by SMEs.

The categorization has been done for purposes of suggesting Lean construction tools for use by SMEs based on their peculiar characteristics. Table 2 is based on the categorization as indicated above.

Table 2. Lean tools

Researchers such as Rose et al. [35] and [36] have affirmed tools such as 5S, Kanban SMED, Kaizen, Increased Visualization, Last Planner, Daily Huddle Meetings, First Run Studies, Poka-Yoke, Andon that are least costly and can be implemented by SMEs. A review of the literature on the other Lean construction tools (5 Why's, Concurrent Engineering, Choosing By Advantages, A3 report, Ishikawa Diagram, LBMS, Heijunka, VSM, Action Learning) shows these are not capital-intensive and therefore within the reach of SMEs. The authors will subsequently validate these other Lean construction tools. A major issue that acts as a challenge to SMEs is finance. Considering the poverty levels in India, the authors suggest the Lean tools that require less monetary investments to be implemented by construction SMEs as can be seen in Table 2. These SMEs can implement without having to invest

Table 1 Identified lean practices

Lean tools	Brief description	References
Last planner system	The Last Planner™ System (LPS™) provides a regimented process of achieving reliable workflow on simple and complex construction projects. This system was created in order to improve the predictability and reliability of construction production	Salem et al. [36]
Increased visualization	It is about communicating key information effectively to the workforce through posting various signs and labels around the construction site. This includes signs related to safety, schedule and quality	Salem et al. [36]
Daily huddle meetings	Daily huddles are for communication, not only for managers to talk to employees but also for employees to express themselves and learn from each other	Salem et al. [36]
5S process	5S is a basic method for clean-up and organization of the workplace	Hafey [18], Tezel et al. [42]
5 Why's	It works by asking once why an effect happened, and to the response of that question, ask again, why it happened. Same procedure is repeated until asking five times why it happened and by the end of the process, the answer is the root cause	Fuenzalida et al. [16]
Concurrent engineering	A simultaneous engineering that attempts to optimize the design of a project and its construction process by the integration of design, fabrication, construction and erection activities	

(continued)

Table 1 (continued)

Lean tools	Brief description	References
Choosing by advantages (CBA)	CBA is a value-based multi-criteria decision-analysis system that supports sound decision-making based on the comparisons among the advantages of alternatives	Suhr [40],Arroyo et al. [4]; Parrish and Tommelein [31], Kpamma et al. [20]
Building information modelling (BIM)	A virtual process that encompasses all aspects, disciplines, and systems of a facility within a single, virtual model. Some new concepts and BIM applications have been developed for different purposes in the construction industry, such as 4D, 5D, 6D and 7D dimensions	Lee [21], Harris and AlvesHarris[19], Dave et al. [11], Abou-Ibrahim and Hamzeh [1]
Kaizen	Kaizen is a Japanese word for improvement. This Lean construction tool involves looking at some task in the field and finding out how to do it better, more efficiently, safer and quicker.	Pasquire and Connolly [32]
Poka-Yoke	In building a culture of stopping to fix problems, poka-yoke is one of the Lean tools that help the employees to detect the defects and halt the process. It is synonymous with fail safe for quality and safety	Shang [39], Tommelein [43]
A3 report	An A3 is an orderly document that aids thinking. A3 reports are so named because they fit on one side of an A3 size paper. The A3 report is a way of representing an action course, in which goals, methodology, agents involved and others are included. The document is for problem solving, proposing action or project status reporting	Fuenzalida et al. [16], Gupta et al. [17]

(continued)

Table 1 (continued)

Lean tools	Brief description	References
The Ishikawa diagram	The Ishikawa diagram is a representation of a cause–effect analysis that is carried out for any type of result	Fuenzalida et al. [16]
Location based management system (LBMS)	The Location-Based Management System (LBMS) provides a much-needed spatial element to planning	Kenley and Seppänen (2010), Dave et al. [12], Frandson et al. [15]
Andon	A visual control tool which shows the operation status and signalize the occurrence of abnormalities	
Heijunka	Leveling the work flow of a production system and balancing or distributing load and capacity	
Value stream mapping	Systemic view of the production process (of the value flow), identification of real problems and wastes and proposition of improvements	Murguia et al. [26]
Material Kanban cards	It is used as a material process flow technique for wastes and proposition of improvements. the pull replenishment logic system	
Six minute exchange die (SMED)	SMED practices in project management can be seen as a method for fast tracking the project schedule	
Action learning	The core idea behind Action Learning is to create small, mutually supportive groups (known as SETs) of people who band together to solve real problems or difficulties which are not solved in current best practice	Davey et al. [13]

so much and as they grow they can implement tools such as BIM which requires some investments in software and hardware. Construction SMEs can also implement soft aspects of Lean construction such as employee involvement, teamwork, reward and recognition, and communication, which will not require huge monetary investment. In the same vein SMEs can implement some aspects of BIM such as 3D, which does not require substantial investments. SMEs can still benefit from fragmented or

Table 2 El-Nino related variables

Criteria lean tools	A	B	C
Last planner system	×	×	
Increased visualization	×	×	
Daily huddle meetings	×	×	
First run studies	×	×	
5S process	×	×	
5 Why's	×	×	
Concurrent engineering	×	×	
Choosing by advantages (CBA)	×	×	
Building information modelling (BIM)			×
Kaizen	×	×	
Poka-Yoke	×	×	
A3 Report	×	×	
The Ishikawa diagram	×	×	
Location-based management system (LBMS)	×	×	
Andon	×	×	
Heijunka	×	×	
Value stream mapping (VSM)	×	×	
Material Kanban cards	×	×	
Six minute exchange die (SMED)	×	×	
Action learning	×	×	

isolated use of Lean tools, but the benefits will not come close to the implementation of the full system. Since SMEs lack capacity to fully adopt Lean construction, there is a clear need to build on their capacity to fully adopt Lean construction philosophy.

3.1 Strategies: Adopting Lean Practices Within Construction SMEs in India

A change in mind-set is required before embarking on the implementation.

Professional bodies such as the Indian Institute of Architects (IIA), the Indian Institution of Surveyors (IIS) and the Institution of Engineers (IE) should expose their members to the concept of Lean thinking through their continuing professional development programs. There could also be collaboration with leading institutional proponents of Lean construction, like the Lean Construction Institute (LCI), to offer special training for Indian contractors on the strategies of applying the Lean thinking concept within the industry. Furthermore, there should be a deliberate government policy to implement Lean, especially within public sector construction works. There is the need for research and teaching to be reinforced into the formal and vocational training of students pursuing similar professions in building.

4 Conclusions

The global economy is changing and becoming much more competitive. The benefits of implementing Lean are substantial while the cost of not being able to meet project goals may be very significant. The primary objective of this paper was to propose Lean construction tools that can be implemented within construction SMEs setup. This is against the backdrop of SMEs lack of the needed resources to implement Lean construction. Through a review of literature, this paper has proposed Lean construction tools that can be implemented within the construction SMEs setup. The study found that SMEs can implement some soft and hard aspects of Lean such as the last planner, 5S, and teamwork, which will not require having to invest huge sum of monies. This is an important finding as some authors have argued that SMEs do not have the capacity to implement Lean construction. This is a preliminary work on an ongoing PhD which aims to develop a Lean implementation framework to enhance the performance of construction SMEs. The findings in this study will be validated through a nationwide survey, case studies and interviews in India.

References

1. Abou-Ibrahim, H., & Hamzeh, F. (2016). BIM: a TFV perspective to manage design using the LOD concept. In Proc. 24th Ann. Conf. of the Int'l. Group for Lean Construction, Boston, MA, USA, Sect. 4 (pp. 3–12).
2. Achanga, P., Shehab, E., Roy, R., & Nelder, G. (2006). Critical success factors for Lean implementation within SMEs. *Journal of Manufacturing Technology Management, 17*(4), 460–471.
3. Ackah, J., & Vuvor, S. (2011). The challenges faced by small & medium enterprises (SMEs) in obtaining Credit in Ghana. A thesis submitted to Blekinge Teknista Hogskola for the award of Master's Degree.
4. Arroyo, P., Tommelein, I., & Ballard, G. (2013). Using choosing by advantages to select ceiling tile from a global sustainable perspective. In Proc. 21st Ann. Conf. Int'l. Group for Lean Construction (IGLC), Fortaleza, Brazil.
5. Atuahene, B. T. (2016). Organizational culture in the Ghanaian Construction Industry, A thesis submitted to the Department of Building Technology, College of Art and Built Environment, KNUST in partial fulfilment of the requirements for the degree of Master of Philosophy
6. Ayarkwa, J., Agyekum, K., Adinyira, E., & Osei-Asibey, D. (2012). Perspectives for the implementation of lean construction in the Ghanaian Construction Industry. *Journal of Construction Project Management and Innovation, 2*(2), 345–359.
7. Bauchet, J., & Morduch, J. (2013). Is micro too small? Microcredit vs. SME finance. *World Development, 43*(C), 288–297.
8. Berroir, F., Harbouche, L., & Boton, C. (2015). Top down vs. Bottom up approaches regarding the implementation of lean construction through a French case study. In Proc. 23rd Ann. Conf. of the Int'l. Group for Lean Construction. Perth, Australia, July 29–31 (pp. 73–82).

9. Bhamu, J., & Sangwan, K. S. (2014). Lean manufacturing: literature review and research issues. *International Journal of Operations & Production Management, 34*(7), 876–940.

10. Bhuiyan, N., & Baghel, A. (2005). An overview of continuous improvement: from the past to the present. *Management Decision, 43*(5), 761–771.

11. Dave, B., Kubler, S., Pikas, E., Holmström, J., Singh, V., Främling, K., Koskela, L., & Peltokorpi, A. (2015). Intelligent products: shifting the production control logic in construction (with Lean and BIM). In Proc. 23rd Ann. Conf. of the Int'l. Group for Lean Construction, Perth, Australia, July 29–31 (pp. 341–350).

12. Dave, B., Seppänen, O., & Modrich, R. (2016). Modelling information flows between last planner and location based management system. In Proc. 24th Ann. Conf. of the Int'l. Group for Lean Construction, Boston, MA, USA, Sect. 6 (pp. 63–72).

13. Davey, C. L., Powell, J. A., Cooper, I., & Hirota, E. (2000). Innovation and culture change within a medium-sized construction company: success through the process of action learning. In: Proc. IGLC-8, 8th Conf. of Int. Group for Lean Construction, Brigthon, UK.

14. Dowlatshahi, S., & Taham, F. (2009). The development of a conceptual framework for just-in-time implementation in SMEs. *Production Planning and Control, 20*(7), 611–621.

15. Frandson, A. G., Seppänen, O., & Tommelein, I. D. (2015). Comparison between locations based management and Takt Time Planning. In Proc. 23rd Ann. Conf. of the Int'l. Group for Lean Construction, 28–31 July, Perth, Australia (pp. 3–12).

16. Fuenzalida, C. Fischer, B., Arroyo, P., & Salvarierra, J. L. (2016). Evaluating environmental impacts of construction operation before and after the implementation of lean tools. In: Proc. 24th Ann. Conf. of the Int'l. Group for Lean Construction, Boston, USA, Sect. 10 (pp. 3–12).

17. Gupta, A. P., Tommelein I. D., & Blume, K. (2009). Framework for using A3s to develop shared understanding on projects people, culture and change. In Proc. 17th Annual Conference of the International Group for Lean Construction (IGLC 17), 15–17 July, Taipei, Taiwan.

18. Hafey, B. (2010). Lean safety. Transforming your safety culture with lean management. CRC Press, Taylor & Francis Group (p. 163).

19. Harris, B. N., & Alves, T. C. L. (2016). Building information modeling: a report from the field. In Proc. 24th Ann. Conf. of the Int'l. Group for Lean Construction, Boston, MA, USA, Sect. 5 (pp. 13–22).

20. Kpamma, Z. E., Adjei-Kumi, T., Ayarkwa, J., & Adinyira, E. (2014). Enhancing user-involvement through a multi-criteria decision aid: a lean design research agenda. In Proc. 22nd annual conference of the International Group for Lean Construction (IGLC 22), Oslo, Norway.

21. Lee, C. (2008). BIM: Changing the AEC industry. In PMI global congress. Denver, Colorado, USA: Project Management Institute.

22. Lee, C. Y. (1996). The applicability of just-in-time manufacturing to small manufacturing firms: an analysis. *International Journal of Management, 13*(2), 249–258.

23. Lee, G. L., & Oakes, I. (1995). The 'Pros' and 'Cons' of total quality management for Smaller-Firms in manufacturing: some experiences down the supply chain. *Total Quality Management, 6*(4), 413–426.

24. Mazanai, M. (2012). Impact of Just-In-Time (JIT) inventory system on efficiency, quality and flexibility among manufacturing sector, Small and Medium Enterprise (SMEs) in South Africa. *African Journal of Business Management, 6*(17), 5786–5791.

25. Miller, C., Packham, G., & Thomas, B. (2002). 'Harmonization between main contractors and subcontractors: A prerequisite for lean construction?' Journal of Construction Research, 3(1): 67–82.

26. Murguia, D., Brioso, X., & Pimentel, A. (2016). Applying lean techniques to improve performance in the finishing phase of a residential building. In Proc. 24th Ann. Conf. of the Int'l. Group for Lean Construction, Boston, USA, Sect. 2 (pp. 43–52).

27. Ofori, G., & Toor, S. R. (2012). Leadership development for construction SMEs.' In Working Paper Proceedings, Engineering Project Organizations Conference, Rheden, Netherlands, July 10–12, 2012.

28. Ormsby, J. G., McDaniel, S. W., & Gresham, A. B. (1994). Behavioural considerations for small businesses and JIT. *Journal of Business and Entrepreneurship, 6*(1), 51–58.
29. Oviatt, B., & McDougall, P. (1994). Toward a theory of international new venture. *Journal of International Business Studies, 25*(1), 45–64.
30. Panizzolo, R., Garengo, P., Sharma, M., & Gore, A. (2012). Lean manufacturing in developing countries: evidence from Indian SMEs. *Production Planning & Control, 23*(10–11), 769–788.
31. Parrish, K., & Tommelein, I. (2009). Making design decisions using choosing by advantages. In Proc. 17th Annual Conference of the International Group for Lean Construction (IGLC 17), 15–17 July, Taipei, Taiwan (pp. 501–510).
32. Pasquire, C. L., & Connolly, G. E. (2002). Leaner construction through off-site manufacturing. In Proc. for the 10th Annual Conference of the International Group for Lean Construction, Gramado, Brazil.
33. Pérez, C. T., Fernandes, L. L. A., & Costa, D. B. (2016). A Literature review on 4-D BIM for logistics operations and workspace management. In: Proc. 24th Ann. Conf. of the Int'l. Group for Lean Construction, Boston, USA, Sect. 8 (pp. 53–62).
34. Powell, D., & Gran, E. (2012). Applying a BITESIZE lean change methodology to intermodal terminal operations. In Proc. of the POMS 23rd Annual Conference, Chicago, USA.
35. Rose, A. N. M., Deros, B. M., & Abdul-Rahman, M. N. (2013). Lean manufacturing perceptions and actual practice among Malaysian SMES in automotive industry. *International Journal of Automotive and Mechanical Engineering, 7,* 820–829.
36. Salem, O., Genaidy, J. S. A., & Luegring, M. (2005). Site implementation and assessment of lean construction techniques. *Lean Construction Journal, 2*(2), 1–21.
37. Salem, O., Solomon, J., Genaidy, A., & Minkarah, I. (2006). Lean construction: from theory to implementation. *Journal of Management in Engineering, 22*(4), 168–75.
38. Shah, R., & Ward, P. T. (2003). Lean manufacturing: context, practice bundles, and performance. *Journal of Operations Management, 21*(2), 129–149.
39. Shang, G. (2013). The Toyota way model: an implementation framework for large Chinese construction firms. In A Thesis Submitted for the Degree of Doctor of Philosophy at the Department of Building National University of Singapore.
40. Sowards, D. (2004). '5S's that would make any CEO Happy.' Contractor Magazine, May 2004. Suhr, J. (1999). 'The Choosing By Advantages Decision Making System'. Quorum, Wesport, pp. 293.
41. Taleghani, M. (2010). Success and failure issues to lead lean manufacturing implementation. World Academy of Science, Engineering and Technology, 615–618
42. Tezel, A., Aziz, Z., Koskela, L., Tzortzopoulos, P. (2016). Benefits of visual management in the transportation sector. In Proc. 24th Ann. Conf. of the Int'l. Group for Lean Construction, Boston, USA, Sect. 6 (pp. 123–132).
43. Tommelein, I. D., (2008). 'Poka Yoke' or quality by mistake proofing design and construction systems. In Proc. for the 16th Annual Conference of the International Group for Lea Construction, Manchester, UK.
44. Womack, J. P., Jones, D. T., & Roos, D. (1990). *The machine that changed the world: the story of lean production.* New York: Harper Perennial.
45. Wong, Y. C., Wong, K. Y., & Ali, A. (2009). A study on lean manufacturing implementation in the Malaysian electrical and electronics industry. *European Journal of Scientific Research, 38*(4), 521–535.

Reliability Analysis of a System Using Universal Generating Function

Nupur Goyal, Subhi Tyagi, and Mangey Ram

1 Introduction

Reliability is one of the essential parameters for the analysis of the system's performance. Engineering systems generally have two states, that is, working and failed. The systems having two states are called binary systems. Reliability with these types of systems with complex configuration can be evaluated with the help of universal generating function (UGF) technique. This is an effective technique to estimate the reliability of any system having a very complex configuration as well. With UGF, one can also calculate the reliability of the system that has more than two states, which means the system having multiple states. These types of systems are known as multi-state systems (MSS). Many researchers had done work in this field [18, 19].

In the context of UGF technique, Levitin [1] had generalized a linear consecutive *k*-out-of-*r*-from-*n*: *F* system to the multi-state case. Also, the proposed system had multiple failure criteria. The system had multiple states from perfectly working to completely failed. The author had applied the UGF technique for the given system and proposed an algorithm for MSS based on the extension of UGF. Levitin [2] had extended the UGF technique for the MSS where the element's performance distribution also sometimes depends on another element's state or the state of the group of elements. Li and Zio [3] for the reliability assessment of the distributed generation had proposed an approach of analytical multi-state modeling. The UGF technique was applied to the multi-state system's modeling. Ebrahimipour et al. [4]

N. Goyal · S. Tyagi
Department of Mathematics, Graphic Era Deemed to Be University, Dehradun, India

M. Ram (✉)
Department of Mathematics, Computer Science & Engineering, Graphic Era Deemed to be University, Dehradun, India
e-mail: drmrswami@yahoo.com

© The Author(s), under exclusive license to Springer Nature Singapore Pte Ltd. 2021
P. K. Kapur et al. (eds.), *Advances in Interdisciplinary Research in Engineering and Business Management*, Asset Analytics,
https://doi.org/10.1007/978-981-16-0037-1_7

75

had considered a multi-state series-parallel model having subsystems in k-out-of-n type configuration. The authors had solved the model with the UGF technique to allocate the redundancy of the subsystems and to minimize the cost. Destercke and Sallak [5] had discussed and gave the extension of the UGF technique for a multi-state system considering epistemic uncertainties. The given method was used to measure the reliability of the system even if the model had ill-known probabilities and transition rates. Instead of using straightforward and Markov process to access the reliability of the complex multi-states systems, Lisnianski [6] had used the L_z-transform method by using the universal generating operator. This technique is easy to apply and also other indices can be evaluated such as expected performance and availability. For evaluating reliability function, MTTF, and so on, L_z-transform was used. Zhao et al. [7] had proposed a new approach to evaluate the reliability of real engineering systems with the concept of balanced systems. The authors had used the finite Markov-chain embedding approach and UGF technique to evaluate the reliability of such a system and also presented various case studies to demonstrate the new model.

Dhillon and Fashandi [25] presented an overview of most suitable robot safety and various reliability methods, for example, failure mode and effects analysis, fault tree analysis, and Markovian analysis. Manglik and Ram [24] studied the concept of a repairable system, which can fail completely due to the failure of its subsystems and failure of its last components and both the cold standby units connected in parallel with that component. Nagiya and Ram [26] investigated the various reliability characteristics of a satellite communication system which includes the earth station and terrestrial system. Kumar et al. [27] analyzed the reliability of the casting process and obtained the generalized expression using the Markov process. Goyal et al. (2015) proposed a mathematical model of the water-cooling system and simulated the results using the Markov process. They investigated the reliability measures water-cooling system and also seek the cost–benefit analysis in the maintenance of it.

In the context of signature reliability, Samaniego et al. [23] had extended the work in system signature reliability for the system having dynamic reliability settings. The authors also studied two more applications of dynamic signatures in detail. Navarro et al. [21] had extended the work in signature and gave some results for the system having heterogeneous components. The authors gave results for two considered cases, first for components with independent lifetimes and second for components having some specific lifetime dependence. Eryilmaz (2012) had discussed the reliability properties of the m-consecutive-k-out-of-n:F system with overlapping runs. The author had studied the reliability properties of the system via system signature and also evaluate various characteristics of the discussed system. Eryilmaz [9] had shown the usefulness of the system signature for repairable systems. For the computation of reliability indices, the system signature was evaluated for consecutive-k-within-m-out-of-n:F and m-consecutive-k-out-of-n:F systems. The author had also calculated stationary availability, MTTF and rate of occurrence of failure for the systems. Da and Hu [10] defined bivariate signature in sense of order statistics and proposed the formula of bivariate signature of the dual for the computation of a three-state system. The authors had also presented a useful method to evaluate the tail signature of the system. Marichal and Mathonet [12] presented an effective method to compute the signature of the system by the diagonal section of the reliability function via derivatives. Marichal [20] gave a general multilinear expression of a structure-function for the arbitrary semi-coherent system. The expression was given in terms of minimal path and cut sets. Huang et al. [22] had simplified the optimization model of the reliability redundancy allocation problem with the help of survival signature theory.

2 Algorithm for Computing the Signature Using Reliability Function

First, by using Boland [11] the signature of the designed system is obtained using the reliability function of the system.

$$B_a = \frac{1}{\binom{s}{s-a+1}} \sum_{\substack{k \subseteq [s] \\ |k|=s-a+1}} \phi(K) - \frac{1}{\binom{s}{s-1}} \sum_{\substack{K \subseteq [s] \\ |K|=s-1}} \phi(K). \tag{1}$$

The computed reliability polynomial for the system using $K(P) =$

$$\sum_{j=1}^{s} e_j \binom{s}{j} P^j q^{n-j},$$

where $e_i = \sum_{i-s-j+1}^{s} w_i$, $j = 1, 2, \ldots s$.

Then, for the next step, the tail signature is to be estimated, that is, $(p+1)$-tuple $W = (W_0, \ldots, W_p)$ by the formula given below

$$W_a = \sum_{i=a+1}^{s} w_i = \frac{1}{\binom{s}{s-a}} \sum_{|H|=s-a} \phi(K). \tag{2}$$

Then by using Taylor evolution at $w = 1$, the reliability function of the system is to be calculated in the form of the polynomial with the help of the given formula

$$P(w) = w^s K\left(\frac{1}{w}\right). \tag{3}$$

Now, by using Marichal and Mathonet [12] and the Eq. [2] the tail signature of the considered system is to be calculated as follows

$$W_a = \frac{(s-1)!}{s!} d^a P(1), \quad a = 0, 1, \ldots, s. \tag{4}$$

Then, the signature is to be estimated with the help of the tail signature as above

$$w = W_{a-1} - W_a, \quad a = 1, 2, \ldots s. \tag{5}$$

3   The Algorithm to Assess the Expected Lifetime of the System by Using Minimum Signature

For calculating the expected lifetime for the system, first, we have to determine the MTTF of the independent and identically distributed (i.i.d) elements of the bridge structure who have exponentially distributed elements with mean $(\mu = 1)$.

Then by using Navarro and Rubio [13], the estimation of $E(T)$ for i.i.d. elements are being done with the help of the following formula

$$E(T) = \mu \sum_{i=1}^{n} \frac{e_i}{i}. \tag{6}$$

where $e = (e_1, e_2, \ldots, e_n)$ is a vector coefficient. We obtain the estimation with the help of minimal signature.

4   Algorithm for Obtaining the Barlow-Proschan Index for the System

For computing the Barlow-Proschan index of the bridge structure, the reliability function is considered. The elements of the system are i.i.d. By using the formula given by Shapley [16] and Owen [14, 15], the Barlow-Proschan index is calculated as follows:

$$I_{BP}^{(a)} = \int_0^1 (\partial_a K)(w)dw, \quad a = 1, 2, \ldots, n. \tag{7}$$

where K is the reliability function of the system.

5 Algorithm to Determine the Expected Value of the System (Eryilmaz, 2012)

The expected value of the system is evaluated as follows:

$$E(X) = \sum_{i=1}^{n} i w_i, \quad i = 1, 2, \ldots, n. \tag{8}$$

At last, two expressions $E(X)$ and $E(X)/E(T)$ of the system are calculated.

6 Example

In Fig. 1, there are three subsystems A, B and C in which subsystems A and B are connected in parallel. Subsystem B is connected in series with A and B subsystem. In subsystem A, two components are connected in series.

The UGF of all the components can be defined as

$$u_i(z) = p_i z^1 + (1 - p_i) z^0; \quad i = 1, 2, 3, 4.$$

Now A can be calculated as:

$$u_5(z) = \min\{u_1(z), u_2(z)\}$$

$$= p_1 p_2 z^1 + p_1(1 - p_2) z^0 + (1 - p_1) p_2 z^0 + (1 - p_1)(1 - p_2) z^0.$$

Combination of A and B is estimated by the following procedure:

$$u_6(z) = \max\{u_3(z), u_5(z)\}$$

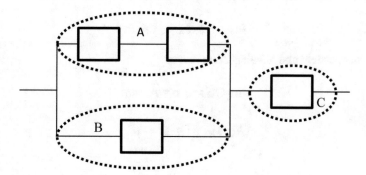

Fig. 1 Block diagram

$$= p_1 p_2 p_3 z^1 + p_1 p_3 (1 - p_2) z^1 + (1 - p_1) p_2 p_3 z^1$$

$$+ p_3 (1 - p_1)(1 - p_2) z^1 + p_1 p_2 (1 - p_3) z^1 + p_2 (1 - p_1)(1 - p_3) z^0$$

$$+ p_1 (1 - p_2)(1 - p_3) z^0 + (1 - p_3)(1 - p_1)(1 - p_2) z^0.$$

The final u-function of the considered system is:

$$u_7(z) = \min\{u_4(z), \, u_6(z)\}$$

$$= p_1 p_2 p_3 p_4 z^1 + p_1 p_3 p_4 (1 - p_2) z^1 + (1 - p_1) p_2 p_3 p_4 z^1 + p_1 p_2 p_4 (1 - p_3) z^1$$

$$+ p_3 p_4 (1 - p_1)(1 - p_2) z^1 + p_4 (1 - p_3)(1 - p_1)(1 - p_2) z^0 + p_1 p_2 p_3 (1 - p_4) z^0$$

$$+ p_1 p_4 (1 - p_2)(1 - p_3) z^0 + p_2 p_4 (1 - p_1)(1 - p_3) z^0 + p_1 p_3 (1 - p_2)(1 - p_4) z^0$$

$$+ p_2 p_3 (1 - p_1)(1 - p_4) z^0 + p_1 p_2 (1 - p_3)(1 - p_4) z^0$$

$$+ p_2 (1 - p_3)(1 - p_1)(1 - p_4) z^0 + (1 - p_1)(1 - p_2)(1 - p_3)(1 - p_4) z^0$$

$$+ p_1 (1 - p_3)(1 - p_4)(1 - p_2) z^0.$$

6.1 Reliability Function

Reliability function for the considered example is obtained as follows:

$$R = p_4 p_3 + p_1 p_2 p_4 - p_1 p_2 p_3 p_4 \tag{9}$$

For independent and identical components

$$p_1 = p_2 = p_3 = p_4 = p$$

$$h(p) = p^2 + p^3 - p^4 \tag{10}$$

Table 1 Tail signature

K	$\overline{S}_k$
0	1
1	¾
2	1/6
3	0
4	0

6.2 Min Signature = (0, 1, 1, -1)

6.3 Expected Time

With the help of minimal signature, the expected time of the considered system is evaluated as:

$$E(t) = \frac{0}{1} + \frac{1}{2} + \frac{1}{3} - \frac{1}{4} = 0.583333.$$

6.4 Tail Signature

Tail signature of the system is calculated with the help of Eq. [4] and shown in Table 1.

$$\overline{S}_k = \frac{(n-k)!}{n!} D^k(f(x)); \qquad k = 0 \ldots n.$$

where $f(x) = x^n h\left(\frac{1}{x}\right)$.

6.5 Signature

After calculating the reliability function using Owen's method, the signature of the considered system is calculated by using Eq. [5] and revealed in Table 2.

$$S_k = \overline{S}_{k-1} - \overline{S}_k.$$

Table 2 Signature

K	S_k
1	¼
2	7/12
3	1/6
4	0

6.6 Expected Cost

The expected cost of the system is:

$$E(X) = \sum_{k=1}^{4} k S_k.$$

The expected cost rate of the considered system comes out to be as:
=23/12.

6.7 Mean Time to Failure

Mean time to failure of the considered system can be evaluated as

$$MTTF = \frac{E(X)}{E(t)} = 3.2857.$$

6.8 Barlow-Proschan Index

Now, with the help of Eq. [7] the Barlow-Proschan index of the considered system is evaluated and shown in Table 3.

Table 3 Barlow-proschan index

K	I_{BP}
1	1/12
2	1/12
3	1/2
4	7/12

7 Conclusion

This paper studies a complex series-parallel system. The elements of the system are considered to be independent and identically distributed, that is, they have the same reliability. By using the universal generating function, reliability function of the proposed system is derived. With the vital examination of reliability function minimum signature, tail signature, signature, expected time, expected cost, mean time to failure and Barlow-Proschan index is calculated. Through overall calculations, we found that the subsystem C has less failure probability and subsystem B has the highest failure probability.

References

1. Levitin, G. (2005). Reliability of linear multistate multiple sliding window systems. *Naval Research Logistics (NRL), 52*(3), 212–223.
2. Levitin, G. (2004). A universal generating function approach for the analysis of multi-state systems with dependent elements. *Reliability Engineering & System Safety, 84*(3), 285–292.
3. Li, Y. F., & Zio, E. (2012). A multi-state model for the reliability assessment of a distributed generation system via universal generating function. *Reliability Engineering & System Safety, 106*, 28–36.
4. Ebrahimipour, V., Sheikhalishahi, M., Shoja, B. M., & Goldansaz, M. (2010, November). A universal generating function approach for redundancy optimization for hot-standby multi-state series parallel k-out-of-n systems. In *2010 Fourth UKSim European Symposium on Computer Modeling and Simulation* (pp. 235–239). IEEE.
5. Destercke, S., & Sallak, M. (2013). An extension of universal generating function in multi-state systems considering epistemic uncertainties. *IEEE Transactions on Reliability, 62*(2), 504–514.
6. Lisnianski, A. (2016, February). Application of extended universal generating function technique to dynamic reliability analysis of a multi-state system. In *2016 Second International Symposium on Stochastic Models in Reliability Engineering, Life Science and Operations Management (SMRLO)* (pp. 1–10). IEEE.
7. Zhao, X., Wu, C., Wang, X., & Sun, J. (2020). Reliability analysis of k-out-of-n: F balanced systems with multiple functional sectors. *Applied Mathematical Modelling*.
8. Eryilmaz, S. (2012). *m*-consecutive-*k*-out-of-*n*: *F* system with overlapping runs: Signature-based reliability analysis. *International Journal of Operational Research, 15*(1), 64–73.
9. Eryilmaz, S. (2014). Computing reliability indices of repairable systems via signature. *Journal of Computational and Applied Mathematics, 260*, 229–235.
10. Da, G., & Hu, T. (2013). On bivariate signatures for systems with independent modules. In *Stochastic orders in reliability and risk* (pp. 143–166). New York, NY: Springer.
11. Boland, P. J. (2001). Signatures of indirect majority systems. *Journal of Applied Probability, 38*(2), 597–603.
12. Marichal, J. L., & Mathonet, P. (2013). Computing system signatures through reliability functions. *Statistics & Probability Letters, 83*(3), 710–717.
13. Navarro, J., & Rubio, R. (2009). Computations of signatures of coherent systems with five components. *Communications in Statistics-Simulation and Computation, 39*(1), 68–84.
14. Owen, G. (1975). Multilinear extensions and the Banzhaf value. *Naval Research Logistics Quarterly, 22*(4), 741–750.
15. Owen, G. (1988). Multilinear extensions of games. *The Shapley Value. Essays in Honor of Lloyd S. Shapley*, 139–151.

16. Shapley, L. S. (1953). A value for *n*-person games. In *Contributions to the Theory of Games*, vol. 2. In *Annals of Mathematics Studies*, vol. 28. *Princeton University Press, Princeton, NJ,* 307–317.
17. Eryilmaz, S. (2012). The number of failed elements in a coherent system with exchangeable elements. *IEEE Transactions on Reliability, 61*(1), 203–207.
18. Ushakov, I. A. (1986). A universal generating function. *Soviet Journal of Computer and Systems Sciences, 24*(5), 118–129.
19. Levitin, G. (2003). Linear multi-state sliding-window systems. *IEEE Transactions on Reliability, 52*(2), 263–269.
20. Marichal, J. L. (2016). Structure functions and minimal path sets. *IEEE Transactions on Reliability, 65*(2), 763–768.
21. Navarro, J., Samaniego, F. J., & Balakrishnan, N. (2011). Signature-based representations for the reliability of systems with heterogeneous components. *Journal of Applied Probability, 48*(3), 856–867.
22. Huang, X., Coolen, F. P., & Coolen-Maturi, T. (2019). A heuristic survival signature based approach for reliability-redundancy allocation. *Reliability Engineering & System Safety, 185,* 511–517.
23. Samaniego, F. J., Balakrishnan, N., & Navarro, J. (2009). Dynamic signatures and their use in comparing the reliability of new and used systems. *Naval Research Logistics (NRL), 56*(6), 577–591.
24. Manglik, M., & Ram, M. (2013). Reliability analysis of a two unit cold standby system using Markov process. *Journal of Reliability and Statistical Studies, 6*(2), 65–80.
25. Dhillon, B. S., & Fashandi, A. R. M. (1997). Safety and reliability assessment techniques in robotics. *Robotica, 15*(6), 701–708.
26. Nagiya, K., & Ram, M. (2013). Reliability characteristics of a satellite communication system including earth station and terrestrial system. *International Journal of Performability Engineering, 9*(6), 667–676.
27. Kumar, A., Varshney, A. K., & Ram, M. (2015). Sensitivity analysis for casting process under stochastic modelling. *International Journal of Industrial Engineering Computations, 6,* 419–432.
28. Goyal, N., Kaushik, A., & Ram, M. (2016). Automotive water cooling system analysis subject to time dependence and failure issues. *International Journal of Manufacturing, Materials, and Mechanical Engineering (IJMMME), 6*(2), 1–22.

Analyzing the Concept of Priority Queue for Scheduling in Cloud Scenario

Rukaiya Naim and Nisha Chaurasia

1 Introduction

Cloud computing provides facilities of accessing, creating, configuring, and customizing the applications, resources over the Internet. The word "cloud" refers to a set of links or something like which is situated at remote place or are able to provide services easily over public and private networks, i.e., WAN, LAN, or VPN. It facilitates shared resources or online storage on demand (as per user requirement), infrastructure, services, and applications in industry as well as in research areas without the need of any infrastructure at the user's system except the input, output devices, and the most important Internet connection [1]. Cloud computing provides platform independency as the software is not required to be installed locally on the physical system and hence it makes the business mobile and collaborative. All these services are controlled by the concept of virtualization, which is a technique that helps to share the demanded part of a resource, application by the help of using pointer, or by assigning a logical method to that memory, etc. Hypervisor acts as controller which monitors everything like CPU and memory, allocation. In cloud, users take advantage of resources (CPU, memory, etc.) by using the criteria of pay-as-you-go pricing model [2].

Billions of people can use the resources of cloud with the help of submitting their tasks to host in cloud system. Task scheduling is a one of the biggest challenges faced by many researchers while applying migration techniques in cloud computing,

R. Naim (✉)
Jaypee Institute of Information Technology, Computer Science & Engineering Department& Engineering Department, Sector 62, Noida, Uttar Pradesh, India
e-mail: ruks2692@gmail.com

N. Chaurasia
Dr. B. R. Ambedkar, National Institute of Technology, Jalandhar, India
e-mail: chaurasianisha21@gmail.com

optimization of tasks, and optimal use of resources allocation must help one decide the optimal number of VMs by which overall cost gets reduced. In cloud scenario, various popular CPU scheduling algorithms are used for task scheduling like FCFS (first come, first serve), RR (round robin), SJF (shortest job first), etc. [3] by which overall execution time of processes could be minimized. FCFS was used as the base scheme on which SJF provides better results [4] stating that FCFS has disadvantages that it demands advance information regarding execution time, waiting time taken by process. Basically, it is better suited for the processes those have short burst time while SJF scheduling algorithm works for the processes which comes simultaneously in the queue, and with the help of preemption concept it show better results. While on comparing these two algorithms with remaining RR [5] algorithm, RR provides better results for the tested scenario. It is shown that RR [6] provides the better response time compared with the other abovementioned algorithms. When large number of processes comes in the queue then the speed of processor goes slow as it totally depends upon the processor load. So, for resolving this issue concept of priority is being introduced which works upon the multilevel queue concept. However, for achieving more optimal makespan, concept of multilevel feedback queue (MLFQ) is introduced in this paper. MLFQ is expanded from RR algorithm where each process is scheduled as per their priority (low, medium, high), and consists of multiple levels of queues (level of queue is also maintained from high to low priority) which uses feedback to determine the priority of given job, the highest priority job is given first preference (i.e., the job is switched toward higher priority queue). Therefore, MLFQ is an outstanding model for a scheme that learns from the past to predict the future and creates a new strategy that reduces the overall total execution time of a process in a system.

The further sections in the presented paper include related work, CloudSim (tool for testing the proposal) [7], proposed methodology, experimental outcomes, and conclusion.

2 Related Work

Noshy, Mostafa, et al. survey, examine, discuss, and compare the methods to recognize their advancement as well as their concerned difficulties. Their work additionally features the open research issues that require further assessment to advance the procedure of live migration for virtual machines. In live migration, pre-copy, post-copy, and hybrid approach are the main approaches [1].

In this paper, Zhang, Weishan, et al. proposed Genetic Algorithms (GAs)-based methodology which is viable in addressing to multi-objective optimization issues. The author performed some fundamental assessments of the proposed methodology which indicates very encouraging outcomes, utilizing one of the traditional genetic algorithms [2].

Beloglazov et al. proposed a novel methodology for any known stationary remaining task at hand and at a given state configuration ideally takes care of the issue of host overload recognition by increasing the average between migration times under the particular Quality of Service (QoS), whose objective is dependent on a Markov chain model. They heuristically adjusted the algorithm to deal with obscure non-stationary workloads utilizing the "Multi-size Sliding Window" remaining workload estimation strategy. From end-to-end simulation with real-world workload traces follows of extra than a thousand Planet Lab VMs, they demonstrated that their methodology beats the best standard algorithm and gives roughly 86% of the performance of the ideal offline algorithm [8].

Raheja et al. state that by using an uncertain set called as Vague-based Multilevel Feedback Queue (VMLFQ) scheduler scheduling will be organized. It keenly handles the vulnerability and characterizes the perfect number of queues and furthermore the perfect size of time quantum for each queue moreover settles down the issue of starvation and separates the execution of VMLFQ scheduler with the other multilevel feedback queue systems utilizing MATLAB [9].

In this paper, another algorithm is being proposed for tackling optimization issues and limiting the response time. Recurrent Neural Network (RNN) algorithm has been used to determine the number of queues as well as the optimized quantum for each queue. Moreover, with the end goal to keep any probable faults in processes such as response time computation, another fault-tolerant methodology has been displayed. The trial results demonstrate that utilizing the Intelligent Multilevel Feedback Queue algorithm (IMLFQ) results in better reaction and waiting time in comparison with other scheduling algorithms [10].

Thombare, Malhar, et al. implemented MLFQ approach for utilizing small burst time for the first queue in the manner which is similar to RR scheduling and used SJF algorithm before RR algorithm from second queue onward resulting in better CPU usage. Dynamic time quantum is additionally utilized which further enhances the efficiency of the scheduling [11].

In this paper, Yang.et al. state that a new task scheduling algorithm Multi-dimensional Quality of Service Genetic Algorithm and Ant Colony Optimization (MQoS-GAAC) is considered as tedious, utilization, security, and dependability in the scheduling procedure. Multi-QoS impediments and Ant Colony Optimization algorithm (ACO) is organized with Genetic Algorithm (GA) for creating the initial pheromone more successfully for ACO, GA is conjured. Through the organized fitness function, four-dimensional QoS targets are evaluated [12].

Azad.et al. state that a task scheduling algorithm utilizes a combination of cultural and ant colony optimization algorithm for minimizing the makespan and energy consumption. The author used distinctive strategies to convey priorities to subtasks which deliver different makespan in a heterogeneous computing system. Under investigation aggregate execution time and energy consumption would be considered. After considering the significance of these two factors, another strategy is exhibited that joins ACO and cultural algorithm for scheduling the tasks and priorities in green computing. Azad, Poopak, et al. concluded that a proposed technique which is a combination of cultural and ant colony optimization algorithm beats the Heuristic-based Earliest Finish Time (HEFT) upward rank algorithm and ACO algorithm on the basis of energy consumption and enhances makespan [13].

Since, the overall objective is to achieve effective and optimal task scheduling without compromising QoS in cloud, the MLFQ in this sense firmly holds its position. The tasks in cloud are expected sensitive in one or the other manner and they need to be arranged on priority basis. The studied algorithms lack in this context and drive MLFQ as the most desirable solution.

3 CloudSim

CloudSim is a system created by the GRIDS research center of college of Melbourne which empowers consistent displaying, simulation, and probing structuring cloud computing frameworks [7].

The distinguishing features offered are as follows.

3.1 CloudSim Individuality

It offers the additional things such as

- Maintaining replica of huge scale cloud computing.
- Offering an independent stage for displaying data center, service broker, task scheduling, and allocation policies.
- Cloud can support VM provisions at two dimensions:

    i.   **Host level**: It is possible to indicate the amount of the general preparing intensity of each center will be allocated to every VM called as VM arrangement allocation.
    ii. **VM level**: For individual task, VM will divide available power in equal amount and within its executable engine this concept is known VM scheduling.

3.2 CloudSim Platform

Following are the components basically offered by CloudSim [7]:

- **Data center**: It deals with the center part (i.e., infrastructure level) which is offered by cloud suppliers and puts the host (homogeneous and heterogeneous) in a set and exemplifies it.
- **Data center broker**: It is in charge of making model for a broker and is responsible for mediating negotiations between the (SaaS) model and service supplier.
- **Cloudlet**: It shows the services of cloud-based applications.
- **VM**: Virtual machine is kept running on cloud host to manage a cloudlet.

Fig. 1 CloudSim components

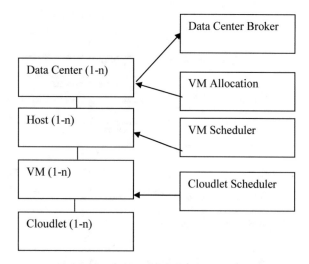

- **VM allocation**: A provisioning policy which is kept running in data center level that distributes VMs to hosts (Fig. 1).

4 Proposed Methodology

In this paper, the (multilevel feedback queue) MLFQ algorithm is used for allocation of incoming tasks toward virtual machines for task scheduling. It has numerous dimensions of queues, and uses feedback to decide the priority of a given task. MLFQ has various inimitable queues, each agreed on an alternate priority level. Whenever a task is going to arrive in a solitary queue and as indicated by needs, choice would be taken by which task will execute first within less time: higher priority will be given first preference. If more than one task comes in a queue with same priority then the concept of (RR—round robin) is used for avoiding starvation by preemption. Here, the processes according to the priority are set by which minimum makespan is attempted to be achieved.

For better outcomes, essential arrangement of laws in MLFQ is as follows:

- **Law 1**: If Priority (C) > Priority (D), C runs (D don't).
- **Law 2**: If Priority (C) = Priority (D), C and D keep running in RR.
- **Law 3**: When a task enters the system, it is set at the most highest priority.
- **Law 4**: If a task goes through a queue and uses up its time allotment while running, its priority is narrowed (i.e., goes down one queue) or if a task surrenders the CPU before the time phase is up, it remains at a similar required level.
- **Law 5**: After some time, phase moves every one of the tasks to the highest point of the queue in the system (Fig. 2).

Different scheduling algorithms analyzed w.r.t. MLFQ are as follows:

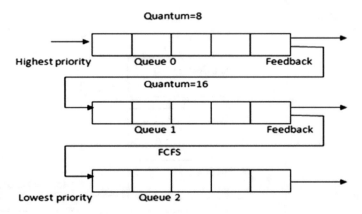

Fig. 2 MLFQ flow diagram

FCFS (First Come First Serve): The main criterion here is arrival time of the process. The mode is non-preemptive (means once you choose a job and give it to CPU it will finish it completely) with lot of processes in Ready Queue (RQ) and one CPU picks one process and gives it to CPU according to criteria of whoever comes first, it will get first preference. Therefore, everything depends on their arrival time. It is the most simplest and easy to implement scheduling algorithm [4].

SJF (Shortest Job First): Here the criteria are burst time and it can be both preemptive and non-preemptive. Whenever a task is to be executed, the preference is given to a task with shortest execution time—choosing it will not stop until it finishes [4].

RR (Round Robin): It is a preemptive version of scheduling, which means when process is still running and wants more execution time, the scheduler pulls it out and schedules it accordingly. The next criterion opted here is arrival time which is the same as in FCFS, and next one is Time Quantum (i.e., the maximum allowance time that process can run once when scheduled). In respect to these scheduling algorithms, the fundamental problem that MLFQ tries to address is to optimize the makespan more effectively. Also, MLFQ attempts to make system feel responsive and interactive toward user [5] (Fig. 3).

5 Experimental Outcomes

5.1 Implementation Results of SJF, FCFS, and MLFQ

In the experiment, different task scheduling algorithms, namely, FCFS, SJF, and MLFQ have been used for analyzing makespan w.r.t. the cloud processes. Twenty cloudlets (processes) on host are sent, and the average is calculated by arrival time and finish time of each allocated process.

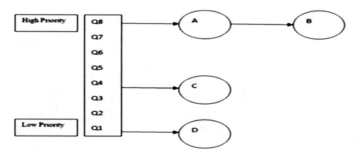

Fig. 3 Priority of processes

Normally, MLFQ maintains distinct number of queues for processes and each assigned with different priority levels. In above figure, every queue has different fixed time slots for 20 processes, for adding process in queue several iterations are to be performed. In the above result, three queues qo, q1, and q2 are used for pushing and popping of 20 processes in it as per their priorities and quant size. In consideration to the levels of queues, MLFQ uses feedback of process to higher priority queue to determine the priority of given task. Highest priority task comes in top most queues where FCFS scheduling is to be performed while rest of the processes comes in q1 and q2 queues where shuffling among processes is done as per priorities to decide which task should run at a given time and with higher priority. If more than one task in a queue may have same priority, then the Round Robin (RR) comes into action which uses the concept of quantum (maximum allowance time) quant = 8 for queue1 and quant = 16 for queue 2 by which minimum makespan could be achieved (Fig. 4).

Fig. 4 Priority of queues   Queue 0

P1	P2	P4	P5	P3	P6	P7

Queue 1

P8	P9	P10

Queue 2

P11	P12	P13	P14	P15	P16	P17	P18	P19	P20

Table 1 Time taken by processes in SJF scheduling

Cloudlets ID	VM ID	Time	Start time	Finish time
1	2	1923.5	0.1	1923.6
2	6	2457.05	0.1	2467.15
3	5	2834.43	0.1	2834.15
4	4	3146.43	0.1	3146.53
5	3	3401.58	0.1	3401.78
6	2	2182.18	1923.6	4105.78
7	6	1951.91	2467.15	4419.06
8	5	1903.96	2834.53	4738.49
9	2	1401.86	4105.78	5507.64
10	3	2416.1	3401.7S	5817.88
11	4	3500.47	3146.53	6647
12	3	874.13	5817.88	6692.01
13	5	2249.62	4738.49	6988.11
14	6	2719.92	4419.06	7138.98
15	2	1890.77	5507.64	7398.41
16	6	419.36	7138.98	7558.34
17	5	953.16	6988.11	7941.27
18	4	1847.14	6647	8494.14
19	6	2141.66	7558.34	9700
20	5	1824.5	7941.27	9765.77

The average makespan value results of FCFS, SJF, and MLFQ are shown as follows. It can be clearly seen that by increasing the number of tasks, the MLFQ takes lesser time compared to SJF and FCFS algorithms. This concludes that MLFQ is better than the SJF and FCFS as (Tables 1, 2, and 3 and Figs. 5, 6, and 7)

1. Time execution of task is maintained.
2. Service Level Agreement (SLA) violation is avoided.
3. Quality of Service (QoS) is maintained.

Table 2 Time taken by processes in FCFS scheduling

Cloudlets ID	VM ID	Time	Start time	Finish time
1	4	1827.96	0.2	1828.16
2	3	2455.02	0.2	2465.22
3	3	2209.55	2455.22	4674.87
4	5	2074.13	0.2	2074.33
5	4	2910.13	1828.16	4738.28
6	3	2831.54	4674.87	7506.41
7	4	1956.85	4738.28	6595.13
8	6	2927.85	0.2	2928.05
9	6	2575.81	2928.05	5503.86
10	4	3266.88	6695.13	9952.01
11	6	2584.45	5503.86	8088.31
12	4	1957.37	9952.01	11919.38
13	3	718.07	7506.41	8224.48
14	3	220.06	8224.48	10424.54
15	3	2050.54	10424.54	12475.08
16	3	2888.51	12475.08	15363.58
17	5	597.29	2074.33	2671.62
18	2	2868.93	0.2	2859.13
19	5	2517.12	2671.52	5188.74
20	3	1279.98	15363.58	16643.57

6 Conclusion

In this paper, comparison of available and dominant CPU scheduling algorithms like FCFS, SJF, RR, etc. is done to find the optimal resource allocation and minimize the total execution time in cloud scenario. While performing experimentation SJF takes lesser time than FCFS. However, for more optimal results, Multilevel Feedback Queue Algorithm (MLFQ) with RR is analyzed in CloudSim providing more optimized makespan results.

Table 3 Time taken by processes in MLFQ scheduling

Cloudlets ID	VM ID	FCFS arrival time	FCFS finish time	MLFQ time
1	4	2	1829	2
2	3	2	2455	2
3	3	2465	4675	2465
4	5	2	2075	2
5	4	1829	4739	1829
6	3	4675	7506	4675
7	4	4739	6695	4739
8	6	2	2929	4773
9	6	2929	5504	4773
10	4	6695	9963	6695
11	6	5503	8089	9782
12	4	9962	11920	9962
13	3	7507	8225	9782
14	3	8225	10425	9782
15	3	10425	12476	10425
16	3	12476	15364	12476
17	5	2075	2672	9782
IB	2	2	2870	9782
19	5	2672	5189	9782
20	3	15364	16644	15364

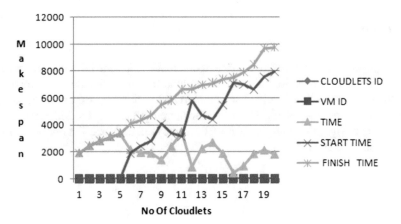

Fig. 5 Makespan using SJF is 4200.90348 s

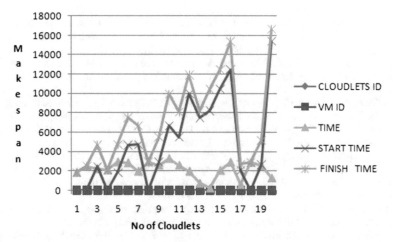

Fig. 6 Makespan using FCFS is 5868.389335 s

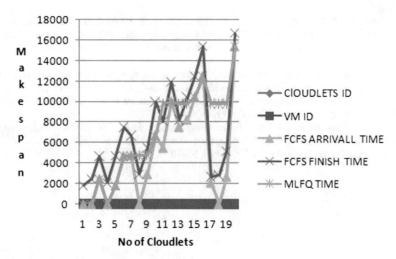

Fig. 7 Makespan using MLFQ is 3398.34 s

References

1. Noshy, M., Ibrahim, A., & Ali, H. A. (2018). Optimization of live virtual machine migration in cloud computing: A survey and future directions. Journal of Network and Computer Applications.
2. Zhang, W., Tan, S., Lu, Q., Liu, X., & Gong, W. (2015). A genetic-algorithm-based approach for task migration in pervasive clouds. *International Journal of Distributed Sensor Networks, 11*(8), 463230.
3. Elmougy, S., Sarhan, S., & Joundy, M. (2017). A novel hybrid of Shortest job first and round Robin with dynamic variable quantum time task scheduling technique. *Journal of Cloud Computing, 6*(1), 12.

4. Kaur, A., Maini, R. DIFFERENT TASK SCHEDULING ALGORITHMS IN CLOUD COMPUTING. International Journal of Latest Trends in Engineering and Technology Vol.(9)Issue(3), pp. 17–2-23DOI: http://dx.doi.org/10.21172/1.93.3.
5. Behzad, S., Fotohi, R., & Effatparvar, M. (2013). Queue based job scheduling algorithm for cloud computing. International Research Journal of Applied and Basic Sciences ISSN, 37853790.
6. Siahaan, A. P. U. (2016). Comparison analysis of CPU scheduling: FCFS, SJF and Round Robin. *International Journal of Engineering Development and Research, 4*(3), 124–132.
7. Calheiros, R. N., Ranjan, R., Beloglazov, A., De Rose, C. A., & Buyya, R. (2011). CloudSim: a toolkit for modeling and simulation of cloud computing environments and evaluation of resource provisioning algorithms. Software: Practice and experience, 41(1), 23–50.
8. Beloglazov, A., & Buyya, R. (2013). Managing overloaded hosts for dynamic consolidation of virtual machines in cloud data centers under quality of service constraints. *IEEE Transactions on Parallel and Distributed Systems, 24*(7), 1366–1379.
9. Raheja, S., Dadhich, R., & Rajpal, S. (2016). Designing of vague logic based multilevel feedback queue scheduler. *Egyptian Informatics Journal, 17*(1), 125–137.
10. EffatParvar, M., Faez, K., EffatParvar, M., Zarei, M., & Safari, S. (2006, October). An Intelligent MLFQ Scheduling Algorithm (IMLFQ) with Fault Tolerant Mechanism. In Intelligent Systems Design and Applications, 2006. ISDA'06. Sixth International Conference on (Vol. 3, pp. 80–85). IEEE.
11. Thombare, M., Sukhwani, R., Shah, P., Chaudhari, S., & Raundale, P. (2016, March). Efficient implementation of Multilevel Feedback Queue Scheduling. In Wireless Communications, Signal Processing and Networking (WiSPNET), International Conference on (pp. 1950–1954). IEEE.
12. Dai, Y., Lou, Y., & Lu, X. (2015, August). A task scheduling algorithm based on genetic algorithm and ant colony optimization algorithm with multi-QoS constraints in cloud computing. In Intelligent Human-Machine Systems and Cybernetics (IHMSC), 2015 7th International Conference on (Vol. 2, pp. 428–431).IEEE.
13. Azad, P., & Navimipour, N. J. (2017). An energy-aware task scheduling in the cloud computing using a hybrid cultural and ant colony optimization algorithm. *International Journal of Cloud Applications and Computing (IJCAC), 7*(4), 20–40.

Data Acquisition from Smart Meters of Electric Sub-station Using Gateway Module

Vishakha Chaudhary, Ranjeeta Singh, and H. K. Verma

1 Introduction

In the modern electric sub-stations of power distribution systems and the recent micro-grids, a number of parameters related to quality of power are measured in addition to the basic measurement of energy. This is achieved by using smart multi-function meters (SMFMs) rather than simple energy meters. SMFMs have the capability to communicate in addition to measuring up to 100 quantities. The importance of SMFM/smart electricity meter and data acquired from them is reported in reference [1]. It describes smart meter as an important element of smart grid and also gives emphasis on data acquisition, its processing, and further interpretation. Use of smart multifunction meters in preference to the conventional digital meters is explained in reference [2]. Reference [3] elaborates the algorithms and techniques used toward smart meter development.

Data from the smart electricity meters need to be acquired remotely and automatically in a central place, which is generally the office or the control room of the utility (distribution company). The data acquisition from the smart meters can be carried out broadly by using either wireless ossr wired communication technologies. Reference [2] reports use two types of wireless communication from smart energy meters, one for short distances and the other for long distances. For short distances, the energy consumption is transmitted with the help of an RF transmitter within the energy meter to a central node. For long distances, the readings obtained in the central

V. Chaudhary (✉) · R. Singh · H. K. Verma
Sharda University, Greater Noida, India
e-mail: ch.vishakha@gmail.com

R. Singh
e-mail: ranjeeta.singh@sharda.ac.in

H. K. Verma
e-mail: hk.verma@sharda.ac.in

© The Author(s), under exclusive license to Springer Nature Singapore Pte Ltd. 2021   97
P. K. Kapur et al. (eds.), *Advances in Interdisciplinary Research in Engineering and Business Management*, Asset Analytics,
https://doi.org/10.1007/978-981-16-0037-1_9

node are transferred to the utility office using a GSM (Global System for Mobile) module.

Wireless communication technologies can work satisfactorily when the data is acquired from the meters installed in the premises of consumers primarily for the purpose of raising electricity bills for payment by consumers. But wireless communication technologies will not work satisfactorily when data is to be acquired from SMFMs installed in a distribution substation, due to the strong electro-magnetic interference (EMI) created by the high voltages (11 or 33 kV) and high currents (typically several hundred amperes) present in the substation. Error-free data communication is both important and critical because it is meant for the operation of the distribution system and future planning. The solution lies in the use of wired communication technologies.

This paper presents a robust solution using wired communication technologies. It reports the successful development of an advanced metering system for acquiring data from SMFMs installed in an electric substation of a university power distribution system and transmitting to its control room. All the SMFMs in a substation are connected in a personal area network (PAN) based on Modbus/RS485 protocols. Values of different parameters acquired from SMFMs through this PAN are transmitted to the control room connected to the campus LAN of the university using a gateway module. The gateway module selected here has the capability to convert between Modbus TCP protocol running on the RS-485 PAN and Modbus ASCII/RTU protocol running on the Ethernet LAN of the university campus. By using this module, data on one network can be transferred to another network.

This paper has been organized as follows: next section describes the proposed solution for acquiring data. Section 3 presents an overview of the SMFM used in this work. Section 4 describes the gateway module used. Section 5 shows connections between SMFMs and the gateway module along with the pin assignments. Section 6 explains the two protocols used in this system. Software used for data acquisition is described in Sect. 7. Section 8 gives the results obtained from testing the system. The last section (Conclusion) presents the outcome of the work along with the scope of future work.

2 Proposed Approach

The approach proposed in this paper is to acquire the data from the smart meters using MB3180 gateway module. Here, the meters are connected in a PAN through daisy chain wiring and the data is acquired by the gateway module through its serial port. The data acquired by the gateway module is shared over the Ethernet LAN forming part of the Intranet of Sharda campus. RS-485 protocol, specifying MAC and physical layers, is used along with Modbus RTU protocol, which specifies the application layer. The master connected to this LAN can easily access this data. This data can be seen on PCs through ModScan32 WinTech software. The approach is

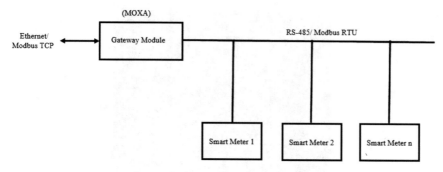

Fig. 1 Data acquisition using gateway module

illustrated in Fig. 1, wherein a gateway is connected to multifunction meters through RS-485/Modbus on one side and to Ethernet/Modbus TCP on other side.

3 Smart Multifunction Meters

Data acquired here is from PROCOM ACE multifunction meter. This meter can communicate and measure up to 100 quantities. It is equipped with customized six-digit, three-row alphanumeric displays [4].

3.1 Features of PROCOM ACE Multifunction Meter

- i.   Measured quantities and parameter can be simultaneously displayed.
- ii.   Auto-scaling of kilo, giga, mega, and decimal point.
- iii. Parameters are password protected so that no one can change them.
- iv.   Modbus communication is done on RS-485.
- v.   Trip function is available.
- vi.   Auto and manual scroll for display can be easily selected.
- vii. Input voltages are Vr, Vy, Vb, Vn.
- viii. Input currents are Ir, Iy, Ib.
- ix.   Input current required is 50 mA to 6 A.
- x.   Maximum withstand voltage is of 1000 V (Fig. 2).

The power terminals are present in the rear panel of meter. It consists of six current terminals, one in and one out. Its three-phase four-wire connections are shown in Fig. 3, wherein the four voltage terminals, two auxiliary power supply terminals, and two terminals of RS-485 communication port can also be seen.

Fig. 2 Smart multifunction meter [4]

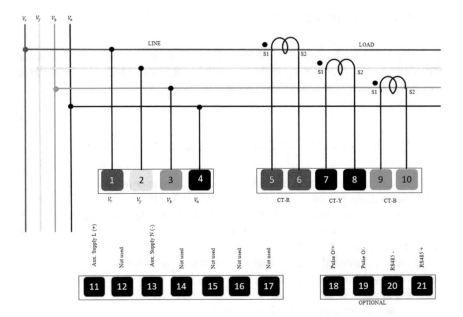

Fig. 3 Wiring diagram of multifunction meters

4 Gateway Module

The gateway used here is MGate MB3180 from Moxa. The MGate MB3180 is a one-port Modbus gateway that converts between Modbus TCP and Modbus ASCII/RTU protocols. It allows Ethernet master to control serial slaves, or serial masters can control Ethernet slaves. Up to 16 TCP masters and 31 serial slaves can be connected simultaneously [5].

4.1 Gateway Features

The MGate MB3180 has the following features:

i.   Integrates with Modbus TCP and Modbus RTU/ASCII and can be used to connect 31 Modbus RTU/ASCII slaves and 16 TCP master/client devices.
ii. Easily configured with Ethernet and can access any Windows and can work up to 100 Mbps. Communication protocol used is RS-485.
iii. It supports serial interface with high speed of 921.6 Kbps.
iv. Transmission distance up to 4 km through multi-mode fiber model and up to 20 km through single-mode fiber models.

4.2 Gateway Specifications

Important specifications of the gateway module are described below:

(a)   **LAN**

- Ethernet can work for speeds of 10/100 Mbps, RJ45 port is also present.
- Magnetic isolation is present with protection of 1.5 kV.

(b)   **Serial Interface**

- Protocols that can be interfaced are RS-232/422/485.
- Number of Ports: 1
- DB9-male, connector is present with 9 pins.
- Signals used here are RS-485 (4-wire) Tx+, Tx−, Rx+, Rx−, GND.
- RS-485 data direction is patent.

(c)   **Serial Communication Parameters**

- Parity none, even, odd, space, mark.
- Data bits used are 7, 8.
- Stop bits used are 1, 2.
- Transmission speed is 50 bps to 921.6 kbps.

(d)   **Power Requirements**

- Power input required is 12 to 48 VDC.
- Power connector (adapter) and power jack are provided.
- Power consumption is 200 m for 12 VDC and 60 mA for 48 VDC.

4.3   Gateway Hardware

Hardware of the module shows two parts (Fig. 4), one is rear panel view and other is front panel view. Rear panel view has two ports, one is for adapter and other is for Ethernet, and there is a reset button present that helps in resetting the device when connected to another device.

In front panel view, there are three LEDs, i.e., one for ready, other for Ethernet, and the last one is for P1, which indicates that the device is receiving data.

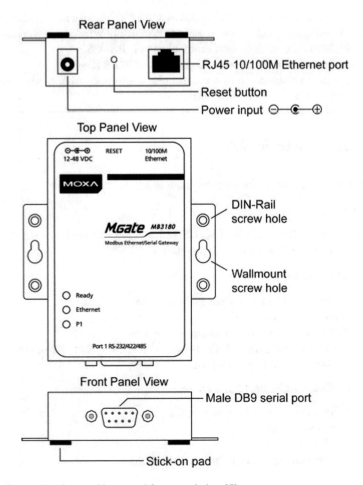

Fig. 4 Gateway hardware with rear and front panel view [5]

4.4   LED Indicators

The LED indicators present on the front panel view of the gateway indicates the different color lights which have different functions and these functions are explained in Fig. 5.

5   Wiring Diagram

The gateway module used is connected to the Ethernet cable through Ethernet port. gateway module, on the other side, is connected to the smart multifunction meter through RS-485 terminals. RS-485 is used along with Modbus RTU for successful communication. The same Ethernet port is connected to PC to give the acquired data. The whole connection diagram is shown in Fig. 6. Software used for taking readings from meters is ModScan32. The settings of the gateway are maintained by MGate Manager Software.

LED Indicators—Three LED indicators are located on the top panel:

Name	Color	Function	
Ready	Red	Steady on:	Power is on and the unit is booting up
		Blinking:	IP conflict exists, or DHCP or BOOTP server is not responding properly
	Green	Steady on:	Power is on and the unit is functioning normally
		Blinking:	Unit has been found by the Location command in MGate Manager
	Off	Power is off or power error condition exists	
Ethernet	Orange	10 Mbps Ethernet connection	
	Green	100 Mbps Ethernet connection	
	Off	Ethernet cable is disconnected or has a short	
P1	Orange	Unit is receiving data from device.	
	Green	Unit is transmitting data to device.	
	Off	No data is being exchanged with device.	

Fig. 5   LED indicators [5]

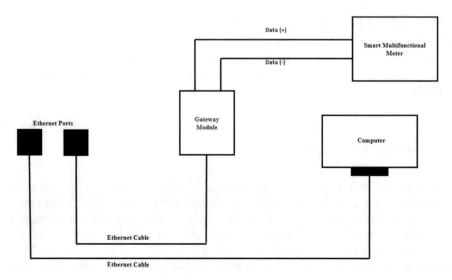

Fig. 6 Gateway connected to the Ethernet and multifunction meters

5.1 Pin Assignments of Ethernet Port and DB9 Connector

Pin assignment is the main part of the data acquisition from the meters. The pin assignment is shown in Fig. 7. Pin number 3 of DB9 connector will be connected to the positive terminal of RS485 port and pin number 4 of DB9 connector will be connected to the negative terminal of RS485 port and pin number 5 will be connected to ground which is optional.

5.2 Daisy Chain Wiring

Daisy chain wiring is used to reduce the complexity in the wiring connection. It requires a smaller number of wires. It gives us the availability of connecting different meters back to back. The different multifunction meters are connected to each other with this type of wiring for reducing the mess in the system. The connection of the meters through daisy chain wiring is shown in Fig. 8.

6 Protocols/Technologies Used

The protocols used here for application layer are Modbus RTU and Modbus TCP/IP. Modbus RTU has been used for transferring the data between the field devices (smart meters) and the gateway module and Modbus TCP/IP for exporting the data from

Ethernet Port (RJ45)

Pin Number	Signals
1	Tx+
2	Tx-
3	Rx+
4	Rx-

DB9 Connector

Pin Number	RS 485 (2 wire)
1	-
2	-
3	Data (+)
4	Data (-)
5	GND
6	-
7	-
8	-
9	-

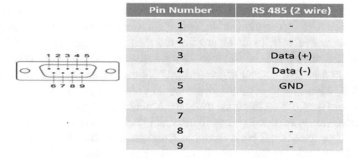

Fig. 7 Pin assignment of Ethernet port and DB9 connector

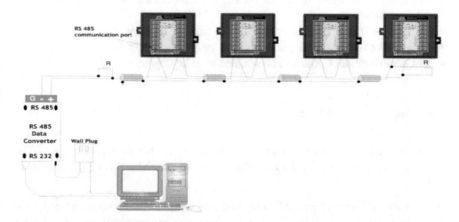

Fig. 8 Daisy chain wiring [4]

the gateway module to master computer present in control room. RS 485 is used for connection between the meters and the gateway modules and specifies MAC and physical layers of the network stack. It is advantageous to use this protocol as it follows the simplest master–slave protocol.

Modbus TCP/IP protocol is the application layer protocol specified by Modbus when data is transmitted on TCP/IP. When TCP/IP protocol suite is used for communication over a local area network (LAN) operating on the Ethernet protocol, the LAN is called as Ethernet TCP/IP network.

7 Data Acquisition Software

Software used for data acquisition is ModScan32 from WinTech. It is a Win32 application designed to operate as a Modbus Master device for accessing data points in a connected PLC (programmable logic controllers)-compatible slave device [6]. It is software which helps us to acquire the meter readings and to display them on computer screen. It is a Windows-based software and can work easily on any Windows-compatible device. It gives us provision of acquiring as many quantities as needed. ModScan32 supports direct serial connections, and connections with modems and network connections using the Modbus/TCP protocol.

Register values are displayed as 16 discrete bits. Data comes in the form of 8 bits once at a time then first 8 bit is stored in 1 register and other 8 bit is stored in next register. Each basic ModScan document represents a series (array) of Modbus data points identified by the following parameters:

- Slave device address—it represents the physical device that is attached to the Modbus network.
- Data type—it represents internal data, i.e., input, coil, or register.
- Data address—it shows the point address present in the device.
- Length—number of points to be scanned or to be displayed.

7.1 Software Settings

Software used for taking readings from meters is ModScan. The settings of the gateway are maintained by MGate Manager Software. The setting process is as follows:

I.   Open MGate Manager Software, then search the "Broadcast Search," check all the settings, and set the Baud rate to 9600 and communication protocol used is RS-485 2-wire.
II. Set the slave ID, i.e., from 1 to 4 (as all meters have different IDs, i.e., 1, 2, 3, 4, resp.).
III. Finally make auto-detection.
IV. Now open the ModScan32 software and set the register to the Holding register.
V.   Set the address from 1001 to 1003 (can differ according to quantity) and connect the system.

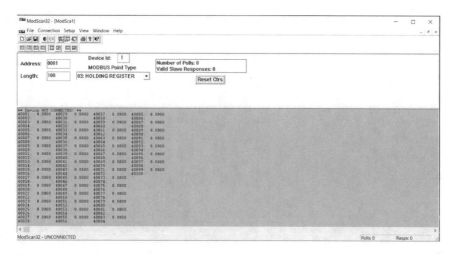

Fig. 9 Software showing error when address is wrong

8 Results

Data acquired from the meters through ModScan32 software have different errors like the address is wrong or the slave ID is wrong. The data acquired can be from minimum one parameter to all the required parameters and there are different cases of data acquisition that are explained below. Device ID indicates the IDs of multifunction meters connected. Address shows the Modbus parameters assigned by the meters.

8.1 Case I: Wrong Address

Address describes Modbus parameter for smart multifunction meters in which different quantities from the smart meters are stored. For meter 1, device ID is 1, and address 1001 is used for import/export purpose.

When parameter address is wrong, i.e., 0001, software will show "Device not Connected," see Fig. 9.

8.2 Case II: Correct Address and Displaying 10 Parameters

Address describes Modbus parameter for smart multifunction meters in which different quantities from the smart meters are stored. If device ID is 1 that means data is acquired from meter 1 and if that address is set to 1003 that means it shows the value of voltage phase 1 to N.

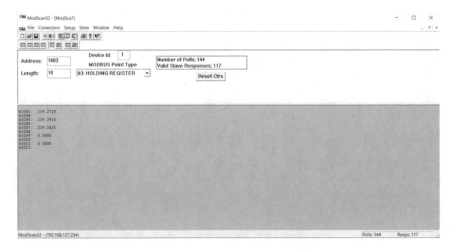

Fig. 10 Readings from meter number 1, displaying 5 parameters only

In Fig. 10, the readings are obtained from meter number 1 and the length taken is 12, which means number of parameters displayed will be 5 only (as 1 Modbus parameter is displayed on 2 addresses).

8.3 Case III: Correct Address and Displaying All Parameters

Address describes Modbus parameter for smart multifunction meters in which different quantities from the smart meters are stored. If device ID is 1 it means that the data is acquired from meter 1 and if that address is set to 1003 it means that it shows the value of voltage phase 1 to N.

In Fig. 11, the readings are obtained from meter number 1 and the length taken is 100, which means number of parameters displayed will be 50. Similarly, data can be acquired from different meters by changing the device ID.

9 Conclusion

Data acquisition from smart multifunction meters, installed in an electric substation of a university power distribution system, has been successfully attempted. All the details pertaining to the approach used, the wired communication protocols/technologies used in the PAN and LAN, the smart meters and the gateway module selected are presented in this paper. The SMFMs were connected to the gateway module through RS485/Modbus RTU PAN. The master (PC) in the control room is connected to the gateway module through Ethernet/Modbus TCP campus

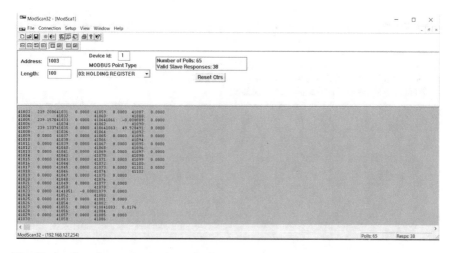

Fig. 11 Readings from meter number 1, displaying 50 parameters

LAN. Software used for data acquisition is ModScan from WinTech, which allows the user to change the values of the parameters to be displayed on PC during the data acquisition. The settings of the gateway can be managed using MGate Software for the gateway module.

The tests performed have given satisfactory results. Three conditions were taken into consideration during testing. First, in case of the wrong address, the system will display error and will not show the readings. Second, we can set the number of parameters to be displayed by setting the length of the parameters. Third, maximum length that can be set is 100 and a maximum of 50 parameters can be displayed, as one Modbus parameter is displayed in two addresses.

The RS485 protocol used for the wired PAN is very robust against the strong EMI present in the substation because of the differential-mode signals used by the protocol and use of Shielded Twisted Pair (STP) cable. Consequently, no communication problems were experienced during experiments.

In regard to the future scope, the readings of the meters can be saved as a database for further processing or keeping history. The database can be used with GUI for good visualization. Monitoring of parameters of interest and giving alarms in case of overload, etc. can also be done.

References

1. Alahakoon, D., Yu, X. (2015). Smart electricity meter data intelligence for future energy systems: a survey. *IEEE Transactions on Industrial Informatics,* 1–12.
2. Zala, H. R., Pandya, V. C. (2014). Energy meter data acquisition system with wireless communication for smart metering application. *International Journal of Engineering Research and Technology,* 1590–1595.

3. Khadar, A., Khan, J. A., & Nagaraj, M. S. (2017). Research advancements towards in existing smart metering over smart grid. *International Journal of Advanced Computer Science and Applications,* 84–92.
4. Procom: ACE Series Multifunction Panel Meter User Manual (2014).
5. Moxa: MGate MB3000 Modbus Gateway User's Manual (2007).
6. WinTech: Modscan32 User Manual. (2012).

RTU Design and Programming for Supervisory Control of Electric Sub-station

Vivek Thakur, Ranjeeta Singh, and H. K. Verma

1 Introduction

One of the major components of smart grid is incorporating automation at every stage of power system, namely, generation, transmission, and distribution. Distribution automation refers to automation of distribution sub-stations, feeders, and major loads. It includes monitoring, control, and protection of distribution sub-station and remote control of feeders. The modern technique of such a distribution control is supervisory control and data acquisition (SCADA). SCADA system consists of a number of remote terminal units (RTUs) and a master terminal unit (MTU). RTU provides interface to different field devices through its digital inputs (DIs) and digital outputs (DOs).

Reference [1] gives design and implementation of a SCADA system for a micro-grid. The design includes front-end processing modules or RTUs, supervisory control or MTU, and communication gateway modules. A controller area network (CAN) bus is used for communication between front-end processing modules and field devices while Ethernet is used to communicate between front-end processing modules and supervisory module. Reference [2] presents the design of a laboratory-based SCADA system. Reference [3] proposes a model to develop the system for controlling the main circuit breaker of sub-station and to collect, display, and store electrical parameters for future reference. The status of sub-station is transferred to the receiver through virtual proxy network (VPN).

V. Thakur (✉) · R. Singh · H. K. Verma
Sharda University, Greater Noida, India
e-mail: vivekt059@gmail.com

R. Singh
e-mail: ranjeeta.singh@sharda.ac.in

H. K. Verma
e-mail: hk.verma@sharda.ac.in

© The Author(s), under exclusive license to Springer Nature Singapore Pte Ltd. 2021   111
P. K. Kapur et al. (eds.), *Advances in Interdisciplinary Research in Engineering and Business Management*, Asset Analytics,
https://doi.org/10.1007/978-981-16-0037-1_10

Power distribution system of the Sharda University is being upgraded and modernized to make it a smart microgrid. This microgrid will be centrally controlled using a SCADA system. This paper presents the design and programming of RTU for supervisory control of one of the four electric sub-stations of the power distribution system of the university. Next section explains the proposed supervisory control system. Section 3 describes Sharda University campus distribution system. Section 4 gives the details of the electric sub-station-1 (ESS-1). Section 5 explains the design of RTU for ESS-1. Section 6 describes the ladder programs of this RTU developed in Allen Bradley MicroLogix 1400. Section 7 presents the result obtained from testing. Last section presents the conclusions and the future scope of the work.

2 Proposed Supervisory Control System

Supervisory control of an electric sub-station of Sharda University consists of a remote terminal unit (RTU), a master terminal unit (MTU), field devices (FDs), MTU-RTU communication system, and RTU-FDs communication system. These components are explained in the following sub-sections.

2.1 Remote Terminal Unit

Figure 1 shows the proposed scheme of installation of 18 RTUs at different places. Two RTUs, RTU-1 in the laboratory (for research and demonstration) and RTU-2 for supervisory control of ESS-1, have been already designed, fabricated, and tested successfully. This paper presents the details of RTU-2 for ESS-1.

2.2 Master Terminal Unit (MTU)

It is located in the control room and its basic functions are (a) to keep on monitoring the whole power distribution system/process and (b) to give control commands as and when necessary. These commands are given by the MTU to the RTU. The RTU converts these control messages into control signals and sends them to the respective field devices, like circuit breakers and tap changers [4].

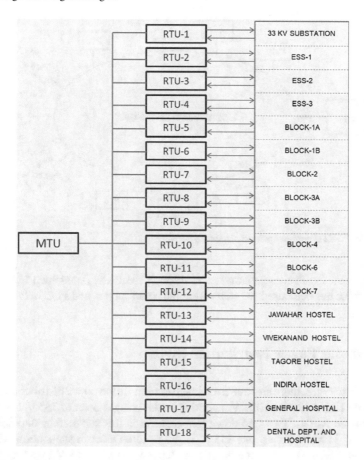

Fig. 1 Proposed scheme

2.3 MTU-RTU Communication System

The system used to communicate between RTU and MTU is LAN. Since Sharda University has its own campus-wide LAN, this network has been used to communicate between MTU and RTUs as shown in Fig. 2.

2.4 RTU-Field Devices Communication System

The RTU communicates with non-smart field devices, like circuit breakers and tap changers, through individual copper cables. On the other hand, the smart multifunction meters in the sub-station are connected in a PAN (personal area network) that terminates on the PLC (programmable logic controller) in the RTU. The RTU

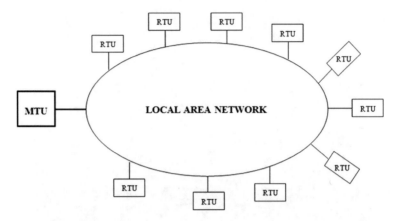

Fig. 2 MTU-RTU communication system [4]

communicates with these smart meters through the PAN. A shielded twisted pair (STP) cable has been used for connecting the smart meters and PLC in PAN.

3 Campus Power Distribution System

As shown in Fig. 3, there are four electric sub-stations in the power distribution system of Sharda University. The 33 kV incoming feeder brings power to 33 kV/11 kV sub-station from NPCL. Two step-down transformers in this sub-station step down the voltage to 11 kV. Further, this 111 V is fed to three electric sub-stations (ESS-1, ESS-2, and ESS-3), where the voltage is stepped down to 415 V by 11 kV/433 V transformers.

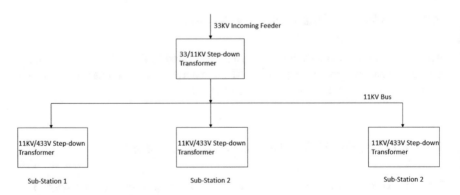

Fig. 3 Campus power distribution system

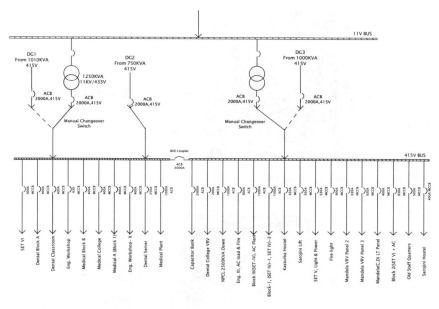

Fig. 4 Single line diagram of ESS-1

4 Details of ESS-1

4.1 Single Line Diagram of ESS-1

Figure 4 shows the single line diagram of Electric Substation-1 (ESS-1) of Sharda University.

4.2 Smart Multi-functions Meters (SMFMs)

In this sub-station, two types of SMFMs are used, Schneider EM 6400 and Schneider EM 6436. Their front panel is shown in Fig. 5. It has eight-segment LED displays, an analog load bar, several LED indicators, and five smart keys for its operation. The meters have capability to communicate and they can measure more than 100 quantities.

The EM6400 and EM6436 power terminals are located on the rear panel as shown in Fig. 6. 14 and details of terminals are provided, seven terminals on each side, six terminals are for current, four terminals are for voltage, two terminals for auxiliary power supply, and two for RS-485 communication port.

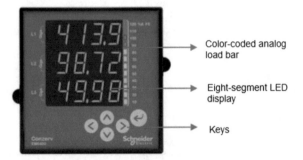

Fig. 5 Smart multi-function meter [5]

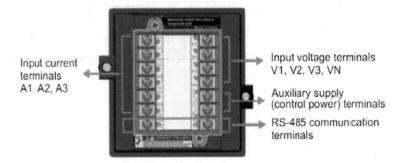

Fig. 6 Rear panel of smart multi-function meters [5]

5 Hardware Design of Remote Terminal Unit (RTU)

A PLC is used as the basic unit of the RTU. The connection diagram of digital inputs (DIs) of the PLC is shown in Fig. 7. One of the important functions of the RTU is to determine the status of two state objects from the status information acquired by it. This information is acquired from the auxiliary coil present in the molded case circuit breakers (MCCBs) and air circuit breakers (ACBs).

Similarly, digital outputs (DOs) are used to send necessary control signals to the MCCBs and ACBs. The PLC used in the sub-station has 12 digital outputs. These DOs are connected through relays (Fig. 8) to the tip coils and motor-based closing mechanisms of the MCCBs and ACBs to open and close them, respectively.

These DIs and DOs of the PLC are not sufficient to connect all the field devices, i.e., MCCBs and ACBs, so two digital input expansion units and four digital output expansion units have been added. Figure 9 shows the connection diagram for these input and output expansion units.

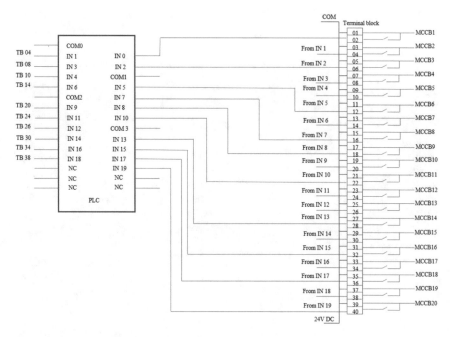

Fig. 7 Connection diagram of digital input terminals of PLC

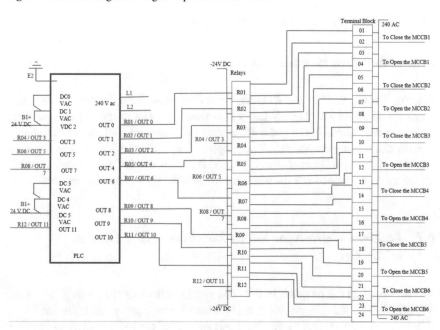

Fig. 8 Connection diagram of digital output terminals of PLC

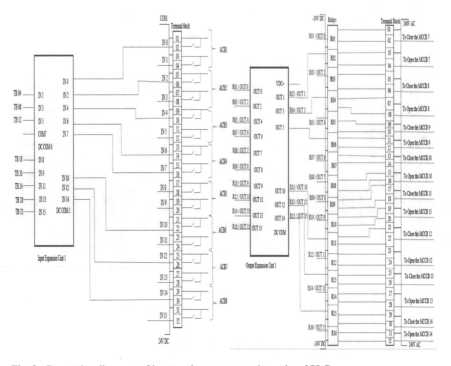

Fig. 9 Connection diagrams of input and output expansion units of PLC

6 Ladder Programming of RTU

Program for the PLC of the RTU has been developed using ladder programming language on Allen Bradley RS Logix 500 Emulator. The program performs three major functions, namely, inputting the status information, outputting the control signals, and reading the SMFMs. The program discussed here is only for verification purpose. It is meant to handle only one SMFM and one MCCB. It can be easily scaled up on the same lines for handling all the SMFMs and circuit breakers.

6.1 Ladder Program for Status Monitoring and Control of MCCB

The first rung of the program in Fig. 10 reads the status of the closing coil of the MCCB. The second rung reads the status of the tripping/opening coil. The third rung is used to give trip signal to the shunt release coil of the MCCB so as to open the circuit breaker and the fourth rung outputs a closing signal to the closing mechanism of the MCCB.

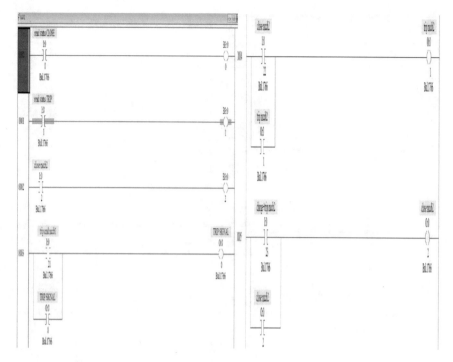

Fig. 10 Ladder program 1

6.2 Ladder Program for Acquiring Data from Smart Meters

A message is sent by the PLC to every meter to send the data every second. This is done by the program in rung 0 of Fig. 11. Rung 1 obtains the value of the frequency sent by the meters and it is stored in N7:15, which is the register used to store integer values but the value we get is in decimals. So, CPW (that is, copy word) is used and this value is stored in F8:0, which is the register used to store floating values.

7 Message Frame Format for Meters

Message format is shown in Fig. 12. This message is sent to the smart meters by the PLC in order to get the value of all electrical parameters from the former. There are three channels available in PLC, channel "0" is "configure for the Modbus communication." Message implies that it will access 100 data from Modbus address 43901 and store it to the N7:1 register of the PLC.

Fig. 11 Ladder program 2

8 Results

Figure 13 shows the values of various parameters as acquired from one SMFM. The data acquired from the meter is stored in F8 register, which is meant to store floating values. For example, value of frequency is stored in F8:0; phase voltages in F8:1, F8:2, and F8:3; line voltages in F8:4, F8:5, and F8:6; and power factor in F8:11.

9 Conclusion and Future Scope

Supervision and control play a very important role in any plant or process for its reliable and effective operation. It enables the operator in the control room of the process to see what is happening in the process. The master unit of supervisory control system continuously monitors the plant through RTUs, generates alarms in case of any abnormal conditions, and necessary control instructions to the plant through RTUs.

General design details of the supervisory control and data acquisition system of Sharda University have been presented in this paper. In particular, design details

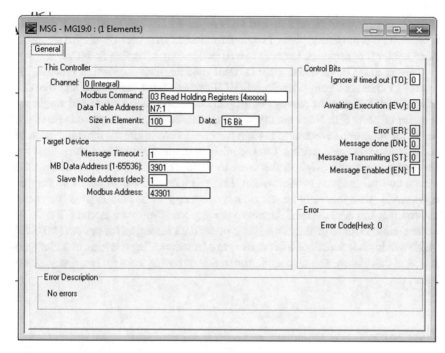

Fig. 12 Message frame format

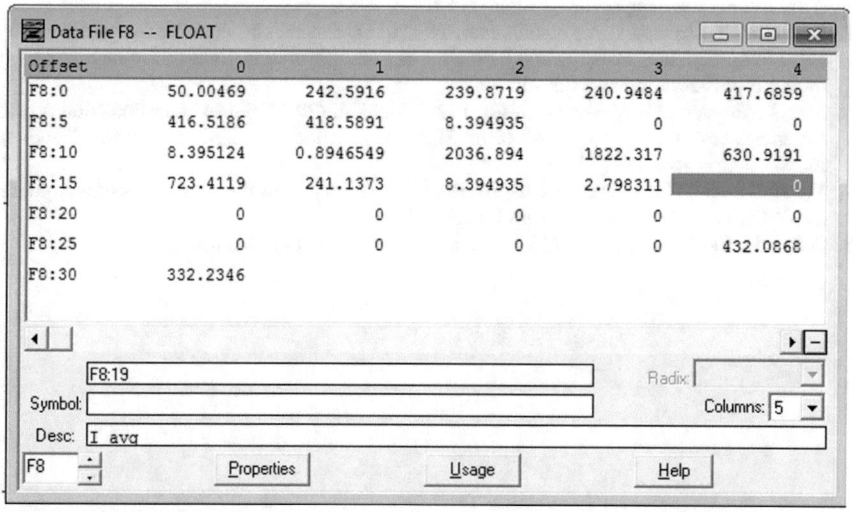

Fig. 13 Floating register of PLC

of the hardware modules of DIs and DOs of the PLC of RTU of ESS-1 and their expansion units have been given. Program for the PLC has been developed and tested using ladder programming language on Allen Bradley MicroLogix 1400.

Various SCADA modules, such as HMI module, data acquisition module, and control module, are being developed. HMI screens have both static and dynamic information. Schematic drawings of all parts of distribution system and single line diagram of each ESS form the static information while the status information of circuit breakers and values of important parameters constitute the dynamic information displayed on these screens. Data acquisition module is used to acquire data from the field devices and displays the same to the operator. Control module provides remote control features to the system. Intervention by the operator is secured by login facilities so that only an authorized operator will be able to give commands and controls the field devices. Another very important software module of SCADA system, which has to be developed, is a meter data management system (MDMS), which will include functions like processing of meter readings; storing of meter readings; and calculating and displaying the trends of power, voltage, frequency, power factor, etc.

References

1. Chen, Y., & Pie, W. (2013). Design and implementation of SCADA system for micro-grid. *Information Technology Journal,* 8049–8057.
2. Rangelov, Y., Avramov, A., & Nikolaev, N. (2015). Design and construction of laboratory SCADA system. In S. Bulgaria (Ed.), *Int. Sc. Conf. on Information Communication and Energy System and Technology* (pp. 300–303).
3. Jose, J., Varghese, C., Abraham, A., Joy, J., & Koilraj, A. (2017). Substation automation system for energy monitoring and control using SCADA. *International Journal of Recent Trends in Engineering & Research*, 32–38.
4. Verma, H. K. (2014). Chapter-2 hardware of supervisory control & data acquisition system. Smart Grid e-monogram at www.profhkverma.info.
5. Schneider Electric: Conzerv EM6400 Series Power Meters User Manual. (2015).

The Missing Link of Job Analysis: A Case Study

Prerna Mathur and Shikha Kapoor

1 Introduction

The online business dictionary [1] has defined "job" as a group of homogenous tasks related by similarity of functions. When performed by an employee in an exchange for pay, a job consists of duties, responsibilities, and tasks which are defined and specific and can be accomplished, quantified, measured, and rated. A job analysis, therefore, has been explained as a process of collecting information of a job by analyzing it from a job's various aspects. These aspects include job aspects such as tasks and responsibilities of a particular job role and details of the required competencies.

The information collected about the job through job analysis helps the organizations to understand the requirements of the job and to understand what employee specifications are required for performing the job. Job analysis, therefore, results in the formulation of two very important job-related documents: the job description and candidate specifications. A job description defined the list of various tasks and responsibilities being performed in a particular job role, whereas the candidate specifications list down the details of a candidate which will be required for performing the job in discussion. These two documents are extremely useful for various human resource functions in the organization. First and foremost, these are the basic requirements for any recruitment-related activity for any job. It helps in clearly communicating the role requirements to the current/prospective employees. It also helps the organization in the selection process for a particular job role by clearly defining the job requirements. In the situation of an ongoing job, these become the basis for conducting the training need analysis, initiatives like capacity building, and defining of key deliverables for

P. Mathur (✉) · S. Kapoor
AIBS, Amity University, Noida, Uttar Pradesh, India
e-mail: prernamathur2019@gmail.com

S. Kapoor
e-mail: skapoor2@amity.edu

© The Author(s), under exclusive license to Springer Nature Singapore Pte Ltd. 2021
P. K. Kapur et al. (eds.), *Advances in Interdisciplinary Research in Engineering and Business Management*, Asset Analytics,
https://doi.org/10.1007/978-981-16-0037-1_11

the performance management system. There are various methods of conducting job analysis which includes methods like questionnaires, interviews, observations, work diary, or log. The selection of the methods depends on the nature of the job and the nature of the participants.

There are various steps involved in conducting job analysis; they can be broadly defined as follows:

- Collecting background and available information about the roles.
- Identifying the representative roles to be included in the study of job analysis.
- Conducting job analysis and gathering relevant data.
- Developing job descriptions.
- Developing job specifications.

The outcome of job analysis lays down the foundation for developing other HR systems and processes. This can be diagrammatically depicted as follows (Fig. 1):

This chapter on job analysis intends to analyze the practice of job analysis in the context of current industry practice.

Job analysis is undertaken by the organizations of the jobs which are existent in nature. The job analysis gives an insight into the different tasks and activities being performed as a part of the roles in an organization. It is sometimes used as a proactive approach, when the organization intends to define and implement a necessary system

Fig. 1 Outcome of job analysis

or process or protocols which make the outcomes of job analysis a pre-requisite for the new system/process/protocols. It is also used as a curative approach, when the organization is able to identify few gaps and feels that the outcomes of job analysis are crucial, the curative approach was implemented. We can understand this in more depth with the help of a case study.

2 Case Study

The organization "A" belonged to the social sector and had the purpose of making a meaningful impact in the rural education of India. However, there was a phase when the organization realized that it was failing to make an impact which it intended to. An internal analysis revealed a problem with delivery of the key result areas of most of the employees. A more detailed look showed that though the tasks were being performed, the deliverables were not making the intended impact.

An HR expert was hired to have a more structured analysis of the problem and for recommendation and implementation of the suggested and approved solution. An introductory and briefing meeting was held between the HR expert and the management of the organization. The discussion gave the HR expert the point of beginning inadequate quality of performance by majority of the employees.

2.1 The Process of Analysis

The HR expert begins with a brainstorming session with the management team to understand their concerns better. The management team was encouraged to think the organizational level problems that they faced which led them to investigate the issue further. The outcomes of the session were as follows:

- The objective of the organization was to make a meaningful impact on the rural education of India. The management further defined it by explaining that this meaningful impact on rural education in India can be achieved by altering the methodologies of teaching. This often required frequent interaction of the employees with the teachers of rural areas. The interaction included discussions about the evolving and innovative methodologies of teaching, the passion, and commitment for "teaching" and ensuring "learning." The impact analysis of the organization revealed that the desired effect was not achieved.
- The teachers were not willing for follow-up discussions.
- The teachers were unwilling to adopt evolving and innovative teaching methodology
- The system was still convinced of a mechanical process of delivery of curriculum, and was not driven by the learning of the students. The passion and commitment were visibly missing even after the interventions of the organization.

It was also noted, discussed, and agreed that all the above measures were greatly dependent upon the interaction, relationship, and mannerism of dealing of the employees with the teachers. This eventually led the organization to further analyze the performance of its employees, since the deliverables of the employees were not making the desirable impact/achieving the expected results, though the tasks and activities were being performed as instructed. It was important for the expert to understand what were the tasks and activities which were being performed of the employees whose role was in discussion. It was also necessary to understand the relations of the tasks and activities being performed to the instructions being provided for performing these tasks and activities. In the absence of a well-defined job description, which could throw some light on the matter, the HR expert recommended job analysis.

2.2   Job Analysis

The process of job analysis was approved for the role of "Trainers" which was an existing role in the organization. This role was the role in discussion and was the link between the organization and the teachers for bringing the intended impact. The following tasks and activities were identified through the interview technique of job analysis:

- Meet the teachers of the schools in designated areas to introduce them to the interventions of the organization.
- Introduce the teachers to the innovative and evolving methodologies of teaching driven by learning.
- Recommend suitable teaching methodologies to enhance the learning experience of the students.
- Conduct capacity building workshops of the teachers for improved teaching methodologies.
- Conduct motivational workshops for teachers for inculcating passion and commitment for teaching.
- Conduct follow-up meetings with the teachers.
- Become the point of contact for any queries/concerns for the teachers.
- Recommend changes in the teacher training content to the designated authority.
- Maintain necessary records and prepare and submit reports.

A job description was developed including the abovementioned tasks and activities. The development of well-defined job descriptions of the role led to the development of the competency maps for the role of "Trainer."

The identified technical and behavioral competencies were as follows (Table 1).

The performance of the employees is greatly dependent on the skill set of the employees, and the achievement of the deliverables with the desired impact of the role is dependent on the fitment of the competencies of the employees as per the requirement of the role. Therefore, it was necessary to understand the talent being hired for

Table 1 Identified technical and behavioral competencies for the role of 'Trainer'

Technical competencies	Behavioral competencies
Pedagogy understanding	Effective communication skills
In-depth understanding of available teaching methodologies	Effective listening skills
Rural education understanding	Excellent interpersonal skills
Expertise in training methodologies	Self-confidence
Proficiency in computer skills	Presentable
Fluency in Hindi and English	Analytical
	Innovative
	Influential
	Organized

the role of trainer. In order to understand the employee performance gaps, it was also essential to understand the performance management system being practiced.

2.3 Process Analysis

2.3.1 Talent Acquisition

The identification of a well-defined job description and related competency maps enabled the HR expert to be clear about the talent required for performing the role effectively. In order to understand the presence/absence of any talent gap, it was essential to understand the skill set of the current manpower. The HR expert reviewed the resumes of the current manpower at the role of the trainer and conducted direct interviews to analyze the current competencies of the trainers in the organization. The HR expert was able to document the following observations:

- Majority of the trainers had a professional experience of either sales or marketing.
- Few of them had undertaken mentoring and coaching of their team members in their past experience.
- Many of them were comfortable with the rural areas.
- All of them were very comfortable with rigorous communication and conducting follow-ups.
- They were all willing to travel extensively.

The comparison of the skill set of the existing manpower with the skill set required for performing the role effectively yielded the following results.
Technical Competencies:

Table 2 Skill set of the existing manpower with the skill set required for performing the role effectively (Technical Competencies)

Skill comparison	
Technical competencies	Current skill status
Pedagogy understanding	Absent
In-depth understanding of available teaching methodologies	Absent
Rural education understanding	Absent
Expertise in training methodologies	Absent
Proficiency in computer skills	Somewhat present
Fluency in Hindi and English	Somewhat present

Table 3 Skill set of the existing manpower with the skill set required for performing the role effectively (Behavioral Competencies)

Skill comparison	
Behavioral competencies	Current skill status
Effective communication skills	Somewhat present
Effective listening skills	Absent
Excellent interpersonal skills	Somewhat present
Self-confidence	Somewhat present
Presentable	Somewhat present
Analytical	Absent
Innovative	Absent
Influential	Absent
Organized	Absent

See Table 2.

Behavioral Competencies:

See Table 3.

Therefore, the comparison clearly showed a talent gap of crucial technical competencies.

2.3.2 Performance Management System

The complete exercise of the HR expert revolved around the lack of impact of the employee's performance. Therefore, it was important to understand the performance management system of the organization. The expert conducted direct meetings with the management, the HR function, and the employees to understand the performance management system, along with reviewing the relevant documents. The focus of the performance management for the role of trainers in organization "A" was found to be on the following deliverables:

- Increasing the outreach of the organization in the rural education sector.
- Enrolling the rural education system in the interventions of the organization.

- Delivery of training sessions to the teachers of the rural schools.
- Maximizing the number of training sessions conducted.
- Ensuring maintenance of required training material as per the requirement of the training sessions.
- Conducting follow-up meetings with the teachers to take feedback and resolve queries.
- Developing and maintaining relevant reports and records.

It was therefore identified that the following deliverables of the role were not getting addressed:

- Introduce the teachers to the innovative and evolving methodologies of teaching driven by learning.
- Recommend suitable teaching methodologies to enhance the learning experience of the students.
- Conduct capacity building workshops of the teachers for improved teaching methodologies.
- Conduct motivational workshops for teachers for inculcating passion and commitment for teaching.
- Recommend changes in the teacher training content to the designated authority.

2.4 Analysis Outcomes and Recommendations

The structured analysis of the issues faced by organization "A" is given in the following:

- Mismatch between the skills and experience hired and the skills and experience required for effective performance of the role resulting in talent gap.
- The employees at the designated role were not able to connect with the targeted audience due to their lack of experience and knowledge in the related field.
- The employees were not able to convince the teachers for the desired interventions due to their limited knowledge of the area of expertise.
- The performance measurement criteria were not all-inclusive.

The HR expert, therefore, offered the following recommendations:

- Redefining the talent acquisition process to acquire the skill set as identified through job analysis for the role of trainer.
- Redefining the performance measurement criteria to be more inclusive of the performance quality.
- Developing and conducting a robust "Train the Trainer" program to build the capacity of the existing trainers.

3 Conclusion

It is evident from the case study illustrated above that the process of job analysis is an inevitable process in an organization. It forms the foundation for developing roles in an organization and enables the organization to function in a structured manner. It also lays the foundation for building various other functions of the human resource function of the organization. If not connected through job analysis, the various functions of HR like talent acquisition, performance management, training, and development can grow in isolation without achieving the desired objective.

Reference

1. http://www.businessdictionary.com/definition/job.html.

Rejection: Major Concern for a Cast Iron Foundry

Amar Wamanrao Kawale and Gajanand Gupta

1 Introduction

This study is based on the data collected from Cast Iron foundry situated at MIDC, Nagpur. It manufactures cast iron fittings used for sewage line. From last few years due to increased use of plastic fittings, the production is decreased to some extent. Though there is no comparison of plastic and cast iron in terms of durability and quality but still many customers preferred plastic due to less pricing. Also there is a high competitiveness among the foundry industry [1]. The customers demanding better and better quality product; on the other hand, the manufacturer tries to reduce operating cost to maintain the profitability [1]. Rejection is one of the major concerns for the cast iron foundry as it affects the customer relationship, company goodwill in the market, moral of the employee, wastage of costly raw material, etc. To keep the rejection level to minimum is of utmost importance to improve the productivity and profitability. There are many reasons for rejection. It may be due to raw material, process, equipment, manpower skill, etc. To identity the exact reasons of the rejection is very essential. Some of the basic techniques that can be used to identify the reasons of the rejection are root cause analysis, fishbone diagram, Pareto method, FMEA, etc. In addition to this, various methods and methodologies such as ISO certification, quality circle, six sigma, total quality management, etc. can be used to improve the quality of the products [6]. In addition to this, various industrial engineering techniques can also use for overall productivity improvement. The various industrial engineering techniques which can be used are work study, time study, proper plant

A. W. Kawale
Jhulelal Institute of Technology, Nagpur, Maharashtra, India
e-mail: amarwkawale@gmail.com

G. Gupta (✉)
Vellore Institute of Technology, Chennai Campus, Tamil Nadu, India
e-mail: gajanandgupta1222@gmail.com

© The Author(s), under exclusive license to Springer Nature Singapore Pte Ltd. 2021   131
P. K. Kapur et al. (eds.), *Advances in Interdisciplinary Research
in Engineering and Business Management*, Asset Analytics,
https://doi.org/10.1007/978-981-16-0037-1_12

layout, systematic movement of man and materials, ergonomics, etc. Total productive maintenance (TPM) also plays vital role in the manufacturing industries. TPM helps in providing breakdown-free manufacturing process by keeping all equipment in better working condition.

2 Literature Survey

Profitability is a major challenge for the industry. To earn profit it is very essential to produce quality product in a best minimum cost. Rejection is the major threat to the industries as it affects productivity and in turn profitability of the company.

Arasu [1] emphasizes importance of cost-effectiveness on the performance of the industry over its competitors. He explains the how rejection badly affect the productivity of the casting industries. He noted that excessive rejection decreases profit, causes wastage of costly raw materials, and demotivates management. He presented design of quality cost system for the foundry which consists of objectives of the system and development of quality cost system. He proposed a methodology to calculate cost of quality analysis, fault tree analysis, and how to implement quality cost system.

According to Damor and Thakkar [2], operational excellence is the key area in which Indian industries need to focus in today's global competition. Foundries and metal working and forming industries are required to implement quality and productivity improvement techniques for their overall improvement. Rejection and rework are the two areas which require serious attention. In his paper, he presented methodology of six sigma and DMAIC by various researchers. According to him, six sigma is one of the best techniques available for the operational excellence.

Borowiecki et al. [3] detected various reasons for the defected castings as slag inclusion, sand hole, porosity, misrun, shrinkage, sand buckle, blow hole, swell, mold shift, and hard spot. Out of this slag inclusion, sand hole, and porosity cause 70% of defected castings and others cause 30%. They used Pareto method for casting defect analysis. After detailed analysis it was found that improper construction of gating system is the main reason for defected castings.

Karnik and Kamble [4] explained the importance of quality on the performance of Indian manufacturing industries after globalization. They noted to restrict rejection to acceptable level of 5–6% is very difficult task in foundry industry. This is because there are many variables which causes rejection which are difficult to control. They suggest general procedure for casting rejection and defect analysis. After preliminary study and problem definition, they used Pareto chart analysis for sorting major and minor defects. It was found that runout and sand drop are the major defects. Using cause-and-effect diagram (Fishbone and Ishikawa diagram), detailed analysis is carried out which reduces rejection percentage from 23 to 3%.

According to Siekanski and Borkowski [5], since casting is a complicated process, there is a high risk of failure during all its process. They implement Pareto Lorenzo's analysis and used Ishikawa diagram to correlate cause and effect of various casting defects. After investigation it was found that displacement, misrun, slaggy, inhomogeneity, shrinkage depression, and hot cracks are the main reasons of rejections. They concluded that noncompliance of technological process and negligence of employee are the main causes of rejection.

Kamble [6] noted the main reason for the defected casting is improper controlling of manufacturing process steps during production. He discussed in detail the various steps in casting manufacturing. He used cause-and-effect diagram developed by Ishikawa to find out the various reasons for the defected casting. On the basis of his study, it will be possible to analyze the defect and remedies for defect-free castings. According to him, it requires continuous efforts to control rejection and then only it is possible to reduce the rejection.

Joshi and Kadam [7] in their paper discussed various processes of castings. According to them, in India, most of the operations in casting industries are manual and hence casualness in human work causes the casting defects. They used Pareto and cause-and-effect diagram to do the analysis of casting defects and to find out remedial action to improve quality and productivity. In their work, first, they did overview of production line and drawn a process flow diagram. Then they emphasized on manual operations like sand preparation, mold preparation, pouring, and shakeout. They used Pareto chart and performed cause-and-effect analysis to find out exact causes and their remedies. They conclude that by following this methodology it is possible to reduce the rejection.

Chokkalingam and Mohamed Nazirudeen [8] proposed a methodology to find and analyze major casting defect in automobile casting industry produced in sand molding process. On the basis of collected data, it was found that rejection percentage was in between 10 and 45%. They performed detailed study of all defects and identified major defects which are mold crush, mold damage, and sand drop. They used cause-and-effect analysis to find out root cause of all major defects. On the basis of their study, they identified various solutions and then selected the best solution. They concluded their working with the findings that by following their methodology the rejection percentage was reduced to 4 from 28%.

3 Methodology

On the basis of collected data we need to solve the problem in a systematic way. Following are the setups which describe the methodology used.

3.1 Data Collection

The company is manufacturing around 95–115 different types of fittings. The daily production capacity is 2400–2800 fittings. As per the customer requirement, the marketing department gives production schedule to production planning and control department (PPC). Based on the availability of the resources the PPC department gives production schedule to shop floor. Sometimes production is also done on the basis of past data to have maximum utilization of resources. Monthly rejection data of last 2 years is collected and tabulated as given in Table 1.

Rejection is the major concern for any industry. Due to rejection there will be increase in production cost and also cost of rework is added. Rejection interrupts dispatch schedule as well as production schedule. Due to delay in delivery there is chance to loose customer and customer goodwill. Due to rejection moral of the employee is also dwindled. Overall rejection creates an unhealthy environment in the industry which impede the growth of the company. For quality enhancement, detailed analysis of rejection is very important. From the data collected, it is found that the percent of rejection is in between 4 and 8% for the period of 10 months as given in Table 1. On the basis of data, the cases of rejection were identified (Fig. 1).

Table 1 Monthly rejection data of last 10 months

S. no.	Month	No. of casting produced	No. of casting rejected	Rejection (%)
1	Apr-2019	60,723	3562	5.87
2	May-2019	58,075	4108	7.07
3	Jun-2019	60,994	3796	6.22
4	Jul-2019	62,679	4498	7.18
5	Aug-2019	65,218	4888	7.49
6	Sep-2019	65,936	4100	6.22
7	Oct-2019	57,313	3289	5.74
8	Nov-2019	61,349	3146	5.13
9	Dec-2019	63,250	3753	5.93
10	Jan-2020	65,936	4850	7.36

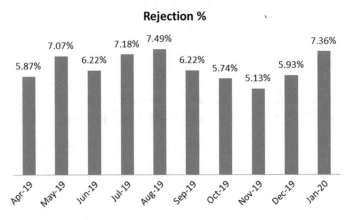

Fig. 1 Graphical representation of monthly rejection percent

3.2   Data Analysis Techniques

On the basis of above data it is necessary to identify the exact causes of the rejections. To do this, the organization must have knowledgable and skill manpower to identify the causes of the rejection. Data collection, data storing, and data retrieval should be the key factors. The main causes of rejection were bifurcated on the basis of different processes which were carried out during the processing of casting. The main processes are pattern making, metal pouring, mold making, core making, and knockout and fettling. Any irregularity in the above process causes defects in castings. Sometimes these defects can be tolerated, sometimes can be reworked, and if not then are rejected. There are various techniques which can be used to do the detailed analysis like FMEA, root cause analysis, Pareto analysis, fishbone diagram, etc.

3.3   Pareto Analysis

Pareto analysis is a creative way of looking at causes of problems because it helps stimulate thinking and organize thoughts. Pareto analysis consists of following procedure to identify the important causes:

a.   Form a frequency of covvurences as a percentage.
b.   Arrange the importance of causes in decreasing order.
c.   Plot a curve with causes on x-axis and cumulative percentage on y-axis.
d.   Plot a bar graph with causes on x-axis and percent frequency on y-axis.
e.   Draw a horizontal dotted line at 80% from the y-axis to intersect the curve.
f.   Draw a vertical dotted line from the point of intersection to the x-axis.
g.   The vertical dotted line separates the important causes (on the left) and insignificant causes (on the right).
h.   Review the chart to ensure that causes for at least 80% of the problems are identified.

As shown in Fig. 2, Pareto analysis is used to find out the cumulative rejection percentage for the respective month (Tables 2, 3, and 4).

From the data analysis, it is found that Misrun, Slag Inclusion, and Sand drop are the three major defects. Since 60 to 70% of the overall rejection is due to above three defects, Pareto analysis can be used to do the rejection analysis. Next step is to find out the root cause for each defect.

4   Cause-And-Effect Diagram

Cause-and-effect diagram represents the correlation between the problem and its causes. It is called as fishbone or Ishikawa diagram (Figs. 3, 4, 5, 6, 7, and 8).

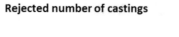

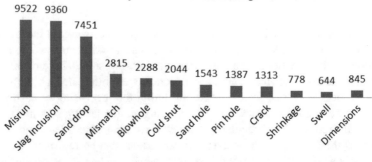

Fig. 2 Pareto chart defect analysis

Table 2 Rejection percentage of castings

Poured Qty	Poured Wt. (in MT)	Rejected Qty (in MT)	Rejected Wt. by Wt.	Rejection %
621,474	2796.633	39,990	179.955	6.43

Table 3 Number of castings ejected due to various defects

Types of rejection	Number of castings rejected
Misrun	9522
Slag inclusion	9360
Sand drop	7451
Blow hole	2815
Mismatch	2288
Coldshut	2044
Sand hole	1543
Pin hole	1387
Crack	1313
Shrinkage	778
Swell	644
Dimensions	845

Table 4 Overall rejection percentage

Poured Qty	Rejected Qty	Rejection (%)
621,474	39,990	6.43

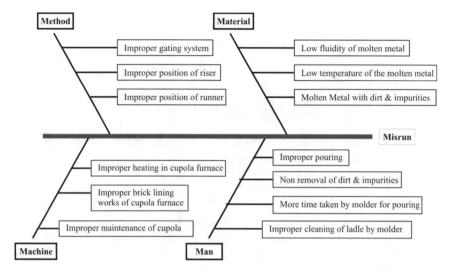

Fig. 3 Cause-and-effect diagram for misrun

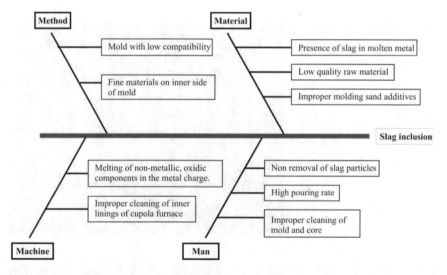

Fig. 4 Cause-and-effect diagram for slag inclusion

5 Analysis of Defects Occurring

5.1 *Misrun*

Misrun is the defect due to the low fluidity of molten metal. As the molder fill ladle
with molten metal and bring it to mold for pouring, the temperature of molten metal

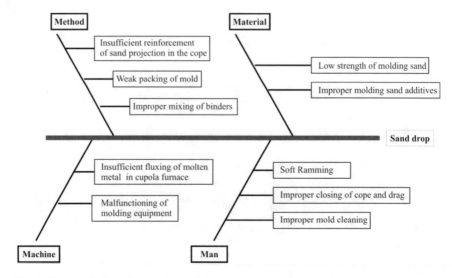

Fig. 5 Cause-and-effect diagram for sand drop

Fig. 6 Casting defect due to
misrun

starts decreasing. Improper gating system and improper positioning of riser also cause misrun defect. Improper heating of molten metal in cupola which is mainly due to improper lining of refractory bricks also causes misrun (Table 5).

5.2 Slag Inclusion

Slag inclusion mainly occurs due to contamination of molten metal with impurities. Low-quality metal charge containing non-metallic materials also forms slag. The

Fig. 7 Defected casting due to slag inclusion

Fig. 8 Defected casting due to sand drop

Table 5 Causes, effect, and remedial action for misrun

Causes for misrun	Effect in misrun	Remedial action
Low fluidity of molten metal	Mold completely not filled with molten metal	Check temperature of molten metal at the time of pouring
Improper design of gating system and riser	It restricts flow of molten metal in all areas of mold cavity	Proper positioning of runner and riser should be provided in the mold
Improper lining of refractory bricks in cupola furnace	It affects the heaing of metal charge inside cupola furnace	Good quality refractory bricks should be used
Improper pouring rate	Pouring rate too high or too low affects flow of molten metal inside mold	Proper pouring rate should be maintained
Improper cupola maintenance	It results in improper heating of metal charge which causes misrun	Proper maintenance activities should be planned

Table 6 Causes, effect, and remedial action for slag inclusion

Causes for slag inclusion	Effect in slag inclusion	Remedial slag inclusion
Slag in molten metal	It gets poured in the mold cavity and remians on the upper surface of the castings	Slag should be removed before pouring
Metal charge containing nonmetalics	During melting of metal charge in the cupola furnace formation of slag occurs	Good quality metal charge should be used in cupola furnace
Improper cleaning of mold and core	Dirt in mold and core causes formation of slag after molten metal is poured into mold cavity	Proper cleaning of mold and core should be done
Improper cleaning of ladle and casting tools	Impurities in ladle and casting tools added to molten metal	Cleaning of ladle and casting tools
Improper runner design	Due to this impurities not get skimmed	Proper runner design

cleaning of mold cavity and core is very important as it causes formation of slag after pouring molten metal into to the mold cavity (Table 6).

5.3   Sand Drop

Sand drop mainly occurs due to low strength of molding sand and improper molding sand additives. Soft ramming also causes sand drop. Sand of mold cavity breaks as the molten metal enters the mold cavity. Other reasons of sand drop are malfunctioning of molding equipment, insufficient fluxing of molten metal in cupola furnace, improper closing of cope and drag, and improper mold cleaning (Table 7).

Table 7 Causes, effect, and remedial action for sand drop

Causes for sand drop	Effect in sand drop	Remedial sand drop
Soft ramming	Due to soft ramming sand falls into mold cavity after pouring of molten metal	Ramming should be done properly
Low strength of molding sand and improper molding sand additives	Sand breaks in mold cavity	Use of good quality of sand and binding materials
Malfunctioning of molding equipment	Poor quality mold caused sand drop	Proper maintenance of molding equipment

6 Conclusion

From the above analysis, it is found that with the help of Pareto chart analysis and fishbone or Ishikawa diagram, it is possible to find out the exact causes of rejections. Since it is completely manual casting operations, it is very essential to evaluate various defects and its causes at different stages of operations, so that the remedial actions can be taken to reduce rejection. In this paper, attempt has been made to propose the systematic approach to find out defect causes and its remedial action to minimize overall rejections. It was found that out of total rejection the major defects were at the initial stages of operations, i.e., at metal charging in cupola furnace, sand mixing with various additives and bindings, gating system of the mold, position of runner and risers, properties of green sand, etc. Any casualness in the initial stages of manual operations will result in production of castings with defects.

The above methodologies prove that with the help of effective analysis of above techniques and processes, rejection can be minimized.

References

1. Arasu, M. (2010). Development of quality cost system in cast iron Foundries. Available online foundryinfo-india.org
2. Damor, Y., & Thakkar, H. R. (2014). A literature review on quality and productivity improvement in foundry industry. *Journal of Emerging Technologies and Innovative Research, 1,* 827–829.
3. Borowiecki, B., Borowiecka, O., Szkodzińka, E. (2011). Casting defects analysis by the Pareto Method. Published quarterly as the organ of the Foundry Commission of the Polish Academy of Sciences 11:33–36.
4. Karnik, S., & Kamble, B. (2018). Diagnostic approach towards analyzing casting defects—an industrial case study. *International Journal of Advanced Research in Science, Engineering and Technology, 5,* 5616–5625.
5. Siekanski, K., & Borkowski, S. (2003). Analysis of foundry defects and preventive activities for quality improvement of castings. *METALURGIJA, 42,* 57–59.
6. Kamble, B. (2016). Analysis of different sand casting defects in a medium scale foundry industry—a review. *ResearchGate, 5,* 1281–1288.
7. Joshi, A., & Kadam, P. (2014). An application of pareto analysis and cause effect diagram for minimization of defects in manual casting process. *International Journal of Mechanical And Production Engineering, 2,* 36–40.
8. Chokkalingam, B., & Mohamed Nazirudeen, S. S. (2009). Analysis of casting defect through defect diagnostic study approach. *Journal of Engineering Tome, 7,* 209–212.

A Chronological Literature Review of Evolution, Concept and Various Aspects of Employee Engagement Worldwide

Anoop Kumar and Shikha Kapoor

1 Introduction

Organisations are working in global perspective now. They have become more dependent on digitalisation and adopting newer technologies. Virtual working environments are growing with the constant need of physical outcomes. The scenario compels industries and organisations to connect their employees for more engagement for better productivity and efficiencies.

Employee engagement can play a major role in current circumstances, where contribution of all employees and contribution by each individual matter. Researches, especially in public sector organisations, are now showing that more employee engagements are resulting into higher organisational performance and increased productivity with improved net profits. Not only the reduced absenteeism rate and higher retention, employee engagement also results in healthy environment within the organisation [1]. Rathi Nandini articulates that in order to achieve superior performance through its people, employee engagement practices required in the fast-changing regime of technological disruptions should be redefined as automations with economic fluctuations changing the ways the businesses operate today [2]. Sahoo and Mishra advocated that not just a workforce but rather a community can be created through implementation of successful employee engagement practices [3]. Employees generate and form an emotional connection with the organisation, if engaged effectively and positively.

A. Kumar (✉)
Amity International Business School, Amity University, Noida, U.P. 201301, India
e-mail: anoop79.ak@gmail.com

S. Kapoor
HR Department, Amity International Business School, Amity University, Noida, U.P. 201301, India
e-mail: skapoor2@amity.edu.in

© The Author(s), under exclusive license to Springer Nature Singapore Pte Ltd. 2021   143
P. K. Kapur et al. (eds.), *Advances in Interdisciplinary Research
in Engineering and Business Management*, Asset Analytics,
https://doi.org/10.1007/978-981-16-0037-1_13

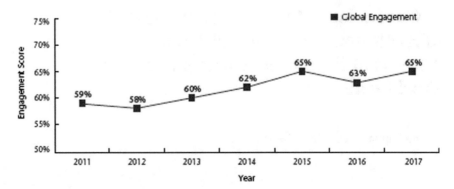

Fig. 1 AON Report, 2018; global trends in employee engagement

According to recent study by AON on 2018 global employee engagement trends as shown in Fig. 1, the employee engagement reached to its apex in 2015, dropped in 2016 and again matched to previous best in 2017. It can be seen as world average of employee engagement scores has increased by 6% over a period of 6 years from 2011 to 2017 with the world average of 65% in 2017 [4].

In current business scenario, more is being expected from the employees than ever before. Employee engagement has been consistently emerging as a valuable solution to cater the requirements of organisations. Although there is a paucity of clear-cut academic literature on employee engagement and little is known how management influence employee engagement.

2 Research Objectives

The objective of this literature review is to critically examine the available academic literature on employee engagement across the organisations. Research aims to identify various available definitions, history and origin, evolution, current scenario and other various aspects of employee engagement. The distinction between employee engagement and related concepts is elaborated. The paper also reviews the relationship between employee engagement and its different constructs.

3 Research Methodology

The study reviewed mostly secondary data available through various online available data, research papers and articles. Major research works from various journals over a larger time period from 1970 to 2018 were reviewed. Approximately, 150 articles were collected and 68 articles (as given in reference) were found relevant to the study. Following structure of methodology for literature review is adopted:

- Identification of relevant literature and potential evidences available.
- Screening of identified literature which is potentially relevant for the study.
- Arranging the screened literature logically and systematically.
- Analysing and reviewing the literature for critical examination.

4 Review of Literature

4.1 Employee Engagement

Employee engagement acts like a barometer to measure the level of association of employees with the organisation. Association levels are actually the level of involvement and commitment exhibited by an employee with its organisation and its value systems. At the same time, organisation also has responsibility towards the employee to develop and nurture employee's engagement.

Employee engagement studies are getting admiration from the professional community and prompted for intensive academic research to unify and integrate various available employee engagement concepts. Flade did an extensive work on engaged workers for 11 countries and calculated the proportions of engaged workers [5]. Accordingly, the percentages in countries were USA: 27%, Chile: 25%, Canada: 24%, New Zealand: 23%, Israel: 20%, Great Britain: 19%, Australia: 18%, France: 12%, Germany: 12%, Japan: 9% and Singapore: 6%. Gallup Inc. published that only 12% workforce in China and Thailand is engaged [6]. Crabtree calculated the ratio between aggressively disengaged-to-engaged workers in Western Europe as 1:0.81 [7]. The same ratio between US and Canada was 1:1.44 and for New Zealand versus Australia was nearly 1:1. Very particularly, for India, the total engaged workforce was approximately 10% only.

From the available data, it can be substantiated that majority of workers across the globe are disengaged. Approximately, 90% of employee population are inferred to be disengaged. Even the 10% of engaged employees are not exhibiting their full potential at workplace.

4.2 Definitions

Kahn defined employee engagement as 'The harnessing of employee's selves to their work roles [8]. People employ and express themselves physically, cognitively and emotionally during role performances'. Harter et al. defined employee engagement as 'An individual's enthusiasm for work with involvement and satisfaction at business unit level' [9]. Schaufeli et al. defined employee engagement 'as a positive, fulfilling, work-related state of mind that is characterized by vigor, dedication, and absorption' [10]. Baumruk defined employee engagement 'as energy or passion, employee exhibits for their work and even employers and subsequently resulting into their intellectual and emotional commitments' [11]. Wellins and Concelman termed employee engagement as an 'Illusive force which motivates employees to higher levels of performance' [12]. Saks defined employee engagement as 'A distinct and unique construct consists of three components namely cognitive, emotional and behavioural associated with individual role performance' [13].

4.3 Origin and History

The initial work on employee engagement was conceptualised by William A. Kahn in his paper titled 'Psychological Conditions of Personal Engagement and Disengagement at Work', which was published in 1990 in 'Academy of Management Journal' [8]. He quoted that at workplace, people can use themselves physically, cognitively and emotionally. These affect both their work and experiences in respective work roles. Kahn also outlined three psychological conditions, namely, meaningfulness, safety and availability which employees bring in or leave out when they engage or disengage [8].

Brown floated the concept of close association of employee engagement with job involvement [14]. Engagement is an antecedent to job involvement. Job involvement may be termed as a cognitive judgement about the needs satisfying abilities of the job thereby depending on one's self-image. Job involvement may also be defined as the extent to which an employee judges the job situation centrally and relate his or her identity with it [15]. Employee engagement has strong relationship with the emotions of employees about work experience and how they are being treated in the organisations. These emotions are centrally related to drive the bottom-line success of any organisation [3].

4.4 Evolution

Literature review shows that employee engagement concept is considered to be in its nascent stage till first decade of new century. According to a report prepared by Melcrum Publishing on the basis of a survey of more than 1000 human resource practitioners, 74% of the people started concentrating on employee engagement concept only after year 2000 [16]. Woodruffe specifies that employee engagement is a concept far from commitment and motivation concepts and has its own identity [17]. Two important concepts of employee engagement are organisational citizenship behaviour and employee commitment discussed below.

Organisational Citizenship Behaviour (OCB): Rafferty et al. articulate that employee engagement has two important precursors, namely, organisational citizenship behaviour (OCB) and employee commitment [18]. During the extensive review

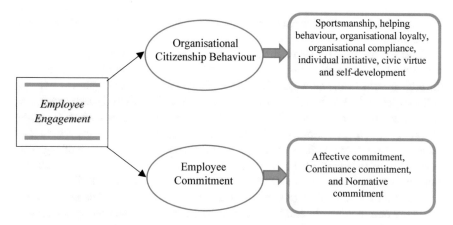

Fig. 2 Author's self-creation. Employee engagement with OCB and EC

of Organisational Citizenship behaviour (OCB), Barkworth identifies key characteristics of OCB as employee's discretionary behaviour, i.e. his choice of 'performing' or 'not performing'. He showcased 7 out of 30 different traits of OCB as sportsmanship, helping behaviour, organisational loyalty, organisational compliance, individual initiative, civic virtue and self-development [19]. Therefore, the employee going an extra mile for organisation has a strong relation of employee engagement with organisation citizenship behaviour.

Employee Commitment (EC): Employee commitment has multidimensional nature and has deep route through employee engagement. Employee engagement concept will remain incomplete if employee commitment is not touched. Tamkin [20] elaborated three types of commitments defined in model given by Allen and Meyer [21]: (a) affective commitment (emotional attachment employee has with organisation), (b) continuance commitment (related to cost incurred to organisation if an employee leaves) and (c) normative commitment (employee's moral obligation with organisation).

A range of factors affect employee's commitment and these factors differ from employee to employee. Silverman signifies the close relation between employee engagement and affective commitment, which shows the employee's willingness to go extra mile for betterment of the organisation and feels satisfied from their peers and work [22]. Author has tried to show important relation of employee engagement with OCB and EC in Fig. 2.

4.5 Relevance

Vlcek proposed that due to inadequate personal standards, ambitious employees sometimes contradict with the organisational gaols and become the cause of high turnover [23]. Therefore, the engagement proves to be more effective, if the activities and interests of employees are in alignment with the organisation goals.

After the work of Harter et al. relating the employee engagement with business outcomes [9], it became imperative for researchers to correctly position and rank the engagement in the network of different antecedents and consequences with job attitudes. There is dearth of literature having relative weight analysis and regressions, etc. to quantitatively correlate engagement with various variables which increases the interest of current researchers in the topic making it more significant in today's business world. Loch outlined that alignment is the clear understanding of organisational goals by the employee [24]. Macky and Boxall also clarify that it is important to generate systematic role links to achieve high employee involvement [25]. Goodridge suggested the key of getting through the tough times is to evolve confidence, i.e. based on timely communication and built on mutual respect, to create successful leaders with strategic responses [26].

Khetarpal further elaborated that communication affects various factors like performance feedback, motivation, job instruction, interpersonal relations, group efforts, etc. which in turn are responsible for overall organisational climate [27]. Sand, Cangemi and Ingram envisaged that various employee development activities like training, recognition of achievements, and encouragement of efforts create value for the organisation [28]. Job-related training helps employees to become proficient in their work and increases sense of responsibility. Peers also pay respect to recognised employees.

4.6   Various Surveys and Current Scenario

Various consulting firms and organisations have conducted many employee opinion surveys and generated their own strategy and frameworks depicting their approaches and perspectives for employee engagement. Also, consulting firms indicated different business outcomes of employee engagement. Some of the firms and their areas of publications are listed in Table 1.

Across the organisations, employee engagement has emerged as a buzzword and being also called for research interest in various researchers and practitioners studying the HR practices. The prime reasons behind this craze were.

- Many research findings indicate a strong relationship between employee's performance outcome variables and employee engagement [9] and
- Employee engagement has been seen as very positive psychological state of motivation.

Researchers have described engaged employees as highly energetic individuals, always buoyant to perform, exhibit consistent work involvements and experience emotions of challenge, enthusiasm, excitement, passion, significance and pride in their job. They work full heartedly, invest their willing efforts with full intensity and

Table 1 Various surveys and business outcomes

S. No.	Consulting and survey firms	Business outcomes
1	Development Dimensions International [29]; Gallup [30]; Hewitt Research Associates [31]; International Survey Research [32]; Towers Perri— International Survey Research [32]; Wellins, Bernthal and Phelps [33]	Sales and revenue growth
2	Corporate Executive Board [34]; Kenexa [35]	Individual productivity
3	Gallup [30]; HRA [31]; TP-ISR [32]	Financial performance
4	HRA [31]; TP-ISR [32]	Cost of goods sold
5	Harter et al. [9]; Luthans and Peterson [36]	Managerial effectiveness
6	Gallup [37]	Reduced accident rates in organisations
7	CEB [34]; DDI [29]; TP-ISR [32]	Reduced turnover
8	DDI [29]; Gallup [37]	Reduced absenteeism
9	DDI [29]	Reductions in quality errors
10	Tritch [38]	Reduced malpractice costs

get inspiration from their work [39, 40]. According to Schwartz, employee engagement has been reflected as common notion and primary source of organisational sustainability in all levels [41]. Therefore, all socially responsible organisations are emphasising more and more on employee engagement practices nowadays to sustain their positions in business market.

Wagner and Harter showed that engaged employees scored 34% higher on customer satisfaction ratings as compared to 12% with their colleagues [42]. Although new studies and employee engagement strategies emerged over the last 20–25 years, only working definitions have been offered and tested for validity. Moreover, research related to employee engagement has been continually demanding and pressure being felt by organisations to retain their intellectual talent and employee traits like commitment, contribution and competence [43] to survive in hyper-volatile business environment in twenty-first century [44].

4.7 Distinction of Employee Engagement from Other Concepts

It is worth appreciable from literature review that employee engagement is sharing both aspects of employee commitment and organisational citizenship behaviour, but engagement surely implies more than these concepts. Kahn also defined employee disengagement as 'the uncoupling of selves from work-roles' [8]. Till early 2000,

Kahn's defined domains of employee engagement were only widely accepted empirical literature available. Although Maslach et al. only reintroduce the concept of engagement with their research on burnout concept [45]. According to their research, employee engagement is characterised by high levels of pleasure and activation and is a persistent positive state. Accordingly, the concept of employee engagement is taken as positive antithesis to burnout. Also, burnout was further categorised into three factors such as (a) attaining high energy from the zone of emotional exhaustion, (b) strong involvement from the stage of depersonalization and (c) a sense of efficacy from the reduced sense.

As a major work after Kahn, Harter et al. analysed 7939 business units of varied fields on employee engagement during a survey conducted at Gallup organisation [9]. They conceptualised the employee engagement as an individual's involvement and satisfaction of employees parallel to their enthusiasm at work. This explanation acted as a catalyst for different researchers and academicians to measure and calculate engagement at business unit level. This became the first of its kind to link between employee engagement and organisation profits. Further, followed by article published by Harter et al. [9], other big international consultants like TP-ISR [32] and CLC deduced profitability links with employee engagement [34]. However, they also could not arrive on single common definition and concept.

After Harter's work, several researchers and practitioners contributed to engagement literature and tried to connect the determinants of engagement with its consequences. Rafferty et al. quoted that engagement supports mutual relationship between organisation and its employees [18]. However, Saks proved to be the first researcher to test antecedents and consequences of employee engagement [13].

With the publication of '*Effective Practice Guidelines Series*' on commitment and employee engagement by the Society for Human Resource Management [46], many professional societies started taking interest in the concept. Two years after the SHRM study [46], The American Society for Training and Development (ASTD) published a study with Dale Carnegie Training on employee engagement [47], which proved to be the first work to add training and learning perspective to the employee engagement concept. ASTD provided a significant developmental link to the research community with its data collection of 776 HR and learning executives across the world. ASTD linked the employee engagement with the employees who are emotionally and mentally invested in their work and contribution to organisation success.

4.8   Variables and Constructs of Employee Engagement

Famous cited researcher Kahn defined the personal engagement and disengagement psychological conditions at work. According to Kahn, employees express themselves physically, cognitively and emotionally during their role performances. Kahn denoted physical engagement as the actual labour exhibit by the employee, cognitive engagement as knowledge and perception held by employee about work environment and, lastly, emotional engagement as employee's attitudes and feelings for working conditions. Kahn also coined two critical components of employee engagement: (a) attention and (b) absorption [8].

Rothbard further elaborated the concept of 'attention' as amount of time and cognitive ability employee spends thinking about his/her job role. He further interpreted 'absorption' as employee being engrossed in his/her role and one's intensity and focus on a role [48]. Maslach et al. proposed the engagement as positive antithesis to job burnout [45]. It is a persistent affirmative stage of employee with high levels of pleasure and activation in their job roles. Harter's work became the first of its kind to give new perspective to employee engagement correlating it with organisational profit [9].

According to Hulin, employees may perceive work as monotonous and probably only levels of employee happiness can predict about levels of employee engagement both on individual and organisational levels [49]. Baumruk proposed three primary constructs to quantify employee engagement such as (a) Say, (b) Stay and (c) Strive [11]. The construct 'Say' is related to the employees who passionately advocates their workplace positively with their colleagues and customers. 'Stay' is referred to employees with intense desire to remain with their organisation, even getting better opportunity to work elsewhere. Third construct 'Strive' indicates employees who work above and beyond routine, exhibiting extra effort and producing extraordinary results for organisations and customers.

Corporate Leadership Council highlighted new model for employee engagement which focussed on business outcomes [34]. Model proposed that employee performance and retention is dependent on a balance between emotional and rational engagement. Council identified 25 drivers of employee engagement [34]. This definition focussed on business outcomes and helped organisations to measure tangible benefits through measuring employee retention.

Robinson et al. discussed the other constructs of employee engagement as organisational commitment and organisational citizenship behaviour [50]. Literature review also suggests other various dimensions of employee engagement as career, cooperation, communication, fair treatment, health and safety at job, equal opportunities, friendliness, pay and befits, job satisfaction, training and development, role of immediate management and performance and appraisal. Saks advocated social exchange theory and debated employee engagement as a distinct and unique construct with cognitive, emotional and behavioural elements associated with employee's role performance on job [13].

Heintzman et al. proposed that employee satisfaction and employee commitment are two strong drivers for employee engagement [51]. Employee satisfaction may be termed as the level of happiness an employee assigns to job/work, organisation [52]. Employee commitment is the pride felt by employee for organisation, desire to serve and intend to remain with organisation. Richman [53] and Shaw [16] argued that engaged employees have characteristics of job involvement and attachment. These characteristics increase employee's contribution in decision-making of organisation which in turn affects the employee's well-being and organisation's performance.

Richman also coined a three-tier model of employee engagement, where first tier includes safety-, benefits- and compensation-related elements and known as threshold factors. Second tier involves skill development and rewards and known as enablers. Third tier pertains to commitment drivers like job satisfaction, diversity, communication, managerial effectiveness, work-life support and career advancement [53]. Woodruffe categorised the employee's needs as (a) role engagement, (b) job satisfaction and (c) compensation package. He outlined that vertical and horizontal communication is very important for employee engagement, and same is closely dependent on employee's needs and satisfaction [17].

In its publication of 'Employee engagement, a review of current research and its implications' in year 2006, Conference Board analysed and reported 12 major studies by top research firms like Blessing White, Gallup, Towers Perrin and the Corporate Leadership Council [54]. This report collectively outlined 26 key drivers related to employee engagement. Conference Board also outlined eight key drivers as trust and integrity, nature of work, line of sight between employee and organisa-tion's performance, career growth opportunities, pride about organisation, employee development, team members and relationship with one's manager [55].

Scarlett contended for categorisation of drivers of employee engagement into engagement drivers and disengagement drivers [56]. Whereas freedom, advance-ment, job assignment, work and recognition and personal growth were under the dimensions of engagement drivers and communication, benefits, company pride, compensation, fellow associates, manager and work conditions were considered to be disengagement drivers. By measuring the levels of these drivers, the level of employee engagement can be measured.

Macey and Schneider floated the concept of trait engagement (job design attributes), state engagement (presence of transformational leader who directly affects state engagement) and behavioural engagement (presence of transformational leader who indirectly affects behavioural engagement) as the main constructs of employee engagement [57].

Macey and Schneider further enhanced the conceptual research on employee engagement and specified that three factors trait engagement, state engagement and behavioural engagement are separate constructs of employee engagement [57]. They further elaborated that trait engagement is affected by job design attributes, state engagement has direct effect in presence of a transformational leader and at the same time behavioural engagement is indirectly affected in presence of a transformational leader. Anchor asserts that a positive and engaged employee provides a tremendous competitive advantage [58]. While it was previously thought that job success, rewards and happiness are the basic tenets, Anchor specified that it is the happiness which drives successful employees [58]. Cummings and Worley identified four key elements of job involvement, namely, power, information, knowledge and skill, and rewards [59].

Different models, approaches and theories emerged from literature review suggest different variables of engagement. Most of these variables have positive conse-quences with employee engagement, therefore making engagement a popular concept for research. Table 2 summarises the variables and constructs highlighted through various research works across the world in the field of employee engagement.

Table 2 Variable and constructs identified from major works under different studies

Author, Year	Major work	Variables and constructs
Kahn, 1990	Environmental conditions and contextual factors [8]	Organisation norms, social support, job characteristics
Kahn, 1990	Need satisfying approach [8]	Meaningfulness, safety and availability
Schaufeli et. al., 2002	Work engagement approach [10]	Job resources and job demands
JD-R Model	Individual unit and organisational level variables [60]	Job resources and job demands
Bakker et al., 2007	Job demands [61]	Physical, social and organisational facets of job
Bakker et al., 2007	Job resources [61]	Psychological, physical, social or organisational facets of the job that decrease demands, help with realising work goals or motivate learning
Leiter and Maslach, 2004	Burnout antithesis [62]	Strong environment required for different areas of work life
Harter et al., 2002	Satisfaction-engagement approach [9]	Span of control, optimism and positive leadership
Saks, 2006	Antecedents and consequences of employee engagement [13]	Job characteristics, perceived organisation support, rewards and recognition, procedural justice and distributive justice
Hackman et al., 1975	Job Characteristics Theory (JCT) linked to motivation and situational factors [63]	Autonomy, feedback, skill variety, task significance and task identity are resources that catalyses the motivational states and these resources predict motivation, job satisfaction, and turnover and decreased absenteeism
Hobfoll, 1989	Conservation of Resources Theory (CRT) linked to motivation and situational factors [64]	To motivate employees both job resources and the aspiration to preserve and proliferate those resources play an important role
Shraga, 2007	A link between personality variables and vigour [65]	Vigour could be predicted by openness and extraversion Job characteristics is an antecedent to vigour. Job enrichment (job identity, job significance, supervisor feedback and skill utilisation) and perceived control all were positively related with vigour

(continued)

Table 2 (continued)

Author, Year	Major work	Variables and constructs
Avery, McKay and Wilson, 2007	Engagement and job satisfaction [66]	Linked with self-categorisation and social identity

4.9   Outcomes of Employee Engagement

Kahn also suggested that engagement creates both types of outcomes for individual and organisation level [8]. Individual feels happy and elated in their work while organisational outcomes are related to productivity and growth of organisation. Maslach et al. characterised employee engagement with high levels of pleasure and activation, while burnout antithesis approach characterises burnout as low levels of pleasure and activation [45]. Harter et al. outlined that engagement has positive consequences of customer satisfaction, productivity, safety, profitability, turnover and loyalty [9]. It was widely accepted that employee engagement and business outcomes have strong connection. Much of empirical research done on employee engagement identifies the various variables like job resources, co-worker support, extra-role performance, job security and decision-making [44].

Schaufeli et al. experienced engagement and burnout as contrary psychological states and defined both as diverse concepts. His workplace engagement approach was separate from burnout concept as engagement is described as positive, fulfilling, work-related state of mind, which is characterised by three vital factors, namely, vigour, dedication and absorption [10].

Innumerable reasons support that engaged employees perform better and more likely to stay with organisation bringing positive consequences to themselves as well as their organisation. Schaufeli et al. envisaged that positive work outcomes are results of positive experiences and emotions of employees [40]. Engaged employees have a greater attachment to their organisation and thereby have a lower tendency to leave their organisation. Schaufeli also quoted for relationship between engagement and organisational outcomes like organisational commitment and turnover intentions defining in-role and extra-role behaviours, customer service ratings and academic performance. Schaufeli noted that engaged employee has greater belongingness with their organisation and has a low tendency to leave the organisation. Many studies examined different consequences having the factors like organisational commitment with performance under JD-R model of employee engagement [60].

Various employee performance outcome variables were identified as intention to turnover and discretionary efforts [67], and overall performance [60]. Kevin K. Claypool examines on how happiness in the workplace affects the employee engagement for organisational success [68]. Researcher found out that passionate employees will ensure that they are productive in the workplace. Happiness is positively correlated with employee engagement and outlined as an important factor for engagement. Organisations can benefit in productivity and efficiency by ensuring the happiness of its workforce.

5 Conclusions

Organisations are striving for higher productivity and efficiencies in current business scenario with more automation and lean human resource. Increasing employee engagement at work place is need of time. Literature review reveals that different authors have floated different definitions and there is lack of mutual agreement. Kahn has also defined engagement as multifaceted [8]. Study shows that the employee engagement concept has been widely accepted as a protuberant driver for organisational success, although no major work could be carried out on employee engagement from 1990 to 2000. More than academic researchers, significant work on employee engagement is done by different survey and consultancy organisations. Literature reveals a connection between employee engagement and organisational productivity. Cumulatively, employee engagement may be defined as the positive, fulfilling and work-related state of mind of employees in organisations. These states of mind are characterised by vigour, dedication and absorption. Organisational citizenship behaviour and employee commitment are other two major dimensions related to employee engagement. Some of literature also reveals that happy employees are more engaged and put their discretionary efforts on jobs.

Literature shows strong connection of employee engagement with similar constructs. Some of the key variables and constructs like organisation norms, social support, job characteristics, meaningfulness, safety, perceived organisation support, rewards and recognition, etc. have been identified. Also, researchers were able to find some of the drivers like business outcomes, CLC [34]; employee satisfaction and employee commitment [51]; trust and integrity, nature of work, line of sight between employee and organisation's performance, career growth opportunities, pride about organisation, employee development, team members, relationship with one's manager [55]; and vigour, dedication and absorption [10]. However, scope of research on employee engagement persists continuously.

References

1. Vashanti, P. S. (2016). An exploratory study on employee engagement in A PSU. *International Journal of Management and Commerce Innovations, 3*(2), 539–546. ISSN 2348–7585 (Online). Retrieved from www.researchpublish.com.
2. Nandini, R. (2016). Retrieved January 02, 2019, from https://www.peoplematters.in/article/str ategic-hr/new-face-hr-psus-13022.
3. Sahoo, C. K., & Mishra, S. (2012). A framework towards employee engagement: The PSU experience. *ASCI Journal of Management, 42*(1), 92–110.
4. AON Report. (2018). Global trends in Employee Engagement. Retrieved December 25, 2018, from https://www.aon.com/2018global-employee-engagement-trends/index.html.
5. Flade, P. (2003). Great Britain's workforce lacks inspiration. *Gallup Management Journal, 11,* 1–3.
6. Gallup Inc. (2010). Employee engagement: What's your engagement ratio? Retrieved from https://www.gallup.com/consulting/121535/Employee-Engagement-Overview-Brochure. aspx.

7. Crabtree, S. (2011). What employees worldwide have in common? *Gallup Management Journal.* Retrieved from https://gmj.gallup.com/content/149405/Employees-Worldwide-Com mon.aspx
8. Kahn, W. A. (1990). Psychological conditions of personal engagement and disengagement at work. *Academy of Management Journal, 33*(4), 692–724.
9. Harter, J. K., Schmidt, F. L., & Hayes, T. L. (2002). Business-unit-level relationship between employee satisfaction, employee engagement, and business outcomes: A meta-analysis. *Journal of Applied Psychology, 87*(2), 268.
10. Schaufeli, W. B., Salanova, M., González-Romá, V., & Bakker, A. B. (2002). The measurement of engagement and burnout: A two sample confirmatory factor analytic approach. *Journal of Happiness Studies, 3*(1), 71–92.
11. Baumruk, R. (2004). The missing link: The role of employee engagement in business success. *Workspan, 47,* 48–52.
12. Wellins, R., & Concelman, J. (2005). Creating a culture for engagement. Workforce Performance Solutions. Retrieved August 1, 2005, from www.WPSmag.com.
13. Saks, A. M. (2006). Antecedents and consequences of employee engagement. *Journal of Managerial Psychology, 21*(7), 600–619.
14. Brown, S. P. (1996). A meta-analysis and review of organizational research on job involvement. *Psychological Bulletin, 120,* 235–255.
15. Lawler, E. E., III., & Hall, D. T. (1970). Relationship of job characteristics to job involvement, satisfaction, and intrinsic motivation. *Journal of Applied Psychology, 54,* 305–312.
16. Shaw, K., & Bastock, A. (2005). *Employee engagement, how to build a high-performance workforce.* London: Melcrum Publishing.
17. Woodruffe, C. (2006). Employee engagement. *British Journal of Administrative Management, 50,* 289.
18. Rafferty, A.N., et al. (2005). What makes a good employer? Geneva: International Council of Nurses. Retrieved from https://www.icn.ch/global/Issue3employer.pdf.
19. Barkworth, R. (2004). Secondments: A review of current research: A background paper for IES research network members. Institute for Employment Studies.
20. Tamkin, P. (2005). The contribution of skills to business performance.
21. Allen, N. J., & Meyer, J. P. (1990). The measurement and antecedents of affective, continuance and normative commitment to the organization. *Journal of Occupational Psychology, 63*(1), 1–18.
22. Silverman, M. (2004). Non-financial recognition: The most effective of rewards. Institute for Employment Studies Research Network.
23. Vlcek, D. J., Jr. (1987). Organization strategy: Decentralization: What works and what doesn't. *Journal of Business Strategy, 8*(2), 71–74.
24. Loch, C. H. (2008). Mobilizing an R&D organization through strategy cascading. *Research-Technology Management, 51*(5), 18–26.
25. Macky, K., & Boxall, P. (2008). Employee experiences of high-performance work systems: An analysis of sectoral, occupational, organisational and employee variables. *New Zealand Journal of Employment Relations (Online), 33*(1), 1.
26. Goodridge, M. (2008). A strategic approach to managing a downturn. *Strategic HR Review, 8*(1), 28–33.
27. Khetarpal, V. (2010). Role of interpersonal communication in creating conducive organisational climate. *ASBM Journal of Management, 3*(1/2), 77.
28. Sand, T., Cangemi, J., & Ingram, J. (2011). Say again? What do associates really want at work? *Organization Development Journal, 29*(2), 101.
29. Development Dimensions International 04, 16, 2014. Retrieved from https://www.ddi.com/pdf/ddi_employeeengagement_wp.pdf.
30. Gallup, . (2007). *Financial services company: Employee and customer engagement.* Princeton, NJ: Author.
31. Hewitt Associates, L., 2004 (HRA). Research brief: Employee engagement higher at double digit growth companies. www.hewitt.com.

32. International Survey Research. (2004). Engaged employees drive the bottom line. Retrieved from https://twrcc.co.za/Engaged%20employees%20drive%20the%20bottom%20line.pdf.
33. Wellins, R. S., Bernthal, P., & Phelps, M. (2005). Employee engagement: The key to realizing competitive advantage. Development Dimensions International, 1–30.
34. Council, C. L. (2004). *Driving performance and retention through employee engagement.* Washington, DC: Corporate Executive Board.
35. Kenexa (2008). The impact of employee engagement. 2008 Work Trends Annual Report. Retrieved from https://www.kenexa.com/getattachment/8c36e336-3935-4406-8b7b-777f1a faa57d/The-Impact-of-Employee-Engagement.aspx.
36. Luthans, F., & Peterson, S. J. (2002). Employee engagement and manager self-efficacy. *Journal of Management Development, 21*(5), 376–387.
37. Gallup. . (2004). *International manufacturing firm: Employee engagement.* Princeton, NJ: Author.
38. Tritch, T. (2003). Engagement drives results at new century. *Gallup Management Journal, 4.*
39. Bakker, A. B., & Demerouti, E. (2008). Towards a model of work engagement. *Career Development International, 13*(3), 209–223.
40. Schaufeli, W. B., & Bakker, A. B. (2004). Job demands, job resources, and their relationship with burnout and engagement: A multi-sample study. *Journal of Organizational Behavior, 25*(3), 293–315.
41. Schwartz, T. (2011). What it takes to be a great employer. Retrieved from https://blogs.hbr.org/schwartz/2011/01/what-it-takes-to-be-a-great-em.html.
42. Wagner, R., & Harter, J. K. (2006). *12: The great elements of managing.* Washington, DC: The Gallup Organization.
43. Crabtree, S. (2005). Engagement keeps the doctor away. *Gallup Management Journal, 13,* 1–4.
44. Bakker, A. B., & Schaufeli, W. B. (2008). Positive organizational behavior: Engaged employees in flourishing organizations. *Journal of Organizational Behavior, 29*(2), 147–154.
45. Maslach, C., Schaufeli, W. B., & Leiter, M. P. (2001). Job burnout. *Annual Review of Psychology, 52*(1), 397–422.
46. Vance, R. J. (2006). *Employee engagement and commitment: A guide to understanding, measuring, and increasing engagement in your organization.* Alexandria, VA: The SHRM Foundation.
47. Czarnowsky, M. (2008). *Learning's role in employee engagement: An ASTD research study.* Alexandria, VA: American Society for Training and Development.
48. Rothbard, N. P. (2001). Enriching or depleting? The dynamics of engagement in work and family roles. *Administrative Science Quarterly, 46*(4), 655–684.
49. Hulin, C. L. (2002). Lessons from industrial and organizational psychology. In J. M. Brett & F. Drasgow (Eds.), *The psychology of work: Theoretically based empirical research* (pp. 3–22). Mahwah, NJ: Erlbaum.
50. Robinson, D., Perryman, S., & Hayday, S. (2004). The drivers of employee engagement. Report-Institute for Employment Studies.
51. Heintzman, R., & Marson, B. (2005). People, service and trust: Is there a public sector service value chain? *International Review of Administrative Sciences, 71*(4), 549–575.
52. Peters, M. (2007, November 23). Employee engagement: A research snapshot. (pp. 1–2) Retrieved from https://www.cio.gov.bc.ca/local/cio/kis/pdfs/employee_engagement.pdf.
53. Richman, A. (2006). Everyone wants an engaged workforce how can you create it. *Workspan, 49*(1), 36–39.
54. Gibbons, J. M. (2006). Employee engagement: A review of current research and its implications. Conference Board.
55. Soldati, P. (2007). Employee engagement: What exactly is it. Management Issues.
56. Scarlett, K. (2007). What is employee engagement? Retrieved from https://www.scarlettsurv eys.com/papers-and-studies/white-papers/what-isemployee-engagement.
57. Macey, W. H., & Schneider, B. (2008). Engaged in engagement: We are delighted we did it. *Industrial and Organizational Psychology, 1*(1), 76–83.

58. Anchor, S. (2010). *The Happiness Advantage: The seven principles of positive psychology that fuel success and performance at work.* New York, NY: Broadway Print.
59. Cummings, T., & Worley, C. (2014). Organization development and change. Cengage learning.
60. Rich, B. L., Lepine, J. A., & Crawford, E. R. (2010). Job engagement: Antecedents and effects on job performance. *Academy of Management Journal, 53*(3), 617–635.
61. Bakker, A. B., Hakanen, J. J., Demerouti, E., & Xanthopoulou, D. (2007). Job resources boost work engagement, particularly when job demands are high. *Journal of Educational Psychology, 99*(2), 274.
62. Leiter, M. P., & Maslach, C. (2004). Areas of worklife: A structured approach to organizational predictors of job burnout. In P. L. Perrewe & D. C. Ganster (Eds.), *Emotional and physiological processes and positive intervention strategies, research in occupational stress and well being, 3* (pp. 91–134). Oxford: Elsevier.
63. Hackman, J. R., & Oldham, G. R. (1975). Development of the job diagnostic survey. *Journal of Applied Psychology, 60*(2), 159.
64. Hobfoll, S. E. (1989). Conservation of resources: A new attempt at conceptualizing stress. *American Psychologist, 44*(3), 513.
65. Shraga, O., & Shirom, A. (2007). *Work characteristics and job satisfaction as predictors of work-related vigor*. Presentation at the Symposium on Engagement at the annual meeting of the Society for Industrial and Organizational Psychology, New York, NY.
66. Avery, D. R., McKay, P. F., & Wilson, D. C. (2007). Engaging the aging workforce: The relationship between perceived age similarity, satisfaction with coworkers, and employee engagement. *Journal of Applied Psychology, 92*(6), 1542.
67. Shuck, B., Reio, T. G., Jr., & Rocco, T. S. (2011). Employee engagement: An examination of antecedent and outcome variables. *Human Resource Development International, 14*(4), 427–445.
68. Claypool, K. K. (2017). Organizational success: how the presence of happiness in the Workplace affects employee engagement that leads to organizational success. Pepperdine University. Graduate School of Education and Psychology. ProQuest 10269214, Published by ProQuest LLC (2017).

Integrated Robust Design Methodology and Dual Response Surface Methodology Approach in Optimization of Powder Coating Process

Preetam Naik and Suraj Rane

1 Introduction

The powder coating process in application is very much similar to painting. In this process, instead of using a wet paint, dry and free-flowing powder is applied electrostatically which is then heat cured to form a protective layer "skin." The surface obtained in this process is usually tougher and reliable than that produced with conventional surface finishing processes. The powder utilized in for this process is thermoplastic or thermosetting polymer that cures when exposed to temperature ranging usually from 170 to 220 °C in a controlled and enclosed atmosphere. The process does not require any addition of solvents and powder wastage is minimal due to the possibility of reuse of recovery powder. Thus, it is cost-effective, efficient, and environment friendly process.

Powder-coated parts generally have quality issues like low or high dry film thickness, dirt/dust/blisters, oil and bubble marks, adhesion failure, picture framing, poor color and texture, opacity, orange peel, pin-holing, gloss issues, etc. to mention a few among them. These defects eventually result in substantial rework of parts, thereby increasing the overall cost of production.

Through this case study, an attempt has been made to demonstrate the robust design concept using Taguchi methodology and apply it to enhance the quality characteristics of the powder-coated sheet metal components. However, owing to the shortcomings of the Taguchi method, DRS methodology was further used to optimize and achieve the critical-to-quality characteristic nearer to the target value. The

P. Naik (✉) · S. Rane
Mechanical Engineering Department, Goa College of Engineering, Farmagudi, Ponda 403401, Goa, India
e-mail: preetamnaik.mech@gmail.com

S. Rane
e-mail: ssr@gec.ac.in

© The Author(s), under exclusive license to Springer Nature Singapore Pte Ltd. 2021   159
P. K. Kapur et al. (eds.), *Advances in Interdisciplinary Research in Engineering and Business Management*, Asset Analytics,
https://doi.org/10.1007/978-981-16-0037-1_14

selected process was investigated for determining the process parameters that govern the quality of powder coating obtained and identifying the best settings for a set of input variables which would optimize the response. The remaining sections of the paper are arranged as follows.

The literature reviewed is presented in Sect. 2. The methodology is discussed in Sect. 3. The problem description and analysis is given in Sect. 4. Section 5 provides the experimentation details. Taguchi experimental design is explained in Sect. 6. Analysis and discussion is explained in Sect. 7. RSM approach is demonstrated with validation in Sect. 8. The conclusions are discussed in Sect. 9.

2 Literature Review

By reviewing literature of the previous works, it is possible to determine the strengths and weaknesses of various methods used in optimization of process parameters. Studying where and how these methods have succeeded and fallen short, efforts can be made in that direction to build upon successes of the past and avoid the shortcomings of the previous works. Literature review here examines how the various Design of Experiments (DOE) and optimization approaches has been applied and expanded.

Authors in [4] attempted to understand the impact of various parameters on the dry film thickness and microstructure obtained. They have optimized the output by varying the two input parameters (gun standoff distance and traverse speed). They found that the standoff distance was the prime factor which was affecting the dry film thickness. Taguchi DOE method was used by [6] to reduce the cost that was incurred in the painting process due to excess paint consumption. Four factors at two levels were considered for the analysis. The outcome of the experiment resulted in a significant lessening of cost.

The researchers in [14] used the Taguchi approach in process of casting of dies to find out the optimal settings which could obtain quality castings by minimizing the porosity formation. Other researchers [21, 12], and [10] through their research explored various Taguchi parameter designs and performance measures which were applied to solve the problems encountered in engineering and fetching proper levels of operating conditions and expenses.

Authors in [22] through their research demonstrated the effect of standoff distance on the job to be coated, the coating deposition was found to be affected by standoff distance for maximum powder coating deposition, and efficiency varies with the velocity of the powder thrown which in turn affects the coating thickness formed. Authors in [23] through their work revealed that the coarse powder gives higher coating efficiency than the ultrafine powder. The spray distribution for ultrafine powder was found even and concave for coarse powder.

Taguchi technique was used to investigate the effect of the powder coating process parameters on the response characteristics [16]. Through validation they found out that there was a good agreement between the estimated and the obtained values in respect to DFT and orange peel within a preferred significance level. The study by authors in [2] reveals the optimum levels of process parameters for casting of ductile iron components. They also found out the correlation between various factors by using RSM methodology.

As per [5], it is possible to find out the optimum value from a set of known parameters and at selected levels using Taguchi DOE method. This method enables us to obtain the result with lesser number of trials but the optimum value obtained could always not be the global optimum. Authors in [1] demonstrated a method solve complex linear equations by using iterative generalized reduced gradient method. They have shown that this linear equations with constraints can be programmed using Excel to find out optimal solutions. Various techniques of Excel solver were demonstrated by authors in [8].

Authors in [11] used a dual response optimization approach that can be achieved using standard nonlinear programming technique, the generalized reduced gradient algorithm. Also the research claims that this method is more flexible and easier to use when compared with other dual response methods. Authors in [17] proposed a more satisfying and substantially simpler optimization procedure by DRS approach. [3] used the Taguchi along with response surface methodology to optimize the power consumption and also the roughness of surface obtained. Through their study [18] predicted out the advantages of RSM over Taguchi. Their study revealed the RSM method gives an optimized output over Taguchi method. Also other researchers [15, 24, 19], and [20] have used the response surface methodology to find optimum solutions.

3 Methodology

3.1 Taguchi Design of Experiments

DOE is a very powerful statistical tool introduced by R. A. Fisher, which can be effectively utilized to validate the correlation between input and output variables. DOE procedure consists of a series of experimentations to calculate undetermined measurements of causes and also along with it the connections between these causes can be predicted within a fewer trials. Taguchi methods, also called robust design methodology, are statistical methods developed by Genichi Taguchi to minimize the sensitivity to noise for any product or process. Taguchi technique makes use of matrix called "orthogonal array" to study the entire parameter space with lesser number of experiments when compared with traditional methods [7]. Orthogonal arrays are then used to conduct a set of experiments and the results of these experiments can be used to analyze and predict the required quality characteristic. Thus, in Taguchi methods, the use of the loss function is used to measure the performance of quality characteristic which tend to deviate from the mean value or desired value. This value is transformed further into a single measure called as Signal-to-Noise (S\N) ratio. The

Fig. 1 Major steps in
Taguchi approach [9]

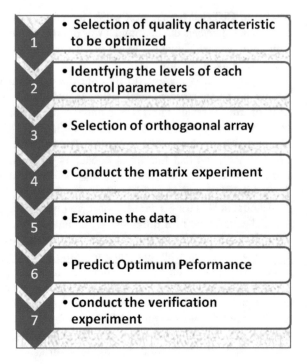

1 • Selection of quality characteristic to be optimized

2 • Identfying the levels of each control parameters

3 • Selection of orthogaonal array

4 • Conduct the matrix experiment

5 • Examine the data

6 • Predict Optimum Peformance

7 • Conduct the verification experiment

motive of Taguchi approach is to maximize the S/N ratio by selecting appropriately the design parameters which will help in getting closer to the desired value or to reduce variation in the selected product's CTQ characteristics. There are three forms of Signal-to-Noise (S/N) ratio that are of common interest for optimization of static problems and depending upon the problem we have to select the one best suited for a particular type of application. They are named as smaller-the-better, larger-the-better, and nominal-the-best [9]. The major steps to complete an effective designed experiment are systematically explained in seven steps as a flow diagram and the same is depicted with the help of Fig. 1.

3.2 Response Surface Methodology

Response Surface Methodology (RSM) is one among the most frequently used statistical and mathematical technique for achieving the target value of the response variable. RSM approach can help to answer the following types of questions [13]:

- How a particular response is affected by a known set of input variables over the required region of interest?
- To what levels the inputs should be controlled, to give a product simultaneously with satisfying desired requirements?

- What input value will yield a maximum for a specific response, and what is the nature of response surface close to the optimum?

The RSM approach stresses mainly on optimizing the estimated mean by identifying the best possible parameter settings for a given set of input or design variables that would optimize the response. Using the polynomial model as given in Eq. 1 the mean is estimated.

$$y_\mu = a_0 + \sum_{i=1}^{n}(a_i) * (x_i) + \sum_{i=1}^{n}(a_{ii}) * (x_i)^2 + \sum_{i<j}^{n}(a_{ij}) * (x_i) * (x_j) \qquad (1)$$

where "y_μ" is the estimated value of the mean of the response parameter y and x_i, $i = 1,2,3,4 \ldots$, n, are the preparatory variables. The optimum values of xi's, which would bring the estimated mean near the target are then determined. In our case of Dual Response Surface Methodology (DRSM) approach, along with the polynomial model for estimating the mean of the response variable, another model for estimating the signal-to-noise ratio of the process is also required and is shown in Eq. 2.

$$y_{\frac{s}{n}} = b_0 + \sum_{i=1}^{n}(b_i) * (x_i) + \sum_{i=1}^{n}(b_{ii}) * (x_i)^2 + \sum_{i<j}^{n}(b_{ij}) * (x_i) * (x_j) \qquad (2)$$

where "$y_{s/n}$" is the estimated value of the S/N ratio of the response parameter "y." Then both the responses (mean and S/N ratios) are simultaneously optimized under a known set of constraints.

4 Scope of Work

The study relates to a manufacturing unit, supplying precision sheet metal fabricated enclosures and structures to healthcare industries. The company was witnessing frequent in-house and customer rejections. When analyzed it was found that a major share of rejections were due to quality issues which occurred from powder coating department, thus affecting quality and delivery performance. The history records of powder coating were collected and analyzed for last 1 year and had shown that high and low dry film thicknesses were the two major contributors for rejection in powder coating process. The analysis is systematically depicted using Pareto analysis and is shown in Fig. 2. It was, therefore, decided to study the powder coating process with an objective of identifying robust parameter settings and for controlling part-to-part variation occurring in the process.

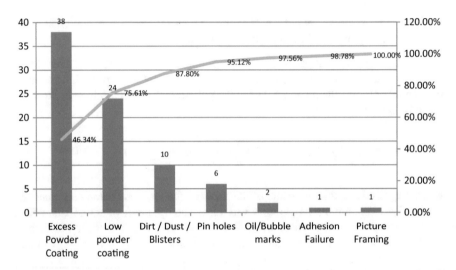

Fig. 2 Pareto analysis for current problem statement

4.1 Parameter Selections

In order to make an analysis, it is necessary to identify the potential contributors of the two major defects as mentioned in earlier section. The potential causes that can have an effect on the thickness of dry film obtained are displayed in Fig. 3. and were identified by a detailed brainstorming session with assistance of the field experts and are well organized using a cause-and-effect diagram.

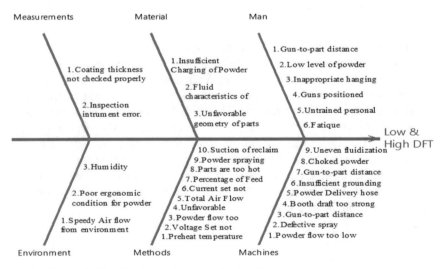

Fig. 3 Cause-and-effect diagram for low- and high-DFT issues

Table 1 Process parameters and their ranges

Sr. No.	Process parameter	Range	Units
1	High voltage	60–100	Kv
2	Current limitation	40–100	μA
3	Total air flow	2–5	m³/h
4	Feed air	40–80	%

Once these causes are identified, robust design method is used to optimize the powder coating operation. With reference from the cause-and-effect diagram while taking into account the prior knowledge of the electrostatic powder coating process the following factors were considered to be most important and necessary to control for optimization and are attributed to four parameters along with their suggested range as specified in Table 1.

5 Experimental Setup

The experimental setup involves a manual-operated semi-automatic electrostatic powder coating machine with a corona gun and is depicted in Fig. 4. Corona-styled powder coating guns charge the powder as it leaves the nozzle while passing through an electric field that is created by an electrode that forms a part of a corona gun. The negatively charged powder particles blown with the help of compressed air get

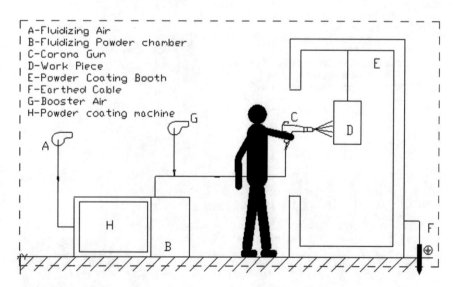

Fig. 4 Powder coating system setup

attracted to the grounded product and allow the powder to temporally accumulate to the product surface thus forming the base of powder coating process.

In the present case study, sheet metal samples of size 120 mm × 120 mm × 2 mm are used for the experimentation purpose. The samples to be studied are passed through a pre-defined and established sequence of operations which involves a nine tank pre-treatment process, electrostatic powder spraying, baking, and curing.

During the baking process, temperature inside the oven is raised up to 220 °C where powder begins to melt, flow, and chemically react to form a cross-linked higher molecular weight polymer in a network-like structure under the influence of intense heat. Thus, a tough continuous film of protective surface is formed upon cooling. The self-explanatory flow diagram of powder coating process is shown in Fig. 5.

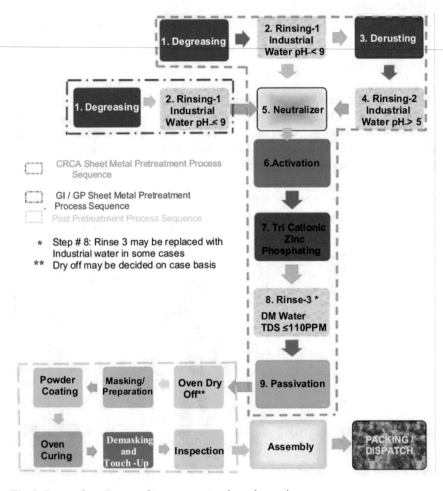

Fig. 5 Process flow diagram of pre-treatment and powder coating

Table 2 Process parameters with their values

Factor	Process parameter	Levels				
		L1	L2	L3	L4	L5
A	High voltage (Kv)	60	70	80	90	100
B	Current limitation (μA)	40	55	70	85	100
C	Total air flow (m^3/h)	2	2.75	3.5	4.25	5
D	Feed air (%)	40	50	60	70	80

6 Experimental Data

In order to get a satisfactory result, five levels selected to each process parameter identified with their corresponding values are represented with the help of Table 2.

In this study, nominal-the-better principle is considered to optimize the DFT in powder coating process. The corresponding loss function is expressed in Eq. 3.

$$S/N_{nominal\ the\ best} = 10 * \log_{10}\left(\frac{\bar{Y}^2}{s^2}\right) \tag{3}$$

The orthogonal array and results obtained from each of the 25 trials are reported in Table 3. The S/N ratios were computed using Eq. 3. The average values of DFT for each parameter at different levels are calculated and plotted in Fig. 6.

7 Data Analysis

7.1 Analysis of S/N Ratio

Taguchi method stresses importance on studying the response variation using the signal-to-noise (S/N) ratio, resulting in minimization of quality characteristic variation due to uncontrollable parameter. The dry film thickness was considered as the quality characteristic of choice with the concept of "nominal-is-better."

As seen from Fig. 7, the peak points obtained from these graphs can be considered to be the optimal combination with their values being high voltage of 60 kV, current limitation of 55 μA, total air flow of 5.0 m^3/h, and feed air of 80%.

7.2 Analysis of Variance

ANOVA is a statistically decision-making tool to detect differences in the average performance and helps testing the significance of all main factors [7, 18]. This method

Table 3 Experimental results obtained

High voltage (Kv)	Current limitation (μA)	Total air flow (m³/h)	Feed air (%)	S\N ratio	Mean
60	40	2	40	16.69	23.83
60	55	2.75	50	26.25	62.83
60	70	3.5	60	26.09	80.33
60	85	4.25	70	29.18	86.67
60	100	5	80	29.73	92.33
70	40	2.75	60	23.03	56.00
70	55	3.5	70	27.33	87.00
70	70	4.25	80	28.80	105.17
70	85	5	40	24.83	61.83
70	100	2	50	22.49	50.83
80	40	3.5	80	29.05	99.33
80	55	4.25	40	22.18	46.50
80	70	5	50	27.13	83.50
80	85	2	60	20.76	39.50
80	100	2.75	70	26.10	70.67
90	40	4.25	50	25.52	59.33
90	55	5	60	30.29	91.17
90	70	2	70	23.30	53.67
90	85	2.75	80	25.33	63.67
90	100	3.5	40	22.55	36.67
100	40	5	70	29.03	90.17
100	55	2	80	23.38	51.67
100	70	2.75	40	15.94	19.00
100	85	3.5	50	25.43	57.17
100	100	4.25	60	27.04	70.67

is widely used to evaluate which design parameters significantly affect the quality characteristic in an orthogonal experiment. Hence, ANOVA was performed to understand the significance of process variables affecting dry film thickness and the same is achieved by separating the total variability, which is measured by the sum of the squared deviations from the total mean into contributions by each of the process parameters. The parameter current limitation with a P-value of 0.734 is found to be more significant than high voltage with a value of 0.123 as seen in Table 4.

Table 5 presents the ANOVA of S/N ratio. From this table, it is also clear that the parameters such as high voltage and current limitation significantly affect the variability of dry film thickness with P-value of 0.568 and 0.421, respectively.

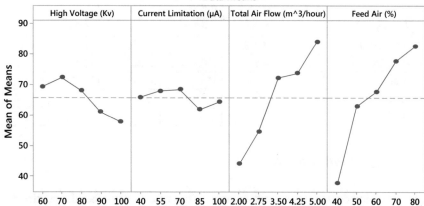

Fig. 6 Main effect plot for means

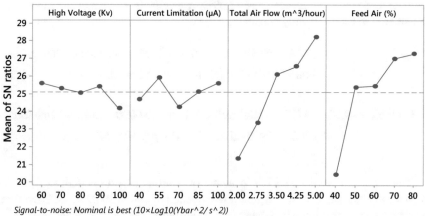

Fig. 7 Main effect plot for S/N ratios

Table 4 Analysis of variance for dry film thickness (means)

Source	DF	Seq SS	Adj SS	Adj MS	F	P
High voltage	4	726.7	726.7	181.68	2.53	0.123
Current limitation	4	145.2	145.2	36.3	0.5	0.734
Total air flow	4	5170.7	5170.7	1292.68	17.98	0
Feed air	4	6129.9	6129.9	1532.48	21.31	0
Residual error	8	575.2	575.2	71.91		
Total	24	12747.8				

Table 5 Analysis of variance for dry film thickness (S/N data)

Source	DF	Seq SS	Adj SS	Adj MS	F	P
High voltage	4	6.259	6.259	1.565	0.78	0.568
Current limitation	4	8.773	8.773	2.193	1.1	0.421
Total air flow	4	150.5	150.501	37.625	18.8	0
Feed air	4	150.756	150.756	37.689	18.83	0
Residual error	8	16.012	16.012	2.001		
Total	24	332.301				

7.3 Regression Analysis

The regression equation is a mathematical model which establishes the relationship between the independent and the response variable [13]. The regression equation for means and S/N ratio is calculated based on all the possible combinations from the available data using the Minitab software and was found out to be as mentioned in Eqs. 4 and 5. The R sq (adj) value of the two regression models, mean and S/N ratio, were found to be 86.46% and 75.69%, respectively, and hence the models are considered to be sufficiently adequate.

$$\text{Mean} = -11.8 - 0.342 * \text{High Voltage (Kv)} - 0.0604 * \text{Current Limitation } (\mu\text{A})$$
$$+ 13.20 * \text{Total Air Flow (m}^3/\text{h)} + 1.046 * \text{Feed Air (\%)} \tag{4}$$

$$\text{S/N Ratio} = 9.72 - 0.0275 * \text{High Voltage (Kv)} + 0.0070 * \text{Current Limitation}$$
$$(\mu\text{A}) + 2.263 * \text{Total Air Flow (m}^3/\text{h)} + 0.1527 * \text{Feed Air (\%)} \tag{5}$$

The predicted value of DFT is around 114 microns which is obtained from the regression equation against a required specification of 100 ± 20 microns. Therefore, a scope for optimization with cost saving opportunity. Hence, dual RSM is used.

8 Optimization by DRSM Approach and Confirmation Experiment

To minimize the cost occurring in the powder coating operation, it was suggested by the top management of the selected company that the dry film thickness of 100 microns would be ideal for coating panels considering its suitability for use in medical industries. Hence, the dry film thickness target of 100 ± 5 microns was chosen. With these conditions known the optimization problem becomes as shown in Eq. 6 with a goal of quality improvement effort that can be stated as attempting to maximize the signal-to-noise ratios.

$$\text{Maximize S/N ratio} = 9.72 - 0.0275 * \text{High Voltage (Kv)} +$$
$$0.0070 * \text{Current Limitation } (\mu A) +$$
$$2.263 * \text{Total Air Flow } (m^3/h) + 0.1527 * \text{Feed Air } (\%) \qquad (6)$$

Subjected to the following constraints:

1.

$$95 \leq -11.8 - 0.342 * \text{High Voltage (Kv)} - 0.0604 * \text{Current Limitation } (\mu A)$$
$$+ 13.20 * \text{Total Air Flow } (m^3/h) + 1.046 * \text{Feed Air } (\%) \leq 105$$

2.

$$60 \leq \text{High Voltage (Kv)} \leq 100$$

3.

$$40 \leq \text{Current Limitation } (\mu A) \leq 100$$

4.

$$2 \leq \text{Total Air Flow } (m^3/h) \leq 5$$

5.

$$40 \leq 1.046 \text{Feed Air } (\%) \leq 80$$

where all constraints considered are integers.

This integer programming problem can easily be solved by using one of the most commonly known robust nonlinear method named as Generalized Reduced Gradient (GRG) algorithm method. The results obtained using this method are displayed in Table 6 which predicts the optimum parameter settings which would give the DFT near to 105 microns, and hence very close to the optimum target value of 100 microns.

In order to validate the results obtained, ten confirmation experiments were conducted for each of the response variables. The results obtained by maintaining optimal process parameters by dual RSM approach are interpreted in Table 7. It is to be pointed out that the outputs match the customer requirements and are within a reasonable degree of variation.

Table 6 Optimal solution

Parameters	Optimal value
High voltage (Kv)	78
Current limitation (μA)	100
Total air flow (m^3/h)	5
Feed air (%)	80
Result	105

Table 7 Results of confirmation experiments

Job no.	1	2	3	4	5	6	7	8	9	10	Mean	Std. Dev.
DFT obtained	100	99	105	100	104	100	105	103	103	98	102	2.58

9   Conclusions

The combination of Taguchi and dual response methodology, namely, Generalized Reduced Gradient (GRG) technique has resulted in a solution, which led to reduction of defects prevailing in powder coating operation. The Taguchi approach of selecting design parameter levels that maximizes the S/N ratio was used to find out the combination which gives the best results. The key requirement of the case being optimization, hence DRS method was further used. To obtain the optimum conditions using DRS method, it was necessary to formulate an objective function with a known set of constraints.

Regression analysis was done on the experimental runs to obtain equations for means and the S/N ratio. The equation obtained as S/N ratio was then used as objective function for maximization while maintaining the constraints. This integer programming problem was formulated and solved using Microsoft Excel. Based on the trials taken from the case study, the following conclusions were drawn:

- Taguchi method was capable of finding the best value at any pre-specified level. Value estimated from Taguchi revealed an output of 114 μm. To be more cost-effective, a further optimization is required.
- From broad literature review, the strengths of the DRS method were identified.
- The study revealed that the optimal combination estimated from DRS method would yield dry film thickness of 105 μm which is very close to a target value of 100 μm.
- Confirmation experimental run was finally carried out at a condition closest to the optimum one which is obtained from GRG method.

- It has been noted that the optimum condition obtained by integrating Taguchi and dual response methodology gives a result of 102 μm and is very near to the target value of 100 μm.
- DRS method was thus able to find out the parameters which could yield optimum results within or beyond the level of specified factors.
- The optimal settings hence can updated in the process control plan to formalize and document the system of control that shall be utilized.

References

1. Brown, A. M. (2006). A non-linear regression analysis program for describing electrophysio-logical data with multiple functions using Microsoft Excel. *Computer Methods and Programs in Biomedicine, 82*(1), 51–57.
2. Johnson Santhosh, A., & Lakshmanan, A. R. (2016). Investigation of ductile iron casting process parameters using Taguchi approach and response surface methodology. *Overseas The foundry, China The Foundry, 13*(5) September.
3. Aggarwal, A., Singh, H., Kumar, P., & Singh, M. (2008). Optimizing power consumption for CNC turned parts using response surface methodology and Taguchi technique—A comparative analysis. *Journal of Materials Processing Technology, 200*(1–3), 373–384.
4. Iyer, B., Nalbalwar, S., & Pawade, R. (Eds.). (2016). ICCASP/ICMMD. *Advances in Intelligent Systems Research, 137*, 71–76. © 2017.
5. Su, C. T., & Yeh, C. J. (2011). Optimization of the Cu wire bonding process for IC assembly using Taguchi methods. *Microelectronics Reliability, 51*, 53–59.
6. Bharathi, D., Baskaran, R. (2014). Improvement of painting process in steel structure using Taguchi's method of experimental design. *International Journal of Mechanical and Industrial Technology, 2*(1), 1–9. ISSN 2348-7593.
7. Montgomery, D. C., & Runger, G. C. (2014). *Applied statistics and probability for engineers* (6th ed.). Singapore: Wiley Singapore.
8. Fylstra, D., Lasdon, L., Watson, J., & Waren, A. (1998). Design and use of the Microsoft Excel Solver. *Interfaces, 28*(5), 29–55.
9. Taguchi,G. (1986). Introduction to quality engineering: Designing quality into products and processes. Asian Productivity Organization, American Supplier Institute, Dearborn, MI, USA.
10. Huang, Y. M., & Shiau, C. S. (2009). An optimal tolerance allocation model for assemblies with consideration of manufacturing cost, quality loss and reliability index. *Assembly Automation, 29*(3), 220–229.
11. Jeong, I. J., Kim, K. J., & Lin, D. K. J. (2010). Bayesian analysis for weighted mean-squared error in dual response surface optimization. *Quality and Reliability Engineering, 26*(5), 417–430.
12. Jailani, H. S., Rajadurai, A., Mohan, B., Kumar, A. S., & Sornakumar, T. (2009). Multiresponse optimisation of sintering parameters of Al–Si alloy/fly ash composite using Taguchi method and grey relational analysis. *International Journal of Advanced Manufacturing Technology, 45*(3–4), 362.
13. Boby, J. (2015). A dual response surface optimization methodology for achieving uniform coating thickness in powder coating process. *International Journal of Industrial Engineering Computations, 6*, 469–480.
14. Apparao,K. C., & Birru, A. K. (2016). Optimization of Die casting process based on Taguchi approach. In *5th International Conference of Materials Processing and Characterization* (Vol. 4, No. 2, pp. 1852–1859).

15. Yoo, K. S., Eom, Y. S., Park, J. Y., Im, M. G., & Han, S. Y. (2011). Reliability-based topology optimization using successive standard response surface method. *Finite Elements in Analysis and Design, 47,* 843–849.
16. Banubakode, M. M., Gangal, A. C., & Shirbhate, A. D. (2013). Taguchi method for improving powder coating process—A case study. *International Journal of Management, Information Technology and Engineering, 1*(3), 167–180.
17. Costa, N. R. P. (2010). Multiple response optimisation: methods and results. *International Journal of Industrial and Systems Engineering, 5*(4), 442.
18. Ross, P. J. (1988). *Taguchi techniques for quality engineering.* Mcgraw-Hill Book Company.
19. Raykundaliya, D. P., & Shanubhogue, A. (2015). Comparison study: Taguchi methodology vis.-a-vis. response surface methodology through a case study of accelerated failure in spin-on-filter. *International Advanced Research Journal in Science, Engineering and Technology, 2*(3), March 2015.
20. Neseli, S., Yaldiz, S., & Türkes, E. (2011). Optimization of tool geometry parameters for turning operations based on the response surface methodology. *Measurement, 44,* 580–587.
21. Kumar, V., & Khamba, J. S. (2008). Statistical analysis of experimental parameters in ultrasonic machining of tungsten carbide using the Taguchi approach. *Journal of the American Ceramic Society, 91*(1), 92–96.
22. Li, W. Y., Zhang, C., Guo, X. P., Zhang, G., Liao, H. L., Li, C. J., et al. (2008). Effect of standoff distance on coating deposition characteristics in cold spraying. *Materials and Design, 29*(2), 297–304.
23. Meng, X. (2009). Influences of different powders on the characteristics of particle charging and deposition in powder coating processes. *Journal of Electrostatics, 67*(4), 663–671.
24. Wang, Y., Huang, C., & Xu, T. (2010). Optimization of electrodialysis with bipolar membranes by using response surface methodology. *Journal of Membrane Science, 362,* 249–254.

Data Mining Techniques and Its Application in Civil Engineering—A Review

Priyanka Singh

1 Introduction

Data mining is the technique of discovering styles of massive data units regarding strategies at the intersection system mastering, facts and database systems. It is the workout of analyzing massive pre-cutting-edge databases in order to generate new records. Data mining is the assessment step of the "information discovery in database" [1, 2]. It is an interdisciplinary subfield of computer science and statics with a preferred aim to extract facts from records set and rework the facts into an understandable shape for further use. This method is likewise used in corporation programs along with market segmentation, fraud detection, and credit danger evaluation further to many different programs [3]. It has the capacity to locate pattern saved inner records and is now taken into consideration as a catalyst for growing business enterprise procedure by using keeping off failure sample. There is a frequent collection of data on each day basis in construction and manufacturing companies while operations. The main intention of data mining is to extract data from the pre-existing dataset for future use in the logical structure. It is now considered as a catalyst for increasing business process by avoiding failure pattern [4, 5].

1.1 Methods and Approach

Some of the applications which are being utilized in the field of civil engineering are as follows [6–9]:

P. Singh (✉)
Department of Civil Engineering, Amity School of Engineering & Technology, Amity University, Noida, Uttar Pradesh, India
e-mail: priyanka24978@gmail.com

© The Author(s), under exclusive license to Springer Nature Singapore Pte Ltd. 2021   175
P. K. Kapur et al. (eds.), *Advances in Interdisciplinary Research in Engineering and Business Management*, Asset Analytics,
https://doi.org/10.1007/978-981-16-0037-1_15

i.   Predicting strength based on the concrete mixture, soil, and several other variables.

ii.   Monitoring infrastructure health of sub- and super-structure on the data received from the sensors.

iii.   Traffic Engineering: Data sensing, analysis, and mining can be used to facilitate decision-making and intelligent transportation systems.

iv.   Spatial data mining to find out the best possible location for construction.

v.   Water resources engineering: In water resources engineering, data mining is used to identify chemical trends in water quality samples.

vi.   Construction Management: It is used to estimate maintenance cost and quality of the construction.

vii.   Environmental Engineering: It could be used to study environmental and natural resource science.

viii. Highway and Transportation Engineering: Traffic and pavement management can be done with the application of data mining in the field of highway and transportation engineering.

ix.   Hydraulics and Water Power Engineering: The prediction of dangerous risks in Hydraulics and Water Power Engineering can be achieved by data mining techniques.

x.   Materials Science and Engineering: Data mining techniques can be applied in many domains such as astronomy, bioinformatics, chemistry, materials science, climate, fusion, and combustion.

xi.   Geotechnical Engineering: Data mining can be explored to formulate several complex geotechnical engineering problems. The complex analysis of geotechnical behavior is due to multivariable soil and rock responses.

xii.   Earthquake Engineering: Application of data mining would be extremely helpful in the prediction of natural geological calamities like flood, tornado, hurricane, volcanic eruption, earthquake, heat-wave, or landslides.

xiii. Coastal and Harbor Engineering: By data mining, the different wave behaviors at the coastal areas can be studied, which will be useful for designing harbor engineering.

xiv.   Tunnel Engineering: The variation in the pattern of geological data, stress–strain data of supporting structures and the deformation data of the surrounding rocks can be interpreted by the data mining technique.

xv.   Surveying and Geo-Spatial Engineering: The use of new and developing technologies such as GPS, satellite imagery, laser mapping, and fast computing to create complex layers of interconnected geographic information can be facilitated by data mining technique.

xvi.   Geomatics: Application of data mining is utilized in delivering spatially referenced information by the data of deposit collection, processing, interpreting, storing of mining, and extraction plant.

xvii. Geosciences: By the application of data mining, petro-physical data, logging data, seismic data, and geological data can be addressed.

xviii. Remote Sensing: Data mining is utilized to extract aerial remote sensing imagery for automatic land-cover classification.

xix.   Geographical Information systems: Data mining can automatically extract knowledge from raw data of Geographical Information Systems to gain spatial analysis result.

The field of construction as well as other industries related to it makes the big statistics technology utilize the large quantity of information received and saved through advanced computing systems. Information comes from anywhere: computers, structures, human beings, sensors, and any device that generates records. In the construction field, every structure that has been erected embodies huge quantities of data [6, 10, 11]. However, managers and civil engineers need to explore the answers hidden in unstructured information. Big data technology gives civil engineers the strength to utilize unstructured information, which is tough to accumulate and analyze manually in any beneficial manner. In fact, all records are useless without correct assessment and verification [10, 12].

2   Application of Data Mining Technology on Construction Project Cost Control System

With the development of construction management, it is essential that the costing-related tasks are done correctly. The innovation regarding data mining process in the development of the cost control framework to resolve the deficiencies in conventional management is highly beneficial [13–15].

Secondarily, the main focus is on direct expenses on items rather than on overhead cost by cost management programs. Rational cost and cost of the quantity of items are often neglected because much more importance is given to financing [16]. There is great redundancy of data in the information system thus making it different to determine variation in the information. Usually issues like selection of important data from huge arise. Hence Data Mining Technology comes as a solution to the problem [17].

The process of sorting the important data from large data storage and extracting relatable patterns to solve problems with ease is facilitated with DM [18]. The user must have adequate knowledge to discover such databases. As a result, Data Mining Technology grows with great importance in the field of multiple costing parameters.

Data Mining Technologies plays a great role in Cost Management. Several goals that are achieved are as follows:-

- Rapid or quick schedule
- Reduction in cost
- Availability of good quality data values
- On-time targets.

Manual recording of data in a common system did not have the provision of accessibility which lead to the problem of missing data. As a result, data were classified into three dimensions:

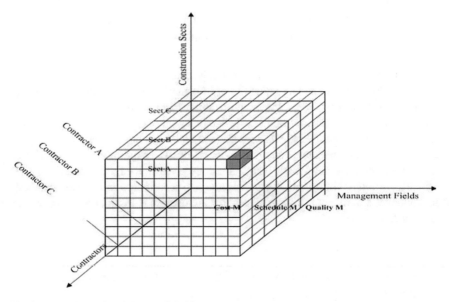

Fig. 1 Three-dimensional data model [20]

- Management field
- Contractors field
- Construction sector.

Initially the data were in one dimension. Now 3D projection and storage of data information has overcome the issue of missing data and has made it easier [19]. Figure 1 illustrates the three-dimensional data model of data information.

The Management field in 3D model can further be separated into three sections (i) cost management, (ii) schedule management, and (iii) quality management. Figure 2 depicts the picture of data dividing in three-dimensional model. Hence special information might be consolidated into a new datum [21].

3 Application of Big Data in Construction Industry

The application of big data in the construction Industry consists of basic three processes:

- Design: Big data analyzes the building design and its modeling. It can also be used to finalize where the building can be constructed and with which material. However, old records can be studied with it to know the probability of risks at the construction site.
- Build: During construction at the site, the data collected from weather, traffic, and other activity is interpreted to decide the stages of construction.

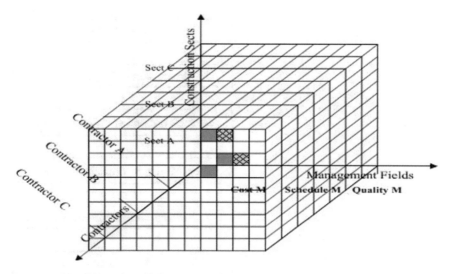

Fig. 2 Data dividing in three-dimensional model [22]

- Operate: The accomplishment of the construction at each level can be tracked by collecting the data from the sensors built into the super and substructures of the buildings. Scheduled maintenance activity can be planned by feeding the data of the sensors in the building information modeling.

The handling of data gets bigger as the size of the data increases. This aspect can be managed by big data analytics. Big data analytics provide inspection alert much before the risk levels [23, 24].

There is a frequent collection of data on daily basis in construction companies for activities and operations. The job overheads increases proportionally with the number of changed orders during the month for high productivity. Projects hidden within the project database can be detected using data mining [25]. The flow chart of the data mining process is shown in Fig. 3.

Data mining serves two objectives, namely Insight and Prediction. Insight prompts in recognizing design and patterns. Prediction gives expectations dependent on information [27]. Large data mining techniques of discrete types, for example, neural networks and conceptual clustering are used in particular domains individually. Data sensing, analysis, and mining can be used to facilitate decision-making and intelligent transportation systems [28].

4   Data Mining in the Field of Transportation Engineering

The study of data mining in the field of Transportation has a great scope. Various problems were acknowledged through data mining such as.

Fig. 3 Flow chart of data
mining process [26]

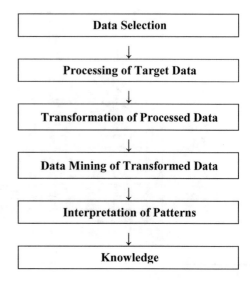

- Productive methodology for dealing with large information with accurate results
- Methods to differ over-fitting
- To check absence and uproarious information
- Build connections between distinct fields
- Better client communication and pre-learning.

Specialized forms of data help to monitor cases like drowsy drivers by sensing their state of driving. It also helps in designing roadways keeping in mind the overtaking speed of vehicles and the rate of accidents [29]. Table 1 illustrates Data Characteristics in GIS Application of transportation engineering.

Table 1 Data characteristics in GIS application of transportation engineering [30]

		Logical	Physical
	Real World	*Legal definitions*	*Actual facilities*
		Route	Highways
		State Trunk Network	Roads
		Country Trunk Network	Interchanges
		Street network	Intersections
		Political Boundary	
	Virtual World	*Data structures*	*Data values*
		Networks	Lines
		Chains	Points
		Links	Poly-lines
		Nodes	Polygons
		Lattices	Attributes

Huge amounts of data are given by PMS for pavement design, the number of vehicles running on that particular pavement, types of materials used to construct that pavement, undulations present, etc. Data provided by GPS helped in tracing the vehicles and to check the inconvenient changing of lanes. In geographic data frameworks (GIS) for transportation, interconnected equipment, programming, information, individuals, and assessment of data are done, which is to be considered as a vital job [31–34].

5 Data Mining in the Field of Environmental Engineering

In the field of environmental engineering, the aspect of data mining can be utilized to the optimum. Data mining techniques in environmental studies provide valuable knowledge and useful patterns which facilitate us to deal with environmental problems, like air pollution, water pollution, etc. [35, 36]. Some of the beneficial aspects of data mining in environmental engineering are as follows:

- Prediction of rainfall based on the parameters of environmental studies.
- Prediction of rainfall and climate change can be predicted on the different patterns obtained by the data mining techniques. [37, 38]
- Ecological systems can be modeled by processing ecological and environmental data [39, 40]
- Better control of air quality and water quality by analyzing environmental data.
- Better waste management and recycling can be accomplished by discovering patterns through data mining [38, 41].

6 Conclusion

This paper shows the various application related to civil engineering where big data plays a very big role in the management of project, maintaining resources, scheduling of jobs, estimation of time and cost involved for civil construction works. Big data analytics along with data mining can make the construction industry fruitful and suggests the methods for improving the health of construction. Data available at various construction sites may be gathered to form big data and the data mining methods like clustering and classification algorithms can be implemented for better analysis, decision–making, and prediction of the new estimates using the past data. Data mining has huge scope in the field of environmental engineering also. It facilitates to have better control over air and water quality. Prediction of the weather forecast and climate change can be optimized.

References

1. Fayyad, U. M., Piatetsky-Shapiro, G., Smyth, P., & Uthurusamy, R. (1996). *Advances in knowledge discovery and data mining*. Cambridge, MA: AAAI Press/The MIT Press.
2. Han, J. &, Kamber, M. (2001). *Data mining: Concepts and techniques*. Higher Education Press.
3. Hall, M. J., Minns, A. W., & Ashrafuzzaman, A. K. M. (2002). The application of data mining techniques for the regionalisation of hydrological variables. *Hydrology and Earth System Sciences, 6,* 685–694.
4. Chau, K. W., & Cao, Y. (2002). The Application of data warehouse and decision support system in construction management. *Automation in Construction, 12,* 213–224.
5. Zhang, J. P., & Wang, H.-J. (2002). Towards 4D management for construction planning and resource utilization. In *The 9 th International Conference on Computing in Civil and Building Engineering*, Taiwan (pp. 1281–1286).
6. Hyperion Software Corp. (1999). The role of OLAP server in a data warehousing solution.
7. Chakrabarti, S. (2002), Mining the web: Statistical analysis of hypertex and semi-structured data. Morgan Kaufmann.
8. . Hong-Yan, L. I, Bu-Ying, C., & Li, D. (2013). Application of data warehouse and data mining in coal information management (Vol. 31, no. 8, pp. 31–32).
9. Forbes,L. H., & Ahmed, S. M. (2003). Construction integration and innovation through lean methods and E-business applications, construction research 2003, Copyright ASCE 2004.
10. Inmon, W. H. (2000). *Building the data warehouse* (2nd ed.). China Machine Press.
11. Bilal M., & Oyedele, O. L. (2016). Big Data Architecture for Construction Waste analytics (CWA): A conceptual framework. *Journal of Building Engineering*, 144–156.
12. Dasu, T., & Johnson, T. (2003). *Exploratory data mining and data cleaning*. Wiley.
13. Hore, A. (2006). Use of IT in managing information and data on construction projects—A perspective for the IRISH construction industry, information technology in construction project management.
14. Zhou, Y., & Ding, L. Y. (2006) International symposium on "Advancement of Construction Management and Real Estate" The CRIOCM 2006.
15. Jin, C. (2017). Real-time damage detection for civil structures using Big Data.
16. Adrians, P., & Zantinge, D. (1996). *Data mining*. England: Addison-Wesley Longman.
17. Cabena, P. (1997). *Discovering data mining: From concept to implementation*. NJ: Prentice Hall.
18. Han, J. (2001). *Data mining: Concepts and techniques*. San Francisco: Morgan Kaufmann Publishers.
19. Hand, D. J., Mannila, H., & Smyth, P. (2001). *Principles of data mining*. Massachusetts: MIT press.
20. Han, J., & Kamber, M. (2006). *Data mining: Concepts and techniques* (2nd ed.). Morgan Kaufmann.
21. Attoh-Okine, N. O. (1997). Rough set application to data-mining principles in pavement management database. *Journal of Computing in Civil Engineering, American Society of Civil Engineers, 11*(4), 231–237.
22. Soibelman, L., & Hyunjoo, K. (2002). Data preparation process for construction knowledge generation through knowledge discovery in databases. *Journal of Computing in Civil Engineering, ASCE, 16*(1), 39–47.
23. Jae-Gil Lee, M. K. (2015), Geospatial Big Data: Challenges and opportunities. *Big Data Research*, 74–81.
24. Ahmad, I., & Ahmed, S. M. (2001). Integration in the construction industry: Information technology as the driving force. In R. L. K. Tiong (Ed.), *Proceedings of the 3rd International Conference on Construction Project Management*. Nan yang Technical University Press.
25. Hu, D. (2005). Research on the systematic framework of computer integrated construction. Huazhong University of Science and Technology.
26. Han, J., & Kamber, M. (2001). *Data mining: Concepts and techniques*. Higher Education Press.

27. Rezania, M., Javadi, A., Giustolisi, O. (2008). An evolutionary-based data mining technique for assessment of civil engineering systems.
28. Shanti, M. A., & Saravanan, K. (2017). Knowledge data map—A framework for the field of data mining and knowledge discovery. *International Journal of Computer Engineering & Technology, 8*(5), 67–77.
29. Barai, S. V., & Reich. (2001). Data mining of experimental data: Neural networks approach. In *Proceedings of 2nd International Conference on Theoretical, Applied Computational and Experimental Mechanics ICTACEM (CD-ROM).*
30. Rabee M. Reffat, John S. Gero, Wei Peng (2004). Using data mining on building maintenance during the building life cycle.
31. Kohavi, R. (2001). Data mining and visualization. In *Sixth Annual Symposium on Frontiers of Engineering* (p.p. 30--40). National Academy Press, D. C.
32. Amado, V. (2000). Expanding the use of pavement management data. In *Transportation Scholars Conference*, University of Missouri.
33. Tan, P., Steinbach, M., & Kumar, V. (2005). *Introduction to data mining.* Addison Wesley.
34. Dzeroski, S. (2003). Environmental applications of data mining. Lecture Notes of Knowledge Technologies, University of Trento.
35. Stojic, A., Stojic, S. S., Reljin, I., Cabarkapa, M., Sostaric, A., Perisic, M., & Mijic, Z. (2016). Comprehensive analysis of PM10 in Belgrade urban area on the basis of long-term measurements. *Environmental Science and Pollution Research, 23*, 10722–10732. https://doi.org/10.1007/s11356-016-6266-4.
36. Gaal, M., Moriondo, M., & Bindi, M. (2012). Modelling the impact of climate change on the Hungarian wine regions using random forest. *Applied Ecology and Environmental Research, 10*, 121–140. https://doi.org/10.15666/aeer/1002_121140.
37. Crimmins, S. M., Dobrowski, S. Z., & Mynsberge, A. R. (2013). Evaluating ensemble forecasts of plant species distributions under climate change. *Ecological Modelling, 266*, 126–130. https://doi.org/10.1016/j.ecolmodel.07.006.
38. Lei, K. S., Wan, F. (2012). Applying ensemble learning techniques to ANFIS for air pollution index prediction in Macau. In *International Symposium on Neural Networks (ISNN'12)*, 11–14 July 2012 (pp. 509–516). Berlin, Heidelberg: Springer.
39. Budka, M., Gabrys, B., & Ravagnan, E. (2010). Robust predictive modelling of water pollution using biomarker data. *Water Research, 44*, 3294–3308. https://doi.org/10.1016/j.watres.2010.03.006.
40. Singh, K. P., Gupta, S., & Rai, P. (2013). Identifying pollution sources and predicting urban air quality using ensemble learning methods. *Atmospheric Environment, 80*, 426–437. https://doi.org/10.1016/j.atmosenv.2013.08.023.
41. Nelson, T. A., Coops, N. C., Wulder, M. A., Perez, L., Fitterer, J., Powers, R., & Fontana, F. (2014). Predicting climate change impacts to the Canadian Boreal forest. *Diversity, 6*, 133–157. https://doi.org/10.3390/d6010133.

Inventory Models for Imperfect Quality Items: A Two-Decade Review

Prerna Gautam, Sumit Maheshwari, Amrina Kausar, and Chandra K. Jaggi

1 Introduction

Inventory management is an important task for the companies to minimize the different types of costs linked with inventory such as ordering cost, holding cost, and purchase cost. Inventory management policies help the companies to reinvest capital, previously wasted on unreasonably planned inventory, in other business areas or take advantage of new market opportunities. The basic EOQ model was presented by Harris [37], which is the realistic model but cannot be applied in different realistic situations. The fundamental of any inventory management surrounds with items. Thus, the practicality of items is an important aspect that should not be ignored while structuring inventory problems. It is not pragmatic to consider the items to be of perfect quality always, as due to various reasons the final item may not withstand the quality characteristic set by the manufacturer. The reason behind this can be due to mishandling, imperfect production process, which may shift from an in-control state to the out-of-control state, accidents during loading/unloading, etc. The extensive research in the area of imperfect quality items was carried out by Porteus

P. Gautam · S. Maheshwari · C. K. Jaggi (✉)
Faculty of Mathematical Sciences, Department of Operational Research, University of Delhi, Delhi 110007, India
e-mail: ckjaggi@yahoo.com

P. Gautam
e-mail: prerna3080@gmail.com

S. Maheshwari
e-mail: sumitduor@gmail.com

A. Kausar
Department of Management Studies, Shaheed Sukhdev College of Business Studies, University of Delhi, Delhi 110089, India
e-mail: amrinakausar@sscbsdu.ac.in

© The Author(s), under exclusive license to Springer Nature Singapore Pte Ltd. 2021
P. K. Kapur et al. (eds.), *Advances in Interdisciplinary Research in Engineering and Business Management*, Asset Analytics,
https://doi.org/10.1007/978-981-16-0037-1_16

[76], Rosenblatt and Lee [79], Lee and Rosenblatt [61], Kim and Hong [58]. Later, Ben-Daya and Hariga [3], Salameh and Jaber [82], Cárdenas-Barrón [23], Yeh et al. [109], Huang [42], Ben-Daya and Rahim [4], Huang [42], Ha and Kim [36], Hsieh and Lee [39], Chen and Lo [12], Sana [85], Khan et al. [52], etc. Kishore et al. [59] investigated the effect of learning in the setup cost for imperfect production systems by using two-way inspection plan. Recently, Khanna et al. [53] developed an integrated model for imperfect production system with maintenance actions and warranty policy.

Considering the sustainable policies in the inventory models Gautam and Khanna [26] developed a sustainable inventory model that considered imperfect production system under integrated decision-making scenario. Recently, Khara et al. [56] considered sustainable recycling for an imperfect production system with acceptance quality level dependent cost and demand.

Initially, the investigation of inventory models with deterioration started by Ghare [29] was then followed by Wee [107], Chaudhari and Chakrabarti [5]. In most of the practical scenarios, many procedural changes and specialized equipment acquisition are used to reduce and control the deterioration rate. However, people often misunderstand deteriorating items with imperfect quality items, both are two different things altogether. The deterioration happens to the items that are already in stock and the imperfect items are those items that are a result of faulty machinery, mishandling while shipment, usage of low-quality raw material for making products, etc. Karlin [47] first addressed the imperfect item in the newsvendor models in random supply to analyze optimal ordering policy. Salameh and Jaber [82] incorporated the concept of imperfect items in the production system. Later, Goyal and Cárdenas-Barrón [32] gave a note on this concept for further improvement of the model. Maddah and Jaber [65] further revisited the model of Salameh and Jaber [82] for imperfect production. Later, Khan et al. [52] proposed a model for deteriorating imperfect quality items accounting credit policies and shortages. Recently, Roy et al. [80] presented a deteriorating inventory model with inspection policy and variable demand. In the literature, there are so many papers for imperfect quality items in the system. This survey paper will cover all the relevant papers related to imperfect quality items in the field of inventory.

The remaining paper is prepared as follows: Section 2 presents the research methodology for performing material collection, descriptive analysis, category selection, and material evaluation. Section 3 explores the content analysis related to this field. Section 4 gives general trends of this field, which highlights the magnificent role of this paper. This section also examines the detailed analysis of imperfect production or imperfect quality items with the content analysis which will give the idea to the readers and as well as practitioners such as the most contributed author, journal, and country of this field. This study may clarify the direction in terms of readers and journals. Section 5 defines the managerial implications and Section 6 concludes this study. Section 7 presents the limitations and future research directions of the study. It is hoped that this study will bring several advantages to the readers, researchers, Editorial Board of the journals, as well as the practitioners who are active in this

Table 1 Past review studies of inventory management based on the different time-span and source

Authors	Title	Specific keywords	The time-span for study	Publisher	Number of papers for review	Source of database
Shekarian et al. [97]	"Fuzzy inventory models: A comprehensive review"	Fuzzy inventory models	Till Aug 2015	Elsevier	210	Scopus
Chaudhary et al. [9]	"State-of-the-art literature review on inventory models for perishable products"	Perishable inventory models	1990–2016	Emerald	419	N.A.
Mosca et al. [68]	"Integrated transportation – inventory models: A review"	Integrated transportation –inventory models	1997–2017	Elsevier	67	Scopus and Google Scholar
Pereira and Costa [75]	"A literature review on lot size with quantity discounts: 1995–2013"	EOQ with incremental discounts	1995–2013	Emerald	49	Scopus and ISI Web of Science
Ahmed and Sultana [1]	"A literature review on inventory modeling with reliability consideration"	Reliability considerations	N.A.	Growing Science	N.A.	N.A.
Khan et al. [49, 50]	"A review of the extensions of a modified EOQ model for imperfect quality items"	Imperfect Quality items	2000–2010	Elsevier	N.A.	N.A.
Present paper	**Inventory models for imperfect quality items: A two-decade review**	**Imperfect quality items in EOQ/EPQ**	**2000–2020**	**Springer**	**70**	**Scopus**

field. Table 1 summarizes the existing review studies based on the search keywords, source of database and time-period.

Table 1 shows that in the past there exist a lot of review papers that considered different aspects of inventory management under different periods with different databases. However, the proposed review paper is based on a recent time-span considering two decades, i.e., from 2000 to 2020. Further, the "Scopus" database is considered for in-depth analysis.

2 Research Methodology

It is a necessary aspect of the literature review of an article. With the help of literature review, it is easy to recognize the theoretical content of the research area. According to Mayring [67], research methodology includes the four essential steps for the systematical review, namely material collection, descriptive analysis, category selection, and material evaluation. This similar approach has been used by Seuring and Müller [96], Gautam et al. [28], Maheshwari et al. [66], and Gao et al. [25] for analyzing the past studies (See Fig. 1).

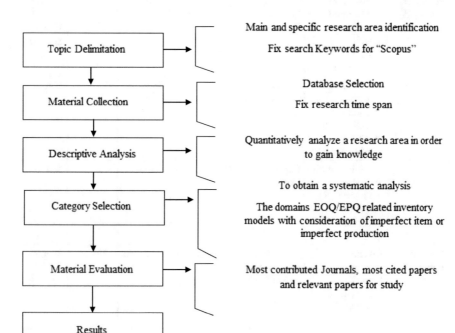

Fig. 1 Summary of the research methodology

2.1 Material Collection

The initial phase of the literature review is the material collection. This study considers only papers written in English and published in Scopus journals with the date range. The web search "Scopus" was used to manage the scientific publication data from 2000 to 2020*, which contains the keywords "EOQ, EPQ, Inventory models, lot size models" with "Imperfect production or imperfect items" for the study. The database of "Scopus" allows individual researchers to present a remarkable work for the effect of the research to its best advantage. For this study, 70 articles have been collected for reviewing and examining the details of various journals, subject, as well as the most cited papers. Figure 2 gives a better understanding of the material collection.

Table 2 shows how the "search" keyword is used in the "Scopus" database. The total number of papers was 1209, efficient enough for the review, but we still carefully compiled these results and removed the repetition of the papers, and the final sample of papers was 640.

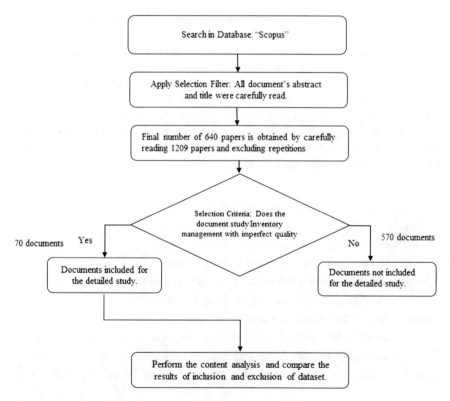

Fig. 2 Data collection and selection for evaluation

Table 2 Search keywords perform on the Scopus database

Search keywords	Number of articles
"Imperfect Production" and "EOQ" OR "EPQ"	87
"Imperfect Item" and "EOQ" OR "EPQ"	64
"Imperfect" and "EOQ" OR "EPQ"	296
"Imperfect Production" and "Lot Size"	93
"Imperfect Item" and "Lot Size"	30
"Imperfect" and "Lot Size"	226
"Imperfect Production" and "Inventory"	279
"Imperfect Item" and "Inventory"	134
Total number of search articles	**1209**
After removing the repletion of articles	**640**

2.2 Descriptive Analysis

The descriptive study and scientific research aim to quantitatively analyze a research area to gain knowledge about how the discipline has evolved. The descriptive analysis gives an idea about the statistical analysis of the study. Here, we are trying to understand the basic concept of quantitative analysis for the data. Figure 3 reveals that relevant articles appear most frequently in IJPE, followed by CIE, AMM, IJPR, and IJSS; all are well-established leading journals of this field based on the search database. Figure 3 shows that the number of papers exploring inventory management for imperfect quality items. It further demonstrates the final number of articles selected for the detailed study, which makes it different from the existing literature, which is based on the Mayring [67] methodology.

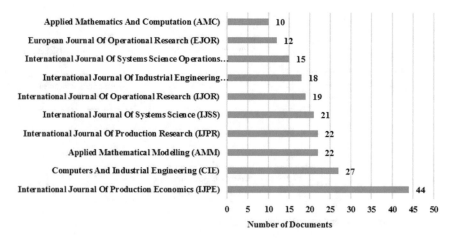

Fig. 3 Number of articles published in different journals (2000–2020*)

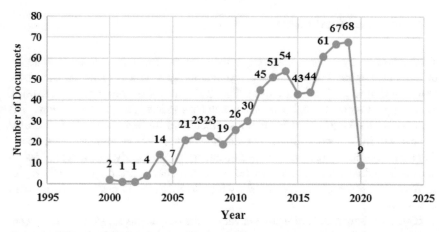

Fig. 4 Distribution of published articles during the year 2000–2020

The distribution of selected journals is shown in Fig. 3, which shows the attention of the researchers and the Editorial Board members. After compiling the result, we have found that IJPE is the most contributed journal in this field for the last two decades and published 44 articles. Further, on this categorization, we cannot ignore the contribution of the other journals. This study considers 10 most contributed journals.

The study of inventory models with imperfect quality items has gained attention exponentially. In 2000, only two articles were published with the consideration of these keywords, and subsequently, in 2001, 2002, 2003, and 2004 the number of published articles are 1, 1, 4, and 14, respectively. In the year 2020, (till Dated—08/03/2020), nine articles have been published. Figure 4 shows the distribution of the numbers of articles published during the year 2000–2020*. Therefore, we can say that the attention of the researcher has been increased exponentially.

2.3 Category Selection

To obtain a systematic analysis, we classify the selected papers into various "categories." We have the major categorization by "links" growth of models with the inclusion of imperfect quality items inventory systems. The final selection for the inclusion of the paper has been done by category selection after carefully reading the title, abstract, and keywords of the papers.

2.4 *Material Evaluation*

This review study ensures that the publications are adequate and appropriate for material evaluation. For the material evaluation, **Appendix A** gives the idea about most contributed publications during 2000–2020* based on the number of citations. This study considers the publications, which have at least 50 or more citations. The key reason for taking at least 50 citations is that there are so many papers that cannot be analyzed in Appendix. But still, we have considered some articles which do not lie in this criterion to perform a scientific analysis of the study. From the collection of data, 51 publications have at least 50 or more citations according to the "Scopus" database. The other relevant documents which do not lie in this category are explored in this study.

If we talk about the classification of data on the subject basis, 18% of articles were purely based on Decision Sciences, 17% in Computer Science, 15% in Mathematics, and 15% articles in the Business, Management and Accounting category from the database of "Scopus." Figure 5 shows the distribution of the published papers based on different subjects with the search category. Some articles are common for many subjects for these keywords. From Fig. 5, we can say that the greatest number of published articles is in the field of Decision Sciences, i.e., 18%, and this classification is based on the total number of papers, i.e., 640.

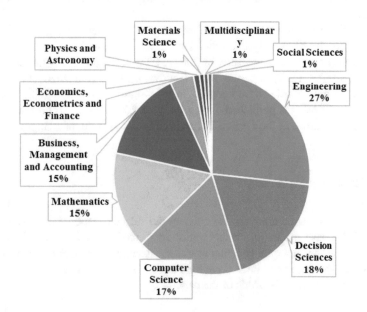

Fig. 5 Distribution of publications with keywords in different subjects

3 Content Analysis

Content analysis is best understood as a broad family of techniques that grants a procedure that can be used to evaluate a large amount of data systematically. It is a crucial stage to recognize various topics covered by the study and to give a clear indication of the areas requiring further investigation. From the content analysis, we observe that a lot of work has been carried out in the field of inventory under the presence of imperfect quality items. Previously, the authors have started to address the impact of imperfect quality items in their basic EOQ/EPQ models and later they proposed some simple solutions like disposing of them or selling them to the secondary markets. A quick tour of inventory models with imperfect quality items under various other realistic conditions are explored as follows.

3.1 EPQ/EOQ Models for Imperfect Items

The key to inventory problem is finding the appropriate levels for holding inventories and ordering/setup sequences and quantities so that the total cost incurred is minimum. The incurrence of imperfect quality items in any ordered/produced lot is inevitable. Thus, the impact of imperfect quality items cannot be avoided while managing the inventory of any organization. Salameh and Jaber [82] put forth a remarkable inventory model that incorporated the effect of imperfect quality items. Goyal and Cárdenas-Barrón [32] developed the simplest approach to obtaining optimal EPQ in the imperfect production environment. Maddah and Jaber [65] revisited the concept of Salameh and Jaber [82] with unreliable supply.

3.2 Supply Chains

The basic essence of the supply chain is the flow of goods from one supply chain player to another. The supply chain players are manufacturers, distributors, retailers, customers, etc. The goods produced or delivered may contain some fraction of defectives, which should be accounted for in the inventory modeling of every player. Working in collaboration is the key to achieve competitive advantages. Various models considered different supply chain players and developed models under similar and distinct distributions of power to the supply chain players. Goyal et al. [34], Chen and Kang [11], Wahab et al. [104], Khan et al. [48], etc. proposed inventory models under collaborative environments.

3.3 Quality Consideration

The quality of goods is the most important parameter for judging the goodwill of any supply chain player. Thus, before selling out to the market an inspection is necessary to make sure that only quality goods are delivered. Despite the emergence of the latest techniques, the field of imperfect quality items holds significant worth for new policies. Management of imperfect quality items is till date a challenging task. The screening procedure is a crucial stage for any business; therefore, it is wise to implement the screening process competently. The initial research that contributed to this area was given by Goyal et al. [33]; they developed an efficient product design to cater to modern production environments. The model where the production process had various stages, and which are followed by a possible screening, was given by Tang [100]. The model with imperfect production process and screening was further studied by Giri and Dohi [31] and Gurnani et al. [35]. Ma et al. [64] elucidated further the topic of imperfect production system for deciding whether and when to apply an inspection on the defectives. Moussawi-Haidar [69], Khedlekar, and Tiwari [57] proposed strategies to manage defectives either through salvaging or reworking.

3.4 Trade Credit Policy

A promotional tool offered by any supply chain player to other participants of the supply chain was to increase the whole profit of the system by facilitating demand. It is the ability to buy the product without paying for it for a certain period, known as the credit period. Chung and Huang [20], Chen and Kang [11], Tsao [102], Jaggi et al. [45], Ouyang et al. [72], Zhou et al. [111], Khanna et al. [54], Sarkar et al. [95], Srivastava et al. [98], etc. used the trade-credit policies in their inventory modeling under the effect of imperfect quality items.

3.5 Environmental Impact

The world is going through the environmental crisis and it calls for effective solutions to reduce the environmental impact of every organization. The imperfect quality items if not treated carefully end up in landfills and waste generation. Thus, effective management techniques are required for dealing with imperfect quality items. Wahab et al. [104], Jauhari and Laksono [46], Gautam et al. [27] proposed defect management techniques to deal with imperfect quality items under integrated and Stackelberg decision instances.

3.6 Rework or Salvage Options

In case the imperfect quality items are produced at the manufacturer's end, he may prefer to send them for the rework process, instead of simply disposing of them at a cheaper price. The reworked items are generally considered as good as new ones and sold at the full selling price. The imperfect quality items can be handled either by reworking them and making them suitable either for primary or secondary market or by disposing of them. Rework and salvage options also play a vital role in the management of imperfect quality items. Chiu [13], Chan et al. [6], Chiu et al. [15], Chiu [106], Cárdenas-Barrón [21], Pasandideh et al. [74], Cárdenas-Barrón [22], Chen [10], Sarkar et al. [90], Moussawi-Haidar et al. [69], Khanna et al. [55], Tayyab et al. [101], Taleizadeh et al. [99] and many other studies incorporated the concept of rework and salvage options in the inventory modeling for imperfect items.

3.7 Learning in the Inspection

Competent supply chain players invest in processes like an inspection to cater only good quality items to their subsequent downstream members. However, due to unavoidable human errors, it sometimes turns out that the inspection process is also imperfect/error-prone and results in Type-I and Type-II misclassification errors. The type-I error results in direct monetary loss to the wholesaler because of the sale of good quality items as bad ones at a cheaper price, whereas the occurrence of Type-II error ends up in penalty and goodwill losses by the sale of bad quality items as good ones by mistake. Type-II error ultimately leads to sales returns. Moreover, learning is the process of gaining knowledge by doing. Learning in inspection describes the ability of the inspectors that comes with performing numerous inspections. Khan et al. [49, 50], Khan et al. [48], Sarkar and Saren [94] and numerous other researchers carried out either inspection or learning in inspection phenomena to showcase a pragmatic inventory scenario.

4 General Trends of Inventory Models Under the Consideration of Imperfect Quality Items

The data has been collected with the help of the "Scopus" database through the topic search during the last two decades, i.e., 2000–2020. The section considered the most contributed country based on the published articles. It has been seen that the authors from India and Taiwan have published more articles in this field. The third most contributed country is Iran, and the fourth is the United States. The consideration of the most contributed country was based on the author's country during the published articles. It has also been seen that there are lots of authors whose parent country was different from the current working country. Therefore, we only consider the author's institutional country/territory and Fig. 6 shows the top 10 most contributed country/territory.

Figure 7 shows the most contributed authors of this field, Chiu, Y.S.P. is the most contributed author of this field during the year 2000–2020. This study considers the author and co-author works. Chiu, Y.S.P. published 30 articles of this field in the last two decades as an author or co-author. The second most contributed author is

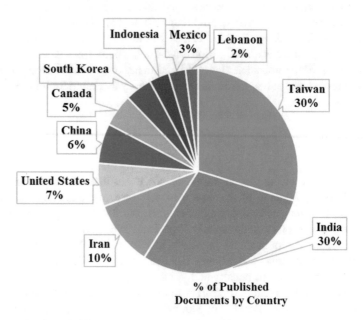

Fig. 6 Contribution of different countries authors in the field of imperfect quality items in the inventory system

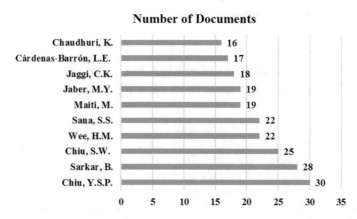

Fig. 7 Most-contributed teamwork in the field of imperfect quality items in the inventory systems

Sarkar, B., the third is Chiu, S.W., and fourth is Wee, H.M., etc. and the top 10 most contributed authors are shown in Fig. 7.

This study gives the idea about the most contributed author, journal, the country as well as others' contribution in this field. This study will be helpful for the academicians, researchers and practitioners, and young learners in the proposed field.

5 Managerial Implications

The inventory management of any organization revolves around the items. Thus, the characteristics of items must be studied to gain insights about the associated risks and profits. Imperfect quality items could be defined as the defective content in an item that has production or design-related defect. Further, the defect content could be because of the weak transport and minor/major accidents while loading and unloading of the items. Without a doubt, imperfect quality items are the part and parcel of any organization. There is a significant need for incorporating them in inventory modeling to cater to the decision-makers with real-time models. Moreover, it is the pressing need to propose solutions on how to deal with imperfect quality items. Through this literature review, it is seen that despite a lot of work that has been done in this field, a lot of research is yet to be explored. For instance, various models in the literature incorporated the effect of imperfect quality items in their models but only a few of the proposed solutions to deal with it. Existing research in the field suggested either disposal of the defectives or reworking, but all the defectives are neither good enough to be reworked nor bad enough to be disposed off. In this direction, some recent articles proposed efficient techniques to overcome such a situation by suggesting strong inspection processes that could bifurcate the defectives into various categories. Not to forget the environmental crisis, imperfect quality items, if disposed of in the environment, may lead to harmful emissions and landfills. Thus, there is an urgent need to minimize the production of imperfect quality items. Further, the organizations must work on recovering these items for bringing them back in the supply chain instead of disposing of them in the environment.

6 Conclusion

In summary, imperfect quality items play a vital role in inventory management. Thus, their effect cannot be avoided while discussing inventory management of any organization. The present review study considered 70 research papers based on inventory modeling for imperfect quality items through the web search on "Scopus." A conceptual research methodology framework has been developed for the different stages of the proposed field. This paper is focused on the original contributions of the authors who studied or summarized the practices of this field. A systematic and crisp review inventory model under the consideration of imperfect quality items based on a recent time is presented. The selected period for the review has a significant and increasing number of publications, through which the latest updates and advancements in the field are well analyzed. The findings and discussions have highlighted research gaps, old methodologies, and new methodologies to catalyze the research development of this field and its system. From this study, it can be concluded that there is a significant research gap to handle the closed-loop supply chain, reverse supply chains, green supply chain, etc. in the existence of imperfect quality items.

7 Limitations and Future Research Directions

The limitations and future research guidelines are given below:

- The findings of this literature review study are focused on the collected data from the "Scopus" with certainly considered inputs in the search option. In the future, the authors can explore other databases like ISI Web of Science, Emerald, Google Scholar, etc. and perform a process-oriented analysis to discuss the recent trends and advancements in this field.
- The period of the data collection is 2000–2020 and it is considered that it is the best representative data for this literature review study. Future works may extend the duration of the search period.
- Based on this review study, we recommend that future research can be implemented for a specific journal, domain, or industry to address imperfect quality items in inventory systems or some other emerging fields.
- Besides, the advancement of the closed-loop supply chain, green supply chain, and reverse supply chain could also be investigated for this research field.

Appendix A: Overview of the Articles with at Least 50+ Scopus Citation

S. no.	Authors	Title	Problem addressed	Scopus citation	Document type
1.	Salameh and Jaber [82]	"Economic production quantity model for items with imperfect quality"	This study extended the EPQ/EOQ model by considering the imperfect quality items.	651	Article

(continued)

(continued)

S. no.	Authors	Title	Problem addressed	Scopus citation	Document type
2.	Goyal and Cárdenas-Barrón [32]	"Note on Economic production quantity model for items with imperfect quality - A practical approach"	This study proposed a practical approach for the EPQ model with consideration of imperfect quality items.	268	Article
3.	Maddah and Jaber [65]	"Economic order quantity for items with imperfect quality: Revisited"	This model analyzed the EOQ model with unreliable supply circumstances in imperfect quality items	199	Article
4.	Chang [7]	"An application of fuzzy sets theory to the EOQ model with imperfect quality items"	This paper developed a fuzzy model for an inventory problem with a 100% screening process of imperfect quality items	177	Article
5.	Cárdenas-Barrón [21]	"Economic production quantity with rework process at a single-stage manufacturing system with planned backorders"	An EPQ inventory model with planned backorders for determining the economic production quantity and the size of backorders for a single product is developed, which is made in a single-stage manufacturing process that generates imperfect quality products and all defective products must be reworked in the same cycle	173	Article
6.	Chiu [13]	"Determining the optimal lot size for the finite production model with random defective rate, the rework process, and backlogging"	This model developed an EPQ model with the reworking of defective items with shortages	168	Article

(continued)

(continued)

S. no.	Authors	Title	Problem addressed	Scopus citation	Document type
7.	Eroglu and Ozdemir [24]	"An economic order quantity model with defective items and shortages"	This paper develops an EOQ model for which each ordered lot contains some defective items. Shortages are allowed	165	Article
8.	Khan et al. [49, 50]	"A review of the extensions of a modified EOQ model for imperfect quality items"	This study reviewed the EOQ-related models with imperfect quality items	160	Review
9.	Sana [83, 84]	"An economic production lot size model in an imperfect production system"	This paper investigates an EPL (Economic Production Lot size) model in an imperfect production system in which the production facility may shift from an "in-control" state to an "out-of-control" state at any random time	140	Article
10.	Sana [83, 84]	"A production-inventory model in an imperfect production process"	The paper develops a model to determine the optimal product reliability and production rate that achieves the biggest total integrated profit for an imperfect manufacturing process	136	Article
11.	Sarkar and Moon [91]	"An EPQ model with inflation in an imperfect production system"	The classical EPQ model is extended with stochastic demand under the effect of inflation	131	Article
12.	Chung and Hou [18]	"An optimal production run time with imperfect production processes and allowable shortages"	This paper develops a model to determine an optimal run time for a deteriorating production system under allowable shortages	130	Article

(continued)

(continued)

S. no.	Authors	Title	Problem addressed	Scopus citation	Document type
13.	Sarkar [88, 89]	"An EOQ model with delay in payments and stock dependent demand in the presence of imperfect production"	This model investigates the retailer's optimal replenishment policy under permissible delay in payment with stock dependent demand within the EOQ framework	118	Article
14.	Khan et al. [48]	"An integrated supply chain model with errors in quality inspection and learning in production"	This paper provides a simple but integrated mathematical model for determining an optimal vendor–buyer inventory policy by accounting for quality inspection errors at the buyer's end and learning in production at the vendor's end	116	Article
15.	Khan et al. [49, 50]	"An economic order quantity (EOQ) for items with imperfect quality and inspection errors"	This paper makes use of Salameh and Jaber [82] and Raouf et al.[77] models to determine an inventory policy for imperfect items received by a buyer	114	Conference Paper
16.	Jaber et al. [43]	"Economic production quantity model for items with imperfect quality subject to learning effects"	This paper extends the work of Salameh and Jaber [82] by assuming the percentage defective per lot reduces according to a learning curve, which was empirically validated by data from the automotive industry	114	Article
17.	Wahab et al. [104]	"EOQ models for a coordinated two-level international supply chain considering imperfect items and environmental impact"	A two-level coordinated supply chain where the shipments from one player to the other contain defectives	113	Article

(continued)

(continued)

S. no.	Authors	Title	Problem addressed	Scopus citation	Document type
18.	Sarkar et al. [90]	"An economic production quantity model with random defective rate, rework process and backorders for a single-stage production system"	This paper revisits the economic production quantity (EPQ) model with a rework process at a single-stage manufacturing system with planned backorders	108	Article
19.	Sarkar [88, 89]	"An inventory model with reliability in an imperfect production process"	The paper analyzes an economic manufacturing quantity (EMQ) model with price and advertising demand pattern in an imperfect production process under the effect of inflation	108	Article
20.	Yoo et al. [110]	"Economic production quantity model with imperfect quality items, two-way imperfect inspection and sales return"	In practice, not only production but inspection processes are often not perfect, thereby generating defects and inspection errors	107	Article
21.	Chan et al. [6]	"A new EPQ model: Integrating lower pricing, rework and reject situations"	This model assumes that items of imperfect quality, not necessarily defective, could be used in another production situation or sold to a particular purchaser at a lower price	107	Article
21.	Ouyang et al. [71]	"Quality improvement, setup cost and lead-time reductions in lot size reorder point models with an imperfect production process"	This paper investigates the lot size, reorder point inventory model involving variable lead time with partial backorders, where the production process is imperfect	97	Article

(continued)

(continued)

S. no.	Authors	Title	Problem addressed	Scopus citation	Document type
22.	Sana et al. [87]	"An imperfect production process in a volume flexible inventory model"	This paper hypothesizes an imperfect production/inventory system where the items produced are a mixture of perfect and imperfect quality	94	Article
23.	Rezaei and Davoodi [78]	"A deterministic, multi-item inventory model with supplier selection and imperfect quality"	This paper considers the scenario of a supply chain with multiple products and multiple suppliers, all of which have limited capacity	90	Article
24.	Goyal et al. [34]	"A simple integrated production policy of an imperfect item for vendor and buyer"	This article develops a simple approach for determining an optimal integrated vendor–buyer inventory policy for an item with imperfect quality	90	Article
25.	Sarkar and Moon [92]	"Improved quality, setup cost reduction, and variable backorder costs in an imperfect production process"	This paper illustrates the relationship between quality improvement, reorder point, and lead time, as affected by backorder rate, in an imperfect production process	85	Article
26.	Liao et al. [62]	"Optimal economic production quantity policy for imperfect process with imperfect repair and maintenance"	This study integrates maintenance and production programs with the EPQ model for an imperfect process involving a deteriorating production system with increasing hazard rate: imperfect repair and rework upon failure (out-of-control state)	82	Article

(continued)

(continued)

S. no.	Authors	Title	Problem addressed	Scopus citation	Document type
27.	Ouyang and Chang [70]	"Optimal production lot with imperfect production process under permissible delay in payments and complete backlogging"	In this study, we explore the effects of the reworking imperfect quality items and trade credit on the EPQ model with imperfect production processes and complete backlogging	80	Article
28.	Chang and Ho [8]	"Exact closed-form solutions for "optimal inventory model for items with imperfect quality and shortage backordering"	This article addressed an inventory problem for items with imperfect quality and shortage backordering, which is the same as Wee et al. [108]	80	Article
29.	Chung et al. [17]	"A two-warehouse inventory model with imperfect quality production processes"	This paper incorporates concepts of the basic two warehouses and imperfect quality to generalize Salameh and Jaber [82]	77	Article
30.	Sarkar et al. [93]	"Optimal reliability, production lot size and safety stock in an imperfect production system"	This paper is concerned with the joint determination of optimal production lot size, safety stock and reliability parameter under the realistic assumptions that the production facility is subject to random breakdowns of machinery system and to change in the variable reliability parameter	74	Article

(continued)

(continued)

S. no.	Authors	Title	Problem addressed	Scopus citation	Document type
31.	Hsu and Hsu [40]	"An EOQ model with imperfect quality items, inspection errors, shortage backordering, and sales returns"	An economic order quantity model with imperfect quality items, inspection errors, shortage backordering, and sales returns is developed	73	Article
32.	Lo et al. [63]	"An integrated production-inventory model with imperfect production processes and Weibull distribution deterioration under inflation"	An integrated production-inventory model from the perspectives of both the manufacturer and the retailer is developed in this study	70	Article
33.	Sarkar and Saren [94]	"Product inspection policy for an imperfect production system with inspection errors and warranty cost"	This paper describes a deteriorating production process that randomly shifts to an out-of-control state from the in-control state. In the case of full inspection policy, the expected total cost together with inspection cost results in higher inventory costs	68	Article
34.	Roy et al. [81]	"An economic order quantity model of imperfect quality items with partial backlogging"	An EOQ model is developed based on some real-life problems where imperfect quality items take place in each ordered lot and partial backordering is done	65	Conference Paper

(continued)

(continued)

S. no.	Authors	Title	Problem addressed	Scopus citation	Document type
35.	Khan et al. [51]	"Economic order quantity model for items with imperfect quality with learning in inspection"	This paper is an extension of Salameh and Jaber [82] for the case where the buyer's inspection process undergoes learning while screening for defective items in a lot	65	Article
36.	Chen and Kang [11]	"Coordination between vendor and buyer considering trade credit and items of imperfect quality"	This paper considers trade credit and imperfect quality in the integrated vendor–buyer model and develops Theorems to determine the optimal solutions of buyer's optimal replenishment period and frequency.	64	Article
37.	Pal et al. [73]	"A mathematical model on EPQ for stochastic demand in an imperfect production system"	In the paper, an EPQ inventory model to determine the optimal buffer inventory is developed	63	Article
38.	Chiu and Chiu [14]	"Mathematical modeling for production system with backlogging and failure in repair"	A production model is developed with failure in the repair process and backlogged shortages	63	Article
39.	Wahab and Jaber [103]	"Economic order quantity model for items with imperfect quality, different holding costs, and learning effects: A note"	An optimal lot-sizing model is developed for imperfect quality items based on Salameh and Jaber [82]	62	Article
40.	Hou [38]	"An EPQ model with setup cost and process quality as functions of capital expenditure"	This paper considers an EPQ model with imperfect production processes, in which the setup cost and process quality are functions of capital expenditure	61	Article

(continued)

(continued)

S. no.	Authors	Title	Problem addressed	Scopus citation	Document type
41.	Bag et al. [2]	"A production inventory model with fuzzy random demand and with flexibility and reliability considerations"	For the first time an EPQ model is developed in the presence of imprecision and uncertainty in demand with flexibility and reliability consideration in the production process	59	Article
42.	Pal et al. [73]	"Maximizing profits for an EPQ model with unreliable machine and rework of random defective items"	This article deals with an EPQ model in an imperfect production system. The production system may undergo in 'out-of-control' state from 'in-control' state, after a certain time that follows a probability density function	57	Article
43.	Chung and Wee [19]	"An integrated production-inventory deteriorating model for pricing policy considering imperfect production, inspection planning and warranty-period- and stock-level-dependent demand"	The purpose of this study is to investigate an integrated production inventory deteriorating model considering the pricing policy, the imperfect production, the inspection planning, the warranty-period and the stock-level-dependent demand with the Weibull deterioration, partial backorder and inflation	57	Article

(continued)

(continued)

S. no.	Authors	Title	Problem addressed	Scopus citation	Document type
44.	Konstantaras et al. [60]	"Economic ordering policy for an item with imperfect quality subject to the in-house inspection"	This article considers a production/inventory system where each lot of items received or produced contains a random proportion of defective units, items of imperfect quality. The purchaser contacts a 100% inspection to identify the perfect (acceptable) quality items	57	Article
45.	Chiu et al. [15]	"Effects of random defective rate and imperfect rework process on economic production quantity model"	This paper considers the EPQ model with a random defective rate and an imperfect rework process	57	Article
46.	Hsu and Yu [41]	"EOQ model for imperfective items under a one-time-only discount"	This paper investigates an inventory model with imperfect quality under a one-time-only discount, where a 100% screening process is performed on the received lot and the defectives are assumed can be sold in a single batch by the end of the screening process.	53	Article
47.	Wang [105]	"The impact of a free-repair warranty policy on EMQ model for imperfect production systems"	In this study, the economic production quantity problem in the presence of imperfect processes for products sold under a free-repair warranty policy is considered	52	Article

(continued)

(continued)

S. no.	Authors	Title	Problem addressed	Scopus citation	Document type
48.	Jaber et al. [44]	"Economic order quantity models for imperfect items with buy and repair options"	This paper revisits this model (Salameh and Jaber, 2000, Int. J. Prod. Econ. 64(1),59–64) and extends it by assuming that a shipment is coming from a distant supplier and, therefore, it is not feasible to replace the imperfect items with an additional order to the same supplier	50	Article
49.	Chiu et al. [16]	"Mathematical modeling for determining the replenishment policy for EMQ model with rework and multiple shipments"	This paper studies the optimal inventory replenishment policy for the economic manufacturing quantity model with rework and multiple shipments. The classic EMQ model assumes a continuous issuing policy and that all items produced are of perfect quality	50	Article
50.	Sana and Chaudhuri [86]	"An EMQ model in an imperfect production process"	A production-inventory model of Giri and Dohi [30] is extended, considering the effect of an imperfect production process subject to random breakdown and variable safety stocks	50	Article

References

1. Ahmed, I., & Sultana, I. (2014). A literature review on inventory modeling with reliability consideration. *International Journal of Industrial Engineering Computations, 5*(1), 169–178.
2. Bag, S., Chakraborty, D., & Roy, A. R. (2009). A production inventory model with fuzzy random demand and with flexibility and reliability considerations. *Computers & Industrial Engineering, 56*(1), 411–416.
3. Ben-Daya, M., & Hariga, M. (2000). Economic lot scheduling problem with imperfect production processes. *Journal of the Operational Research Society, 51*(7), 875–881.
4. Ben-Daya, M., & Rahim, A. (2003). Optimal lot-sizing, quality improvement and inspection errors for multistage production systems. *International Journal of Production Research, 41*(1), 65–79.
5. Chakrabarti, T., & Chaudhuri, K. S. (1997). An EOQ model for deteriorating items with a linear trend in demand and shortages in all cycles. *International Journal of Production Economics, 49*(3), 205–213.
6. Chan, W. M., Ibrahim, R. N., & Lochert, P. B. (2003). A new EPQ model: Integrating lower pricing, rework and reject situations. *Production Planning & Control, 14*(7), 588–595.
7. Chang, H. C. (2004). An application of fuzzy sets theory to the EOQ model with imperfect quality items. *Computers & Operations Research, 31*(12), 2079–2092.
8. Chang, H. C., & Ho, C. H. (2010). Exact closed-form solutions for "optimal inventory model for items with imperfect quality and shortage backordering". *Omega, 38*(3–4), 233–237.
9. Chaudhary, V., Kulshrestha, R., & Routroy, S. (2018). State-of-the-art literature review on inventory models for perishable products. *Journal of Advances in Management Research.*
10. Chen, Y. C. (2013). An optimal production and inspection strategy with preventive maintenance error and rework. *Journal of Manufacturing Systems, 32*(1), 99–106.
11. Chen, L. H., & Kang, F. S. (2010). Coordination between vendor and buyer considering trade credit and items of imperfect quality. *International Journal of Production Economics, 123*(1), 52–61.
12. Chen, C. K., & Lo, C. C. (2006). Optimal production run length for products sold with warranty in an imperfect production system with allowable shortages. *Mathematical and Computer Modelling, 44*(3), 319–331.
13. Chiu, Y. P. (2003). Determining the optimal lot size for the finite production model with random defective rate, the rework process, and backlogging. *Engineering Optimization, 35*(4), 427–437.
14. Chiu, S. W., & Chiu, Y. S. P. (2006). Mathematical modeling for production system with backlogging and failure in repair. *Journal of Scientific & Industrial Research, 65*(6), 499–506.
15. Chiu, S. W., Gong, D. C., & Wee, H. M. (2004). Effects of random defective rate and imperfect rework process on economic production quantity model. *Japan Journal of Industrial and Applied Mathematics, 21*(3), 375.
16. Chiu, Y. S. P., Liu, S. C., Chiu, C. L., & Chang, H. H. (2011). Mathematical modeling for determining the replenishment policy for EMQ model with rework and multiple shipments. *Mathematical and Computer Modelling, 54*(9–10), 2165–2174.
17. Chung, K. J., Her, C. C., & Lin, S. D. (2009). A two-warehouse inventory model with imperfect quality production processes. *Computers & Industrial Engineering, 56*(1), 193–197.
18. Chung, K. J., & Hou, K. L. (2003). An optimal production run time with imperfect production processes and allowable shortages. *Computers & Operations Research, 30*(4), 483–490.
19. Chung, C. J., & Wee, H. M. (2008). An integrated production-inventory deteriorating model for pricing policy considering imperfect production, inspection planning and warranty-period-and stock-level-dependant demand. *International Journal of Systems Science, 39*(8), 823–837.
20. Chung, K. J., & Huang, Y. F. (2006). Retailer's optimal cycle times in the EOQ model with imperfect quality and a permissible credit period. *Quality and Quantity, 40*(1), 59–77.
21. Cárdenas-Barrón, L. E. (2009). Economic production quantity with rework process at a single-stage manufacturing system with planned backorders. *Computers & Industrial Engineering, 57*(3), 1105–1113.

22. Cárdenas-Barrón, L. E. (2012). A complement to "A comprehensive note on: An economic order quantity with imperfect quality and quantity discounts". *Applied Mathematical Modelling, 36*(12), 6338–6340.
23. Cárdenas-Barrón, L. E. (2000). Observation on: "Economic production quantity model for items with imperfect quality" [Int. J. Production Economics 64 (2000) 59–64]. *International Journal of Production Economics, 67*(2), 201.
24. Eroglu, A., & Ozdemir, G. (2007). An economic order quantity model with defective items and shortages. *International Journal of Production Economics, 106*(2), 544–549.
25. Gao, D., Xu, Z., Ruan, Y.Z., & Lu, H. (2017). From a systematic literature review to integrated definition for sustainable supply chain innovation (SSCI). *Journal of Cleaner Production, 142*, 1518–1538.
26. Gautam, P., & Khanna, A. (2018). An imperfect production inventory model with setup cost reduction and carbon emission for an integrated supply chain. *Uncertain Supply Chain Management, 6*(3), 271–286.
27. Gautam, P., Kishore, A., Khanna, A., & Jaggi, C. K. (2019). Strategic defect management for a sustainable green supply chain. *Journal of Cleaner Production, 233*, 226–241.
28. Gautam, P., Maheshwari, S., Kaushal-Deep, S. M., Bhat, A. R., & Jaggi, C. K. (2020). COVID-19: a bibliometric analysis and insights. *International Journal of Mathematical, Engineering and Management Sciences, 5*(6), 1156–1169.
29. Ghare, P. M. (1963). A model for an exponentially decaying inventory. *Journal of Industrial Engineering, 14*, 238–243.
30. Giri, B. C., & Dohi, T. (2005). Computational aspects of an extended EMQ model with variable production rate. *Computers & Operations Research, 32*(12), 3143–3161.
31. Giri, B. C., & Dohi, T. (2007). Inspection scheduling for imperfect production processes under free repair warranty contract. *European Journal of Operational Research, 183*(2007), 238–252.
32. Goyal, S. K., & Cárdenas-Barrón, L. E. (2002). Note on: Economic production quantity model for items with imperfect quality—A practical approach. *International Journal of Production Economics, 77*(1), 85–87.
33. Goyal, S. K., Gunasekaran, A., Martikainen, T., & Yli-Olli, P. (1993). Integrating production and quality control policies: A survey. *European Journal of Operational Research, 69*(1993), 1–13.
34. Goyal, S. K., Huang, C. K., & Chen, K. C. (2003). A simple integrated production policy of an imperfect item for vendor and buyer. *Production Planning & Control, 14*(7), 596–602.
35. Gurnani, H., Drezner, Z., & Akella, R. (1996). Capacity planning under different inspection strategies. *European Journal of Operational Research, 89*(1996), 302–312.
36. Ha, D., & Kim, S. L. (1997). Implementation of JIT purchasing: An integrated approach. *Production Planning & Control, 8*(2), 152–157.
37. Harris, F. W. (1913). How many parts to make at once. *Factory, the Magazine of Management, 10*, 135–136.
38. Hou, K. L. (2007). An EPQ model with setup cost and process quality as functions of capital expenditure. *Applied Mathematical Modelling, 31*(1), 10–17.
39. Hsieh, C. C., & Lee, Z. Z. (2005). Joint determination of production run length and number of standbys in a deteriorating production process. *European Journal of Operational Research, 162*(2), 359–371.
40. Hsu, J. T., & Hsu, L. F. (2013). An EOQ model with imperfect quality items, inspection errors, shortage backordering, and sales returns. *International Journal of Production Economics, 143*(1), 162–170.
41. Hsu, W. K. K., & Yu, H. F. (2009). EOQ model for imperfective items under a one-time-only discount. *Omega, 37*(5), 1018–1026.
42. Huang, C. K. (2002). An integrated vendor-buyer cooperative inventory model for items with imperfect quality. *Production Planning & Control, 13*(4), 355–361.
43. Jaber, M. Y., Goyal, S. K., & Imran, M. (2008). Economic production quantity model for items with imperfect quality subject to learning effects. *International Journal of Production Economics, 115*(1), 143–150.

44. Jaber, M. Y., Zanoni, S., & Zavanella, L. E. (2014). Economic order quantity models for imperfect items with buy and repair options. *International Journal of Production Economics, 155,* 126–131.
45. Jaggi, C. K., Goel, S. K., & Mittal, M. (2013). Credit financing in economic ordering policies for defective items with allowable shortages. *Applied Mathematics and Computation, 219*(10), 5268–5282.
46. Jauhari, W. A., & Laksono, P. W. (2017, November). A joint economic lot-sizing problem with fuzzy demand, defective items and environmental impacts. In *Materials Science and Engineering Conference Series* (Vol. 273, No. 1, p. 012018).
47. Karlin, S. (1958). One stage inventory models with uncertainty. In K. J. Arrow, S. Karlin, & H. (Eds.), *Scarf Studies in the Mathematical Theory of Inventory and Production*, Chap. 8.
48. Khan, M., Jaber, M. Y., & Ahmad, A. R. (2014). An integrated supply chain model with errors in quality inspection and learning in production. *Omega, 42*(1), 16–24.
49. Khan, M., Jaber, M. Y., & Bonney, M. (2011). An economic order quantity (EOQ) for items with imperfect quality and inspection errors. *International Journal of Production Economics, 133*(1), 113–118.
50. Khan, M., Jaber, M. Y., Guiffrida, A. L., & Zolfaghari, S. (2011). A review of the extensions of a modified EOQ model for imperfect quality items. *International Journal of Production Economics, 132*(1), 1–12.
51. Khan, M., Jaber, M. Y., & Wahab, M. I. M. (2010). Economic order quantity model for items with imperfect quality with learning in inspection. *International Journal of Production Economics, 124*(1), 87–96.
52. Khan, M., Jaber, M. Y., Zanoni, S., & Zavanella, L. (2016). Vendor managed inventory with consignment stock agreement for a supply chain with defective items. *Applied Mathematical Modelling, 40*(15–16), 7102–7114.
53. Khanna, A., Gautam, P., Sarkar, B., & Jaggi, C. K. (2020). Integrated vendor–buyer strategies for imperfect production systems with maintenance and warranty policy. *RAIRO-Operations Research, 54*(2), 435–450.
54. Khanna, A., Kishore, A., & Jaggi, C. (2017). Strategic production modeling for defective items with imperfect inspection process, rework, and sales return under two-level trade credit. *International Journal of Industrial Engineering Computations, 8*(1), 85–118.
55. Khanna, A., Mittal, M., Gautam, P., & Jaggi, C. (2016). Credit financing for deteriorating imperfect quality items with allowable shortages. *Decision Science Letters, 5*(1), 45–60.
56. Khara, B., Dey, J. K., & Mondal, S. K. (2020). Sustainable recycling in an imperfect production system with acceptance quality level dependent development cost and demand. *Computers & Industrial Engineering, 142,* 106300.
57. Khedlekar, U. K., & Tiwari, R. K. (2018). Imperfect production model for sensitive demand with shortage. *Reliability: Theory & Applications, 13,* 4–51.
58. Kim, H., & Hong, Y. (1999). An optimal production run length in deteriorating production processes. *International Journal of Production Economics, 58*(2), 183–189.
59. Kishore, A., Gautam, P., Khanna, A., & Jaggi, C. K. (2019). Investigating the effect of learning in set-up cost for imperfect production systems by utilizing two-way inspection plan for rework under screening constraints. *Scientia Iranica.*
60. Konstantaras, I., Goyal, S. K., & Papachristos, S. (2007). Economic ordering policy for an item with imperfect quality subject to the in-house inspection. *International Journal of Systems Science, 38*(6), 473–482.
61. Lee, H. L., & Rosenblatt, M. J. (1987). Simultaneous determination of production cycle and inspection schedules in a production system. *Management Science, 33*(9), 1125–1136.
62. Liao, G. L., Chen, Y. H., & Sheu, S. H. (2009). Optimal economic production quantity policy for imperfect process with imperfect repair and maintenance. *European Journal of Operational Research, 195*(2), 348–357.
63. Lo, S. T., Wee, H. M., & Huang, W. C. (2007). An integrated production-inventory model with imperfect production processes and Weibull distribution deterioration under inflation. *International Journal of Production Economics, 106*(1), 248–260.

64. Ma, W. N., Gong, D. C., & Lin, G. C. (2010). An optimal common production cycle time for imperfect production processes with scrap. *Mathematical and Computer Modelling, 52,* 724–737.
65. Maddah, B., & Jaber, M. Y. (2008). Economic order quantity for items with imperfect quality: Revisited. *International Journal of Production Economics, 112*(2), 808–815.
66. Maheshwari, S., Gautam, P., & Jaggi, C. K. (2020). Role of big data analytics in supply chain management: current trends and future perspectives. *International Journal of Production Research,* 1–26.
67. Mayring, P. (2003). Qualitative Inhaltanalyse – Grundlagen und Techniken [Qualitative content analysis], 8th edn. Weinheim, Germany: Beltz Verlag.
68. Mosca, A., Vidyarthi, N., & Satir, A. (2019). Integrated transportation-inventory models: A review. *Operations Research Perspectives,* 100101.
69. Moussawi-Haidar, L., Salameh, M., & Nasr, W. (2016). Production lot sizing with quality screening and rework. *Applied Mathematical Modelling, 40,* 3242–3256.
70. Ouyang, L. Y., & Chang, C. T. (2013). Optimal production lot with imperfect production process under permissible delay in payments and complete backlogging. *International Journal of Production Economics, 144*(2), 610–617.
71. Ouyang, L. Y., Chen, C. K., & Chang, H. C. (2002). Quality improvement, setup cost and lead-time reductions in lot size reorder point models with an imperfect production process. *Computers & Operations Research, 29*(12), 1701–1717.
72. Ouyang, L. Y., Chuang, C. J., Ho, C. H., & Wu, C. W. (2014). An integrated inventory model with quality improvement and two-part credit policy. *Top, 22*(3), 1042–1061.
73. Pal, B., Sana, S. S., & Chaudhuri, K. (2013). A mathematical model on EPQ for stochastic demand in an imperfect production system. *Journal of Manufacturing Systems, 32*(1), 260–270.
74. Pasandideh, S. H. R., Niaki, S. T. A., & Mirhosseyni, S. S. (2010). A parameter-tuned genetic algorithm to solve multi-product economic production quantity model with defective items, rework, and constrained space. *The International Journal of Advanced Manufacturing Technology, 49*(5–8), 827–837.
75. Pereira, V., & Costa, H. G. (2015). A literature review on lot size with quantity discounts: 1995–2013. *Journal of Modelling in Management.*
76. Porteus, E. L. (1986). Optimal lot sizing, process quality improvement and setup cost reduction. *Operations Research, 34*(1), 137–144.
77. Raouf, A., Jain, J. K., & Sathe, P. T. (1983). A cost-minimization model for multicharacteristic component inspection. *IIE Transactions, 15*(3), 187–194.
78. Rezaei, J., & Davoodi, M. (2008). A deterministic, multi-item inventory model with supplier selection and imperfect quality. *Applied Mathematical Modelling, 32*(10), 2106–2116.
79. Rosenblatt, M. J., & Lee, H. L. (1986). Economic production cycles with imperfect production processes. *IIE Transactions, 18*(1), 48–55.
80. Roy, S. K., Pervin, M., & Weber, G. W. (2020). Imperfection with inspection policy and variable demand under trade-credit: A deteriorating inventory model. *Numerical Algebra, Control & Optimization, 10*(1), 45.
81. Roy, M. D., Sana, S. S., & Chaudhuri, K. (2011). An economic order quantity model of imperfect quality items with partial backlogging. *International Journal of Systems Science, 42*(8), 1409–1419.
82. Salameh, M. K., & Jaber, M. Y. (2000). Economic production quantity model for items with imperfect quality. *International Journal of Production Economics, 64*(1–3), 59–64.
83. Sana, S. S. (2010). A production–inventory model in an imperfect production process. *European Journal of Operational Research, 200*(2), 451–464.
84. Sana, S. S. (2010). An economic production lot size model in an imperfect production system. *European Journal of Operational Research, 201*(1), 158–170.
85. Sana, S. S. (2011). A production-inventory model of imperfect quality products in a three-layer supply chain. *Decision Support Systems, 50*(2), 539–547.

86. Sana, S. S., & Chaudhuri, K. (2010). An EMQ model in an imperfect production process. *International Journal of Systems Science, 41*(6), 635–646.
87. Sana, S. S., Goyal, S. K., & Chaudhuri, K. (2007). An imperfect production process in a volume flexible inventory model. *International Journal of Production Economics, 105*(2), 548–559.
88. Sarkar, B. (2012). An EOQ model with delay in payments and stock dependent demand in the presence of imperfect production. *Applied Mathematics and Computation, 218*(17), 8295–8308.
89. Sarkar, B. (2012). An inventory model with reliability in an imperfect production process. *Applied Mathematics and Computation, 218*(9), 4881–4891.
90. Sarkar, B., Cárdenas-Barrón, L. E., Sarkar, M., & Singh, M. L. (2014). An economic production quantity model with random defective rate, rework process and backorders for a single stage production system. *Journal of Manufacturing Systems, 33*(3), 423–435.
91. Sarkar, B., & Moon, I. (2011). An EPQ model with inflation in an imperfect production system. *Applied Mathematics and Computation, 217*(13), 6159–6167.
92. Sarkar, B., & Moon, I. (2014). Improved quality, setup cost reduction, and variable backorder costs in an imperfect production process. *International Journal of Production Economics, 155,* 204–213.
93. Sarkar, B., Sana, S. S., & Chaudhuri, K. (2010). Optimal reliability, production lot size and safety stock in an imperfect production system. *International Journal of Mathematics in Operational Research, 2*(4), 467–490.
94. Sarkar, B., & Saren, S. (2016). Product inspection policy for an imperfect production system with inspection errors and warranty cost. *European Journal of Operational Research, 248*(1), 263–271.
95. Sarkar, B., Ahmed, W., Choi, S. B., & Tayyab, M. (2018). Sustainable inventory management for environmental impact through partial backordering and multi-trade-credit-period. *Sustainability, 10*(12), 4761.
96. Seuring, S., & Müller, M. (2008). From a literature review to a conceptual framework for sustainable supply chain management. *Journal of Cleaner Production, 16*(15), 1699–1710.
97. Shekarian, E., Kazemi, N., Abdul-Rashid, S. H., & Olugu, E. U. (2017). Fuzzy inventory models: A comprehensive review. *Applied Soft Computing, 55,* 588–562.
98. Srivastava, H. M., Chung, K. J., Liao, J. J., Lin, S. D., & Chuang, S. T. (2019). Some modified mathematical analytic derivations of the annual total relevant cost of the inventory model with two levels of trade credit in the supply chain system. *Mathematical Methods in the Applied Sciences, 42*(11), 3967–3977.
99. Taleizadeh, A. A., Yadegari, M., & Sana, S. S. (2019). Production models of multiple products using a single machine under quality screening and reworking policies. *Journal of Modelling in Management.*
100. Tang, S. (1991). Designing an optimal production system with inspection. *European Journal of Operational Research, 52,* 45–54.
101. Tayyab, M., Sarkar, B., & Yahya, B. N. (2019). Imperfect multi-stage lean manufacturing system with rework under fuzzy demand. *Mathematics, 7*(1), 13.
102. Tsao, Y. C. (2012). Determination of production run time and warranty length under system maintenance and trade credits. *International Journal of Systems Science, 43*(12), 2351–2360.
103. Wahab, M. I. M., & Jaber, M. Y. (2010). Economic order quantity model for items with imperfect quality, different holding costs, and learning effects: A note. *Computers & Industrial Engineering, 58*(1), 186–190.
104. Wahab, M. I. M., Mamun, S. M. H., & Ongkunaruk, P. (2011). EOQ models for a coordinated two-level international supply chain considering imperfect items and environmental impact. *International Journal of Production Economics, 134*(1), 151–158.
105. Wang, H. (2004). The impact of a free-repair warranty policy on EMQ model for imperfect production systems. *Computers & Operations Research, 31*(12), 2021–2035.
106. Wang Chiu, S. (2007). Optimal replenishment policy for imperfect quality EMQ model with rework and backlogging. *Applied Stochastic Models in Business and Industry, 23*(2), 165–178.

107. Wee, H. M. (1995). A deterministic lot-size inventory model for deteriorating items with shortages and a declining market. *Computers & Operations Research, 22*(3), 345–356.
108. Wee, H. M., Yu, J., & Chen, M. C. (2007). Optimal inventory model for items with imperfect quality and shortage backordering. *Omega, 35*(1), 7–11.
109. Yeh, R. H., Ho, W. T., & Tseng, S. T. (2000). Optimal production run length for products sold with warranty. *European Journal of Operational Research, 120*(3), 575–582.
110. Yoo, S. H., Kim, D., & Park, M. S. (2009). Economic production quantity model with imperfect-quality items, two-way imperfect inspection and sales return. *International Journal of Production Economics, 121*(1), 255–265.
111. Zhou, Y., Chen, C., Li, C., & Zhong, Y. (2016). A synergic economic order quantity model with trade credit, shortages, imperfect quality and inspection errors. *Applied Mathematical Modelling, 40*(2), 1012–1028.

Quality Management: Yesterday, Today and Tomorrow

K. Muralidharan and Neha Raval

1 Introduction

The industrial environment has experienced a leap of change over different indus-trial revolutions. The first industrial revolution started during the end of the eigh-teenth century which initiated mechanization of production activities from handmade production tools. The invention of machine tools and steam engines were the major change drivers during the first industrial revolution. With the establishment of the machine tool industry, the second industrial revolution broke in. Improved manufac-turing system by standardization, interchangeability of parts, steel production were prime change drivers of the second industrial revolution dated between 1870 and 1914. With the advent of Internet during 1990, the third industrial revolution began with a bang. Global connectivity through Internet, shift toward renewable energy for power generation, and smart manufacturing system are some of the dimensions of the third industrial revolution. The current emerging industrial scenario with reference to *Integrated Industries* and *Smart manufacturing* is termed as "Industry 4.0". As mentioned by Hermann and Otto [5] major components of Industry 4.0 are Cyber-Physical System, Internet of Things, Smart Factory and Internet of Services, and so on.

With the changes in revolution came the changes in quality of production, services, and delivery mechanism of the organizational activities. Holistic understandings of historical, contemporary, and futuristic purview of quality management (QM) practices require supplementary information regarding peripheral phenomena that

K. Muralidharan (✉) · N. Raval
Department of Statistics, Faculty of Science, The Maharaja Sayajirao University of Baroda, Vadodara 390 002, India
e-mail: muralikustat@gmail.com

N. Raval
e-mail: neha.ravalc@gmail.com

molded and advanced it over a period of time. Among this, the QM practices of the manufacturing segment have experienced monumental shifts over these different revolutions. According to Unyimadu [16], *"The industrial revolution was not merely technical but had impact on values, beliefs, social customs and societies at large".* So everything is changing for the future generation.

We understand that a critical assessment of quality management practices prevailing so far with future generation business practices is not easy but essential to draw commonalities between them. This is effectively done by aligning the technological advancement and socioeconomic condition prevailing during those generations. Each industrial revolution has two components in common: technological advancement and the desire to satisfy customer's needs. The latter of these two leads to the quality movement during different eras. Ever-changing customer demands and escalating technological advancements are putting contemporary organizations under great pressure to dig deep into the quality aspect of their offerings. Quality movement traces its roots back to the craftsmen culture of medieval Europe, where guilds were responsible to lay down strict rules to maintain the quality of their offerings. From craftsmanship to smart manufacturing, quality movement has traveled a lot, and has proved to be a great directional source for businesses to prosper. Hence, this paper proposes a thread of the evolution of quality management theory over different eras and their contribution to the business scenario, and finally its impact on all stakeholders.

This research paper is organized into 7 sections. Section 2 discusses about the origin and its inception of Quality Management practices. Sections 3–6 explain about Quality Management practices considering the technological and socioeconomic changes of each industrial revolution that happened so far. Section 7 proposes discussion with respect to the modern-day business and possible future developments.

2 Journey of Quality Movement

The quality movement can trace its roots back to medieval Europe. Pre-industrial economy is characterized through agriculture-based economy. This agriculture-based economy lead to make the stay people in one place for a longer period and sometimes for generations. The division of labor was low as a very small portion of population was engaged in other crafts except agriculture. As wealth was concentrated in the hands of a mighty few, majorities of people lived in subsistence level with little or no saving. High illiteracy rate and non-hygienic condition resulted in high mortality rate and in turn low population growth. A significant chunk of household income was spent on buying food and the remaining trifling amount enters the market for barter exchange for other products. This limited number of people who were practicing their mastered skills was known as craftsmen. These craftsmen relied on the traditional handmade system to serve their customers without bothering too much on quality.

Due to lower standard of skills, most craftsmen sold their goods locally as each had a tremendous personal stake in meeting customers' needs for quality. If quality needs weren't met, the craftsman can be at the risk of losing customers, which are not easily replaced. Therefore, masters maintained a form of quality control by inspecting goods before sale. These craftsmen across Europe were organized into unions called guilds. These guilds were responsible for developing strict rules for product and service quality. Inspection committees enforced the rules by marking flawless goods with a special mark or symbol. Quality during this era was embodied in fitness for purpose, trueness of material, and beauty of form [15].

The foremost advantage of craftsmen ship was companies gave their full attention to individual customers. The production process was pretty flexible to address customer requirements. High production cost and low productivity were the stumbling blocks of this system. From ice-age to new-age, we have seen all the changes. Everything happened for the benefit of people and the environment. Quality product, quality services, quality designs, quality customer, etc. are the new norms. Quality improvement, quality assurance, quality control, quality council, quality deployment, etc. are the new philosophies. Quality Management is all about these. Let us now understand all this in a bit through various industrial revolutions.

3   First Industrial Revolution (1760 to 1820–1840)

3.1   Technology

The first industrial revolution which dated back from 1760 to mid-nineteenth century lead marginal change from handicraft and agriculture-based economy to a new factory-based economy in Europe and the United States. The invention of machines changed the way in which people made goods. The development of machine tools and the rise of mechanized factory system with dominance in the textile industry are the early development of this era. The use of steam power and water power to perform mechanical work, the new production process for chemical manufacturing and iron production lead to a completely different socioeconomic environment with escalating economic growth in different sectors. Meliorated condition of roads and replacement of horse railways by steam locomotive engines improved average sailing speed by 50% during this period [3]. Mechanization of various industries accelerated productivity and consequently overshadowed craft-based economy [2]. This has increased trust and belief in having machines and technology for production and livelihood.

3.2 Socioeconomic Condition

Increased agricultural productivity and its financial gains laid down many people working on farms and hence they started working in new industrial setup concentrated in cities [14]. This resulted into improved standard of living with a significant increase in workers' wages during the century resulting in improved life expectancy. Though industrialization did little to lower food prizes, efficient transportation system helped in the better food supply in different parts of Britain. Rapid population growth in cities resulted into crowded cities with unhygienic living conditions for many workers. Ordinary working people found ample opportunities in factories with strict conditions of extended working hour's upto 10–12 h a day. Still, people are ready for such earnings. The invention of paper and steam power printing industry resulted in massive printing of newspapers and books which contributed to rising literacy among urban centres. Because of the falling price of commodities, people started spending their money on leisure activities and cities started witnessing consumer culture. All these changes affected the quality of life of people a lot and slowly the perception about the quality of products and processes also changed. Let us see how?

3.3 Quality Management System

The factory system cost many craftsmen their livelihood. Unlike traditional craft system, factory systems were designed based on Adam Smith's principle of division of labor. The ultimate aim of the factory system was to increase productivity. According to the principle of division of labor, if a worker does the entire job then the productivity is far less than when each worker carries out specialized tasks [16]. As a result, the craftsmen who were masters of their trade started working in factories where work was performed based on divided tasks. In place of looking after the whole life cycle of the product, their role was confined to one specific task which contributed to the final finished product. These craftsmen turned factory workers and shop owners turned production supervisors contributed to the decline in employee's sense of empowerment and autonomy in the workplace (ASQ). Quality management under such circumstances was largely left to the workers supplemented by the product inspection by supervisors. Defective products were either reworked or scrapped. A real need for input-process-output quality coordination is warranted.

4 Second Industrial Revolution (1870–1914)

4.1 Technology

With rapid standardization and industrialization, the second industrial revolution started growing rapidly by the end of the nineteenth century. This has contributed to a significant amount of assembly lines and mass production in every business organization. Electrification, mass production of goods and weapons, use of iron, expansion of railways and roads, use of radio and telegraph were the major contributors to the technological advancement of this era. Radio and telegraph changed the way in which people transmit information, do business, and interact with each other. Electrification brought a positive impact on every aspect of production: quality of products, services, and quantity of work with increased speed and lesser operational adjustments are some of them. Many factories relished 30% increase in output just from shifting to electric power.

With the extensive expansion of railroads, the supply of raw material was possible in huge amount and to remote areas. In the beginning of the twentieth century, Henry Ford leveraged the use of electrification by introducing a moving assembly line to his plant. Moving assembly line coupled with the complete and consistent inter-operability of parts changed the face of mass production [17]. The concept of interchangeable parts later adopted by the US department of war resulted in mass production of weapons. The majority of technical advancement of this era are well received by the car manufacturing industry which was later adopted by other industries. Henry Ford demonstrated the efficiency of mass production so much so that he started producing vehicles in huge numbers without allowing customers to wait for the inventory.

4.2 Socioeconomic Condition

Electrification of factories brought along with mechanization has resulted in a dramatic increment in the pace of work to factory workers. The overall health of the workforce declined due to harsh and unhygienic working conditions. Increased mechanization of factories cost many women, craftsmen their jobs which resulted into high unemployment during this period. The rapid increase in urbanization, hard working, and living conditions has forced many class conscious working people to shift their focus to factory-made products, although they cannot afford to, as a matter of prestige. Slowly, mass production improved productivity drastically and hence brought the price of products down. Now many people can afford factory-made products which results in higher demand for the product. Mass production allowed the evolution of consumerism in the US and Europe. Ford's system of mass production was based on the standardization and interchangeability and increased use of machinery leaving all production is carried out with skilled labors and expertise. To tackle high turnover, Ford came up with the improved wage system called 5$ a day. With mass production of cars, he brought cars price so much less that workers of his factory can afford them with their few months' salary. A trust and confidence building has slowly started in quality management.

4.3 Quality Management System

Mass production improved productivity exponentially. Ford reached a peak production volume of 2 million identical vehicle a year in the early 1920s, and he had cut the real cost to the consumer by an additional two-thirds. As explained by Womack et al. [17], to appeal to his target market of the average consumer, Ford had designed unprecedented ease of operation and maintainability into his car. He assumed that his buyers would even be a farmer, so with a modest tool kit and kinds of mechanical skills needed for fixing farm machinery will be easy for then. So, the Model T's owner manual, which was written in question-and-answer form, explained how owners could use simple tools to solve any of the 140 problems likely to crop up with the car. The assembler on Ford's mass-production line had only one repetitive task. The Assembly of different parts was up to newly created professionals—the industrial engineer. The production engineer was responsible for the delivery of parts to the line. Skilled repairmen refurbish the assembler's work and the machine inspector was made responsible to check the quality. The role of the machine inspector was to inspect the production process to ensure quality in the product was maintained. In the mass manufacturing system work that was not done properly was not discovered until the end of the assembly line, where another group of workers was called into play—the rework men, who retained many of the fitters' skills. This resulted into large rework areas in mass manufacturing factories. None of the car was tested on the road before it reaches the market. Quality in mass production was maintained based on the principle of "*Scientific Management*" proposed by Frederick Taylor. This philosophy advocates, dividing process into simple tasks to minimize complexity, so that assignment of simple tasks maximizes efficiency with fewer mistakes [19]. Thus Ford has started its seed for quality revolution in the workplace.

This was also the era of birth of Statistical Quality Control (SQC). Walter A Shewhart-an American Physicist and Statistician, working with bell laboratory in the US invented a schematic tool to examine the quality of output in May 1924. This schematic representation later known as "*Control Charts*", laid down the roots of SQC [11]. Until the invention of control charts, industrial quality was limited to inspecting finished products and removing defective items. Later, statistician and consultant: William Edward Deming inspired by Shewart's work started working on monitoring variation in the process of manufacturing. Deming appreciated the potentiality of Shewhart's work to apply beyond manufacturing to other industrial functions. With the prime focus on improving organizational culture instead of merely statistical tools.

Most of Deming's theories are based on the logic that most product defects resulted from management shortcomings rather than careless workers, and that inspection after the fact was inferior to designing processes that would produce better quality. His philosophy is carried out in four different stages, namely, *Plan–Do–Check–Act*, which describes the basic logic of data (or measurement), based process management. Deming spread his ideas to Japan after World War II. He trained hundreds of engineers and managers in Statistical Process Control (SPC) and taught concepts of quality through his 14 good principles of management. Japanese assimilate this concept so much so that by 1970 Japanese manufacturers gained a major foothold in the US and world market. The association of the Japanese union of Scientists and Engineers (JUSE) is doing all its effort in promoting quality in their processes. The unique method of QM adopted by Japanese manufacturers is known as "*Lean approach*". QM as an improvement tool has further expanded its horizon to other organizational functions in addition to production such as marketing, fiancé, and sales. This augmentation of QM practices gave rise to the field of Total Quality Management (TQM) during the late 1970s and early 1980s, which are still relevant in the modern-day business.

The other experts to join the propagation of quality are Dr. Joseph Juran, who is famous for his "*fitness for use*" (Ref: author's own book titled "Quality Control Handbook") philosophy. In the modern-day quality management, Juran is known as "the architect of quality", and promoted quality planning, quality control, and quality improvement (also called quality trilogy). The quality expert, Philip Crosby is known for his innovative ideas, and is popular for his concept of quality as "*conformance to requirements*" (Ref: author's own book titled "Quality is Free"). Crosby is also responsible to promote the concept of "*zero defects*" and "*Cost of Quality*". Another important quality guru is Armand Feigenbaum. According to him (Ref: Total Quality Control, 1961): "*quality is a customer determination based on the customer's actual experience with the product or service, measured against his or her requirements - stated or unstated, conscious or merely sensed, technically operational or entirely subjective and always representing a moving target in a competitive market.*"

The man who brought quality engineering (QE) and robust design in the modern era is the famous statistician Dr. Genichi Taguchi in the late 1940s. He is a familiar face for Indian scientists and academicians. The methodologies and philosophies developed by Taguchi for optimizing product and process functions are based on engineering concerns, not purely scientific or statistical ones. Another very useful quality improvement tool is the KAIZEN concept introduced by Masaaki Imai in Japan. This concept is treated as the foundation of Japanese management for continuous improvement. When speaking about quality generation, it is worth mentioning the book entitled "*The machine That Changed the World*" by Womack et al. [18]. The book is about the Massachusetts Institute of Technology's $5 million, five-year study on the future of the automobile industry, which eventually resulted in the philosophy of Lean Production System (TPS).

Thus, all quality management principles discussed above have their own relevance and found their own place in improving quality in products and services. Even today, management is finding tremendous applications for improving industrial production and providing sustainable quality services to the customers.

5   Third Industrial Revolution (1960–Mid 20s)

5.1   Technology

With the creation of the Advanced Research Project Agency Network (ARPANET) funded by the US Department of Defence during 1960, the new era of industrialization began. With the advancement of computer systems and adoption of TCP/IP world entered into the new era of World Wide Web (WWW) in 1989–90. Primarily WWW was the "web" of "hypertext documents" that could be viewed through browsers. The Internet reshaped the global communication system with the advancement in web browsers like Internet Explorer, Google Chrome, Firefox, etc. From 1990s to early 2000s was the era when Internet was mainly used for professional communication and exchange. The advancement of emails and e-commerce also started during this time. This was also the time of emergence of the online shopping era with the foundation of *Amazon* in 1994, *ebay* in 1995, and many other players also started appearing in the marketing fray. The online discussion forums, personal websites, and blogs have also started booming. Mobile phone revolution during the beginning of the twenty-first century brought dynamic change in almost all spheres of human life. The popularization of social media which has penetrated information sharing to daily life has brought people at large confined to their room and in isolation. All these developments have brought high awareness of quality in products, processes, and services leaving high competition within the business world. Customer focussed products and services is the norm now.

5.2   Socioeconomic Condition

The major impact that the third industrial revolution brought was a shift of reliance on machines from humans. Penetration of small and smart electronic devices into our everyday lives has begun to abolish the need for human presence in life. Jobs related to computers started paying off vigorously as compared to jobs involving repetitive tasks and human intervention. Increased demand for skilled and knowledgeable labor has put them at the advantageous position with significantly higher pay. This has widened the economic gap between the skilled and unskilled workforce, which is actually not encouraging. Survival of non-skilled labors is really difficult in this changing phase. Let us now see how quality perceptions have changed during this period.

5.3 Quality Management System

This was the era of advancement of QM practices on the Western world front. To recover the major downturn in semiconductor production sector, Motorola first initiated Six Sigma program in their process during 1987. As per the strategy, the Six Sigma program was initiated with the stretched goal of improving product and service quality ten times by 1989 and at least hundred times by 1991. Technically, this goal amounts to producing 3.4 defects per million opportunities (DPMO), and this is possible through the reduction of variation and redundancies in the process. Thus, Motorola mastered this technique and gradually dilated into different industrial sectors. Vogue of Six Sigma is so much that, as reported by American Society for Quality (ASQ) [1], 82% of Fortune 100 companies and more than 53% of Fortune 500 companies that were using Six Sigma approach are highly successful in retaining customer confidence and trust. Many companies realized the potential of Western countries' "Six Sigma" approach and Eastern countries' "Lean approach" and hence started practicing *"Lean Six Sigma"* as an integrated approach of reducing both variation and wastes from the process. As the world started facing extreme climate change during 2011s, many organizations reoriented their quality practices toward environmental sustainability through innovative quality practices such as "Clean Lean", "Green Lean", and "Green Six Sigma". A series of articles on these concepts have come up during the last few years. See Hudson [6], Muralidharan and Ramanathan [12], Muralidharan [13], Lee et al. [10] and references contained there.

6 Fourth Industrial Revolution (Mid 2020 to Future)

6.1 Technology

The current global health crisis COVID-19 pandemic, the emergence of addressing environmental issues has necessitated the advancement in existing digital infrastructure with futuristic technology. Current trending fields of 3D printing, robotics, artificial intelligence (AI), machine learning (ML), Internet of Things (IOT), autonomous vehicles are signaling the fourth industrial revolution at its peak. Automation and connected devices will be the key players of the future era. Self-managed systems will be able to collaborate with each other with transparent data sharing. Industries will be in a position to run their functions without human intervention and a dramatic increase in productivity. High-quality products and transparent services are now in the offing.

With the advent of digital technology, the industrial revolution has also changed rapidly upward. The entire working culture and tradition of manufacturing have contributed to high yield products and services. The relationship between machine and man is improved with the advent of Industry 4.0. Industry 4.0, also known as Smart Manufacturing, is an attempt to aid the complex manufacturing in the recent era. Industry 4.0 is the representation of the current trend of automation technologies in the manufacturing industry, and it mainly includes the Cyber-physical Production systems, Internet of Things, Robotics, Big Data, Augmented Reality, Horizontal and Vertical integration, and Cloud Computing [9, 8]. It is really a revolution, which has no boundaries and barriers.

6.2 Socioeconomic Condition

If there is a leap in the technological developments in future, then the new technology will definitely demand a significant cost of investment in terms of learning and skill acquired for quality maintenance. This will altogether create a new working class with significant knowledge of advanced technology. Monotonous repetitive tasks will be completely replaced by technology which will create a new group of twenty-first century unemployed people [4]. Digital equipment surrounding over lives will be empowered to track information about each aspects of human lives, from health, wealth, political stand to our personal and professional life. This will result in a whole new world driven by data and hence we will end with the new governing religion of humans called "Dataism" [4].

Data will be the most powerful enabler of the twenty-first century. Techno crates will be one of the most empowered groups of the human race holding the power through data supremacy. THE recent COVID-19 global pandemic has brought many uncertainties regarding our future, yet the current trend of remedial measures are directed toward the use of advance technologies such as Augmented Reality (AR), Virtual Reality(VR), AI, and IOT. In the midst of many uncertainties, this pandemic will sure redirect global and local politics, business environment, personal health, and technological advancement. Even sophisticated surveillance system will carry the capacity of optimistic or pessimistic future for human race. Urgency of addressing environmental issues will demand a radical change in our lifestyle from energy generation to conservation to transportation. All this will happen in the future without any delay.

6.3 Quality Management System

One of the promises of Industry 4.0 is the digitalization of quality management. Advanced technology like AI, VR, AR, and ML can be used to improve quality in every function of the business. As explained by Jahn [7] QM in industry 4.0 will be based on the following principles:

- Machines and devices that connect and communicate with each other freely.
- Information transparency and virtualization of real-world situation become frequent.

- The capability of the system to take support of humans in decision-making and solving problems will be common in all digital platforms.
- The ability of the cyber system to become autonomous and make a simple decision on its own will be the demand of time.

Thus, for adopting quality management in future, the companies should require to prioritize their quality areas. This will require (i) determining quality issues with supply chain, (ii) mapping the issues to innovative quality methods, and (iii) apply the most suitable method as per context. Companies with strong quality foundation can further advance their QM practices through adequate technological integration. Such technological advancement will help organizations to work with real-time data which will help them to address quality issues in advance.

7 Discussion

The idea of writing this article is to perceive the growth of quality management over various generations. The advancement in the QM practices over different industrial revolutions in the purview of technological advancement and socioeconomic changes has been discussed in this paper. With the emergence of the factory system and invention of steam machine came the first industrial era and laid down the foundation of QM practices. The necessity of mass production and mechanization was felt in a big way contributing further to the development of quality service and delivery. This brought the importance of standardization and inter-operability and exchangeability of parts to sustain the world-level competition.

The focus on the use of quality management system (QMS) and management information system (MIS) for quality control and quality insurance also started world-wide. This has paved the way for waste reduction and speedy execution of processes through Lean approach which was a major headway in QM practices. The invention of control chart initiated the new sub-branch of QM named SQC, and it boosted the importance of Operations Research (OR) techniques and Statistical Methods for decision-making. Many QM professionals contributed to the advancement of these fields further. As a part of TQM, Deming's PDCA is still relevant and people are adopting it. From 1990s, the entire perception of QM has changed with the advent of digitalization and computing techniques. With computerization, the new industrial revolution brought automated data generation process and database management a common practice in every organization.

Softwares and programs also started contributing to the improvement of quality in every segment of management. The belief in QM practices has further boosted with the introduction of concepts like Six Sigma and Lean Six Sigma, which are now common in many industrial, production, service, and automobile sectors in India. The progress of QM has now percolated into the realization of customer-focussed products and services, which is the real revolution of modern-day quality management. We, as a practitioner of quality feels that for quality to sustain, products and services or anything per se should be subjected to quality audit and accounts before smart devices and technology dominates the scene. In futuristic QM practices, machines are likely to handle the operational part of QM practices and the strategic decision will be left to human intervention.

References

1. ASQ. (2009). Save your company a fortune American Society for Quality.
2. Beckert, Sven. (2014). *Empire of cotton: A global history*. US: Vintage Book Division Penguin Random House. ISBN 978-0-375-71396-5.
3. Coren, M. (2018). *The speed of Europe's 18th- century sailing ships is revamping history's view of the Industrial Revolution*. Quartz. Retrieved 25 March, 2020.
4. Harari, N. Y. (2018). *21 Lessons for the 21st century*: Spiegel & Grau, Jonathan cape.
5. Hermann, T. P., & Otto, B. (2015). Design principles for industrie 4.0 scenarios: A literature review. Technische Universität Dortmund - Fakultät Maschinenbau, *Audi Stiftungslehrstuhl Supply Net Order Management*.
6. Hudson, D. A. (1991). *Business and the environment: Lean, clean, and green*. California: Video Publishing House, Inc.
7. Jahn, N. I. (2019). Quality management and requirements of the fourth technical revolution. *International Scientific Journal "Industry 4.0", 4*(2), 61–63.
8. Korže, S. Z. (2019). From industry 4.0 to tourism 4.0. *Innovative Issues and Approaches in Social Sciences, 12*(3), 29–52.
13. Kumar, N., & Kumar, J. (2019). Efficiency 4.0 For industry 4.0. *Human Technology, 15*(1), February 2019, 55–78.
9. Lee, S. M., Kim, S. T., & Choi, D. (2012). Green supply chain management and organizational performance. *Industrial Management & Data Systems, 112*(8), 1148–1180.
11. Montegomory, D. C. (2003). *Introduction to Statistical Quality Control*. India: Wiley.
10. Muralidharan, K., & Ramanathan, R. (2013). Green statistics: A management perspectives. *Journal of Industrial Engineering, 6*(9), 15–20.
12. Muralidharan, K. (2018). Lean, green and clean: Moving towards sustainable life. *Communiqué, Indian Association for Productivity, Quality and Reliability*, October–December, 40(4), 3–4.
14. Overton,.ark (1996). *Agricultural revolution in England: The transformation of the agrarian economy 1500–1850*. Cambridge University Press.
15. Schaefer, H. (1958). The metamorphosis of the craftsman. *College Art Journal, 17*(3), 266–276.
16. Unyimadu, O. S. (1989). Management and Industrial Revolution in Europe, United States and Japan. *Engineering Management International, 5,* 209–2018.
17. Womack, P. J., Jones, D. T, & Daniel, R. (2007). *The machine that changed the world*. Simon and Schuster.
18. Womack, J. P., Jones, D. T., & Daniel, R. (1991). *The machine that changed the world: The story of lean production*. Harper Perennial.
19. Wood, C. J., & Wood, C. Michael. (2003). *Henry Ford: Critical evaluations in business and management* (Vol. 1). London: Routledge.

Contribution of TQM Towards Organizational Development: A Case Study Conducted at NALCO, Bhubaneswar

Sushree Sangita Ray, Shruti Tripathi, and Sujit Kumar Acharya

1 Introduction

In the era of Liberalization, Privatization, and Globalization (LPG), the leading manufacturing companies are facing cut-throat competition from the leading players of the global world [1, 2]. Many steps and pains have been taken by the MNCs to develop the Industry in such a manner, so as to compete with the global players. Many factors like cost reduction, improvement in quality, and diversified products, safety work measures, and better services are mostly concerned in relation to the implementation of organizational change and development [3]. The manufacturing industries have started realizing the strategic implementation of the above factors which will ultimately lead toward the organizational performance improvement.

Organizational performance and its improvement is a continuous approach which leads towards the implementation of Organizational Development (OD). At the present, manufacturing units are experiencing an incredible amount of change and diversification within it. For successful implementation of OD, several OD interventions have been introduced by many researchers regarding the relationship between or among person, groups, or objects [4]; involvement of change agents in negotiation with clients; participation, involvement, and empowering people at workplace; and these are widely used as OD interventions at present. Considering the present scenario the value of the management aspects related to quality plays a fundamental

S. S. Ray (✉) · S. Tripathi
Amity International Business School, Amity University, Noida, India
e-mail: sushree1973@gmail.com

S. Tripathi
e-mail: stripathi@amity.edu

S. K. Acharya
DDCE, Utkal University, Vani Vihar, Bhubaneswar, Odisha, India
e-mail: sujitkacharya@gmail.com

© The Author(s), under exclusive license to Springer Nature Singapore Pte Ltd. 2021   229
P. K. Kapur et al. (eds.), *Advances in Interdisciplinary Research
in Engineering and Business Management*, Asset Analytics,
https://doi.org/10.1007/978-981-16-0037-1_18

role in the cause of implementation of change and development in the organization. Organization change and development is a practice which gives utmost importance to educate the employees in the implementation of the new quality mechanism, quality process, quality strategy, quality output, and quality services.

Today's knowledge economy reflects tremendous complexities and uncertainty in the business environment but it also provides opportunity to the industrial sector to grow and to expand. Managing in the new changing scenario involves taking strategic decision as well as adopting quality practices which not only relate with the production process but also includes involving quality people, quality decisions, quality process, quality output, and quality services. To provide quality output to the customers these quality and maintenance improvements initiatives are playing a major role. Nowadays, it has compelled the manufacturing industries to adopt quality initiatives to improve their organizational process. Different quality improvement programs and techniques like, "Six Sigma", "Leadership through Quality", "Perfect Design Quality", "Total Quality Control" and "Kizen" leads towards the achievement of TQM in any manufacturing industries [5].

The major objective of Total Quality Management is the continuous improvement in the quality aspects of different sectors of business, such as goods and service, superior–subordinate relationship as well as relationship with the market. Total quality is also a combination of organization culture, people attitude, organization vision, strategy, etc.

Many researchers conducted research on the relationship evolved between TQM and organizational performance, organizational effectiveness, organizational sustainability, and organizational development and organizational excellence. TQM helps in constant enhancement which is recommended as one of the appropriate approach that provide efforts towards organizational growth [6]. The organization which are putting their effort to strengthen their business must adopt TQM practices as a source of their competitive sustainable competitive advantages [7, 8]. TQM is the means by which organization can improve its overall performance and quality [9]. TQM is a management philosophy that helps to enhance the quality and performance of the organization through continuous improvement [10]. So there is a direct impact of TQM on the performance of the organizations [11–13].

To achieve organizational excellence it should focus on its performance [14]. For the achievement of manufacturing excellence TQM excellence is the basic factor [15]. By implementing the TQM practices, organization can sustain as well as it can achieve business excellence [16]. Adoption of TQM practice can add value to the organizational performance, irrespective of the type of organization [17]. TQM is a never-ending process so continuous improvements are required to be implemented in any organization. In this, researchers have tried to find out the connection between TQM and OD.

1.1 Theory

It is a known fact that TQM is indispensable for organizational development. It is the philosophy which seeks to combine the various functions of an organization like marketing, finance, design, engineering and production, and customer service. It aims to do so because it sees meeting the expectations of the customer as foremost and also because it is quality in every aspect that matters most. TQM perceives a firm as a unified process of all the functions mentioned above.

TQM has always focused on satisfaction of the customer and how a management approaches to achieve this, not only in the short term but also in the long term. Its approach is all inclusive. Every employee is single-mindedly devoted to the development of a firm in improving the products or services, processes, and also the work culture.

TQM also believes that every worker irrespective of their position or responsibility is a part of the improvement process. Together everybody contributes to improvement in quality and this is how we can get the effectiveness we are seeking.

The main ideas or principles on which TQM is founded upon are customer orientation (both internal and external), business processes being controlled by statistical tools, maintenance for prevention, participation of the entire management, coordination of all departments, strong leadership, and dedication.

TQM has always focused on consumer gratification through organizational development. TQM views organizational culture as participation of all concerned such as teams involved and their teamwork, the coordination and cooperation that should exist between the teams and units, the creation of relevant information, and learning on continuous basis.

To quote the Guru himself Dr. W. Edwards Deming: "Organizational Development can be best attained through a conversion path". How to bring about a transformation in the entire process that ultimately leads to development. It was realized very early that if we want to thrive in business, compete with our rivals and get an edge over them, we should improve the quality of our goods and services, and we must gratify the needs of our customers in every possible way. That every single member is dedicated to this cause, the management participates whole heartedly and every process is integrated and works towards achieving quality. Only then will a business house reap loyal customer base, positive parol, and positive brand image.

The TQM principle involves every single individual, department, unit, and process of firm. Total means, all areas of function is covered; all employees are covered, as well as it also covers customer, suppliers, and all those who have a stake in a firm. Quality refers to confirming to the requirement of all consumers that means how fit a product is for use as well as the consumer gratification. Management means how effectively we have utilized our resources such as people, tools, raw materials, process, time, capital, and our efforts for continuous development.

The present study is conducted at National Aluminium Company. NALCO which is the first Public sector company in the country to venture into International market. NALCO deals with the industries like mining, metals and thermal power station.

The major products of NALCO are aluminium, alumina and hydrate and zeolite-A. The main units of NALCO are at Damanjodi (Bauxite Plant) and Nalconagar, Angul. NALCO achieved ISO 9001:2000 awards and OHSAS 140,001 for its excellence in production technology and occupational health and safety systems. In the present study, the researchers have identified the plant of NALCO situated at Angul for the purpose of research. The research is mainly based on the perception of the employees irrespective of departments and level of employment. However, the qualification and experience are the two broad factors which were considered for the analysis purpose. Qualification and experience are important variables which are linked with the knowledge on productivity, level of acceptance for implementing change, and improvements. The respondent are the employees of NALCO working at the Angul plant includes both executives and non-executives. The research was conducted at Angul, NALCO for two months, during the period May and June (2018). However, it took another two months to collect all the response sheet distributed among the respondents. The present study is mainly based on the perception of the employees (both executives and non-executives) working in the mines and the plant.

2 Literature Survey

Many studies have been carried out related to contribution of TQM towards organizational performance [18]. These studies include the development of research framework to measure the effectiveness of TQM with regards to performance of the organization [19–22]. There are several research work which have also been conducted to identify the critical success factors of TQM in the manufacturing industry [23–25] in specific countries. These research work highlighted strong customer focus and leadership; quality systems, available information, and MIS are the major contributor towards TQM practices, [26]. Similar studies were also conducted by [27–29] in different national frameworks to find out the impact of TQM from global context.

The globalization and change in market scenario in regards to organizational need, organizational innovation plays a key factor [30]. The research authors presented a comprehensive model for organizational innovation which creates a competitive environment in the global phenomena. A study was conducted in Toyota (Japan), where TQM, Lean Production, and the Toyota Production System were analyzed. A comprehensive model was developed based on three concepts such as creating, defusing, and sustaining organizational innovation. The model suggested that Google Innovation System (GIO) is an open system which is to be involved for developing technical innovation, diversified leadership style, thinking big strategy, and the ability of the firm to implement the change constantly with the support and positive mind set of the senior leaders.

In the traditional context of quality, the term is identified as fitness for use but at present it is applied at all levels, segments, and sphere of organization. These include customer orientation, human capital approach, leadership style, continuous and constant development and improvement, collection and use of information, strategic planning, creating a learning environment to train the employees, team-work, and creating trustworthy employees. To identify quality in such areas [31]. The authors have suggested four stages such as quality inspection, quality control, quality assurance and quality management, and quality development.

TQM is a mechanism by which organization can improve its performance [32]. The researchers have studied to identify that there is a correlation among organizational learning capability, TQM practices, and organizational performance. This study was conducted with the help of a survey form from 90 organization and 270 managers from the textile sector. The hypothesis of the study was proved with the help of hierarchical linear modeling techniques. The study conclude that, TQM practices and organizational learning capability influence the performance of the manufacturing organization, as well as it was proved that TQM practices help to increase the learning capability and performance of the organization.

To sustain in the competitive business environment the organization requires innovative and unique product and practices. Nowadays organizational excellence and organizational performance are two major focuses and to achieve this continuous improvement and development of internal resources, to compete in global market is highly essential [33]. The researchers tried to examine the impact and the relationship between Total Quality Management (TQM), Enterprise Resource Planning (ERP), and Entrepreneurial Orientation (EO) through two models, i.e., Resource-Based View Theory and Knowledge Based View Theory. An attempt was made by the researchers in order to create the base by initiating the organizational excellence. By adapting the mechanism like TQM, ERP, and EO, an organization can improve its performance at par excellence.

Most of the research on TQM studied on Indian framework which have not concentrated on the fundamental issue of relationship between implementation factors and organization development and business performance. These studies mostly address the identification of quality management practices [34] and their benchmarking among Indian industries [35]. Although Mohanty and Lakhe [36] have put their hard work to recognize critical factors for TQM implementation which stands on Leadership, Policy and Strategy, and Customer Focus etc., there is a direct link between TQM and Total Productivity Management (TPM) [37]. TPM is the mechanism applied in TQM practices which includes strategic planning and decision-making, evaluation of output and measuring the quality factors, etc.

Out of different Organizational Development (OD) mechanisms, TQM is an important technique adopted by almost all manufacturing industries. OD plays a vital role for the implementation of organizational change and restructuring. OD is a planned effort of an organization which is broadly managed by outstanding leadership and supportive employees. To examine the psychological health of any organization, employees' perception is mostly considered. The organizational health is maintained by minimizing absenteeism, low production, resistance to change, using inappropriate raw materials, and mechanism. TQM plays a vital role which helps to improve a strategy for planned change, for organizational development and organizational effectiveness.

For organizational development mainly three innovative techniques play a vital role. They are reengineering, benchmarking and empowerment [38]. Reengineering can be explained as the fundamental re-evaluation and restructuring of business methods for the accomplishment of substantial improvement in evaluation. Benchmarking is the method of assessing the performance in comparison with the best organizations, determining how to accomplish the best practices, and utilizing the information as a foundation for own organization's targets, strategies, and implementation. Empowerment means delegating decision-making authority to identify problems and then come up with solutions. Conditions for empowerment—Participation: Encouraging people to improve daily work process and relationship. Innovation: Empowering means encouraging people to try new ideas and take decisions of their own.

3 Methodology

After reviewing the different literature, it seems there is hardly any research conducted on the relationship between TQM and Organizational Development in particular to National Aluminium. In the present competitive environment, the traditional view of TQM which mainly lay on machinery, raw materials, and finished product have shifted to the modern concept of TQM. At present TQM not only based on traditional but also included human capital management, leadership, communication system considering this the researchers have analyzed the importance of TQM and its contribution towards the development of NALCO by understanding the perception of the employees. The employees' perception were analyzed and differentiated on the present topic based on their qualification and experience. The two broad objectives of the paper are 1. To analyze the role of TQM towards the development of an organization in specific to an Aluminium industry. 2. To analyze the opinion of the employees towards the relationship between organizational effectiveness and TQM practices based on their qualification and work experience at NALCO.

3.1 Hypothesis of the Study

On the basis of the objectives, the present study is based on the following two hypotheses:

H0: There is no significant difference of opinion across experience of the employees on effect of TQM on organizational development.
H1: There is no significant difference of opinion across employees' qualification on effect of TQM on organizational development.

After reviewing the different literatures it seems there are hardly any research works that have been conducted in Aluminium industry at NALCO on the present

Table 1 Reliability analysis

Parameters	Cronbach's alpha	Number of items (number of questions under each parameters)
Management	0.837	6
Process	0.752	8
Output and Services	0.826	9

topic. The researchers have also found that very few research works have been made by constructing both independent and dependent variables on the present topic considering the above, and it is an attempt to analyze the Role of TQM towards the development of an organization in specific to an Aluminium industry. However, the views of the employees are collected to study the importance of TQM in the organization. Another objective of the paper is also to analyze the opinion of the employees towards the relationship between organizational effectiveness and TQM practices.

3.2 Reliability

On application of reliability analysis over the dataset containing the information pertaining to management, the Chronbach's Alpha has been found to be 0.837 which is more than 0.7. Hence, the instrument is reliable and paved the way for further statistical analysis (Table 1).

3.3 Sampling

To answer the objectives set for the present research work, an empirical attempt was made by analyzing the perception of the employees towards the effectiveness after implementation of TQM at NALCO. The employees working at refinery plant of Angul were considered. The data analysis is done on the basis of their experience and educational qualification. The total employees on muster roll of NALCO, Angul is around 5600. The sample size of the study has been computed as

$$n_r = \frac{(1.96)^2 \sigma^2}{d^2} \tag{1}$$

$$n_a = \frac{n_r}{1 + \frac{(n_r - 1)}{N}} \tag{2}$$

where Confidence Level $= 95\%$
Standard Deviation $(\sigma) = 1.5$

Margin of Error (d) $= 0.25$.
Sample size $(n_r) = 138.30$.
Adjusted Sample Size $(n_a) = 134.98 \approx 135$.
Hence, the sample size for this study has been taken as 135.

4 Result and Discussion

The researchers have made an attempt to study the effectiveness of the organization
after implementation of TQM at NALCO. The researcher have identified the refinery
plant situated at NALCO Angul, as the employees working in the plant can give better
insight into it. Both primary and secondary sources have been adopted in the research
work. A Self-Administered Questionnaire (SAQ) was constructed during the pilot
survey. The response sheet has two parts, the first part includes two independent
variables (experience and qualification) and the second part includes one dependent
variable (effectiveness after implementation of TQM) which is again subdivided into
three parts such as (management, process, output and services).

For the purpose of collection of data the researchers have adopted Quota sampling
technique. In the present study, there are two demographic variables are considered
to distinguish and to analyze the perception of the employees such as qualification
and experience.

The present SAQ has been identified in three different parts. The first part of the
SAQ, i.e., management includes certain questions relates to the involvement of the
management or the policy-makers towards the implementation of TQM. The second
part is identified as process which required certain basic factors to make successful
implementation of TQM. The third part identifies the outcomes after implementation
of TQM practices. Cronbach's Alpha analysis is used to find out the reliability of the
questions. As per the pilot survey, the Cronbach Alpha have been identified more than
0.7 (for the 6 parameters under management is 0.837,8 parameters for the process
is 0.752, and for 9 parameters of output and services is 0.826).Hence the instrument
is reliable and the applicability for the present research work is justified. After the
pilot survey the samples were drawn from the employees working at NALCO, Angul
refinery. A quota sampling method was adopted as it is suitable for the universe which
is very large for the present sample size which is 135 which is approximately 7% of
the total population working at Angul refinery.

Different statistical tools have been adopted such as Reliability Analysis, Analysis
of Variance (One-Way ANOVA) with Duncan Multiple Range Test (DMRT).The One
Way ANOVA is adopted to find out the variation of opinion between the selected
dependent variables and to determine whether there are any significant differences
between the means of two or more dependent variable. The DMRT is used to compare
the mean of each treatment. The similarity and differentiation of opinion can be
identified among the different variables through DMRT. Quota sampling technique
is adopted. The Likert's Scale is adopted to get distinguished views of the employees

within five options, i.e., strongly agree (SA), agree (A), no opinion (NO), Disagree (D), strongly disagree (SD) (Table 2).

The F-value shown against Management (6.176) is significant at 5% level ($P < 0.05$). This indicates the variation in opinion towards management by different experienced employees is significant.

The F-value shown against process (1.298) is not significant at 5% level ($P > 0.05$). This indicates the opinion towards process by different experienced employees is similar. Hence the hypothesis H0 is rejected in case of management, as there is a difference of opinion among the employees in regards to experience towards TQM on OD.

Table 2 Analysis on variation in opinions by employees of NALCO, Angul of different experience groups on contribution of TQM towards drivers of organizational development

	Sources of variations	Sum of squares	df	Mean square	F	Sig.
Management	Between experience groups	7.270	4	1.818	6.176*	0.000
	Within experience groups	38.256	130	0.294		
	Total	45.526	134			
Process	Between experience groups	1.186	4	0.296	1.298 NS	0.274
	Within experience groups	29.704	130	0.228		
	Total	30.890	134			
Output and Services	Between Experience groups	0.361	4	0.090	0.314 NS	0.868
	Within experience groups	37.432	130	0.288		
	Total	37.794	134			
Total	Between experience groups	0.769	4	0.192	0.912 NS	0.459
	Within experience groups	27.397	130	0.211		
	Total	28.166	134			

N.B: *—Significant at 5% level ($P < 0.05$), NS—Not Significant at 5% level ($P > 0.05$)

The *F*-value shown against Output and Services (0.314) is not significant at 5% level (*P* > 0.05). This indicates the opinion towards Output and Services by different experienced employees is similar. Hence the hypothesis H0 is accepted in case of process.

The *F*-value shown against Total is not significant at 5% level (*P* > 0.05). This indicates the opinions towards total by different experienced employees are similar. Hence the hypothesis H0 is accepted.

This indicates the opinion towards process, output and service and organizational development in total because of contribution of TQM by different experienced groups of employees is similar. The comparison of all probable pairs of mean opinions towards management, process, output and all drivers of organizational development by different experience groups is done with the help of Duncan's Multiple Range Test (DMRT).

In Table 3, and Fig. 1, the letters A and B indicate similar and difference of opinion towards management due to effectiveness of TQM with regards to years of

Table 3 Average opinion by employees of NALCO, Angul of different experience groups on contribution of TQM towards drivers of organizational development

Experience	N	Management		Process		Output		Total	
		Mean	SD	Mean	SD	Mean	SD	Mean	SD
<5 years	22	2.22A	0.99	1.98	0.65	1.83	0.62	1.98	0.64
6–10 years	26	1.49B	0.19	1.94	0.44	1.81	0.44	1.77	0.36
11–15 years	33	1.74B	0.39	1.88	0.35	1.87	0.69	1.84	0.46
16–20 years	14	1.93B	0.39	2.13	0.41	1.68	0.16	1.90	0.31
>20 Years	40	1.94B	0.50	2.10	0.51	1.81	0.47	1.94	0.44
Total	135	1.85	0.58	2.00	0.48	1.81	0.53	1.89	0.46

N.B:- Similar Superscripts over the means indicate their similarity at 5% level (*P* < 0.05)

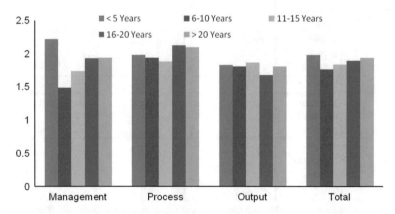

Fig. 1 Average opinion by employees of NALCO, Angul of different experience groups on contribution of TQM towards drivers of organizational development

experience. The table shows that the employees having experience of less than five years in the company have agreed on the contribution of the management on effective implementation of TQM in the organization. On the other hand, employees working for six or more years of experience, have the similar opinions towards effect of TQM on process, output, and all drivers of organizational development.

Table 4 and Fig. 2 presents the average opinion of technical and non-technical employees on contribution of TQM towards drivers of organizational development. The calculated t-values in case of management (1.19), process (1.05), output and services (0.49) and total (0.55) are non-significant at 5% level ($p > 0.05$). This indicates opinions are same in nature by both technical as well as non-technical

Table 4 t-values between employees with technical and non-technical qualification of NALCO, Angul on contribution of TQM towards Drivers of organizational development

	Qualifications	N	Mean	SD	t-value	(Sig.)
Management	Technical	121	1.83	0.61	1.19 NS	0.235
	Non-technical	14	2.02	0.23		
Process	Technical	121	1.98	0.49	1.05 NS	0.295
	Non-technical	14	2.13	0.41		
Output and services	Technical	121	1.82	0.56	0.49NS	0.619
	Non-technical	14	1.75	0.05		
Total	Technical	121	1.88	0.48	0.55 NS	0.583
	Non-technical	14	1.95	0.22		

N.B: NS—Not significant at 5% level ($P > 0.05$) for DF = 133

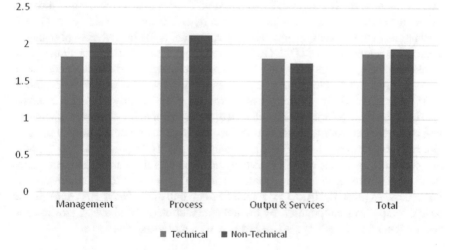

Fig. 2 Average opinion by employees of NALCO, Angul of different experience groups on contribution of TQM towards drivers of organizational development

employees that are agreed towards management, process, output and total on contribution of TQM towards organizational for organizational development. This result
shows that t-value based on the perception of employees on all the parameters are
not significant and all the employees in regards to the qualification are giving similar
opinion towards the contribution of TQM on OD. In such a case, the hypothesis H1
is accepted in all the cases.

5 Conclusion

At Angul refinery of NALCO, there must be a systematic policy to be adopted towards
reduction of accidents. Though NALCO has adopted modern techniques but the work
pressure never reduced. This may be due to demand of the present market. Employees
with six or more years of work experience have valued the TQM process as less
effective. The organization should think of increasing work satisfaction providing
modernized working environment and should produce diversified product. NALCO
has a specific network of customers like other Multinational Corporation but could
not reach to the general public directly. So, NALCO should create a scope for the
same. The researcher says that the employees those who are newly joined or having
less work experience are accepting the involvement of management towards the
implementation of TQM with a motive of organizational development in a positive
sense whereas higher experienced people views differs. NALCO should concentrate
more on implementation of mechanism by which the decision can be taken quickly to
increase intra department relation and cooperation, etc. NALCO should also involve
the experienced people in the decision-making process irrespective of the level for
improvement of quality and implementation of TQM. Considering the qualification
of the perception of employees are similar in regard to the application of TQM and
viewpoints of workforces are low rating as compared to the involvement of management in implementation of TQM. Organization should use statistical techniques in the
decision-making process which will help in research, design and development, views
of consultants and development will give better result for successful implementation
of TQM at NALCO. The environment, surrounding, and quality of work life should
also be given more priority which will lead towards freedom for expansion, openness,
and collaborative working system. Initial expenditure to implement quality policies
and technique are very high. Further, enhancement of TQM practices is an ongoing
and continuing practice. NALCO has adopted all sorts of mechanism and techniques
to create a quality working environment, and quality services to its employees and
customers. At present, NALCO is a loan free and one of the NAVARATANA public
sector company of our country. By the implementation of the above suggestion and
recommendation somehow NALCO can create a motivated task force.

6 Limitation

The study area is very large but it was only restricted to the respondent who are working at NALCO Angul. Only TQM was considered as a parameter of organizational development. Limited budget and time constraints also consider as the major hindrance during collection of data.

7 Future Scope

Considering the limitation, there is a wide scope for the extension of future research work where a comparative study can be done to find out the implication of TQM in OD, and more behavioral aspect can also be studied linking to TQM with OD.

References

1. Nandi, S. N. (1998). *Contribution in implementing quality management in Asian and Pacific firms.* Tokyo: Asian Productivity Organization Publications, pp. 148–183.
2. Sahay, B. S., Saxena, K. B. C., & Kumar, A. (2000). *World class manufacturing—A strategic perspective.* New Delhi: Macmillan India Limited.
3. Chandra, P., & Shastri, T. (1998). Competitiveness of Indian manufacturing: Findings of the 1997 manufacturing futures survey. *Vikalpa, 23,* 25–36.
4. Irwin. Argyris, C. (1970). *Intervention theory and method: A behavioral science view.* Reading, Mass: Addison Wesley; Argyris, C. (1974). Behind the front page. San Francisco: Jossey Bass.
5. Badiger, S. K., & Laxman, R. (2013). Total quality management and organization development. *International Journal of Business and Management Invention, 2,* 34–37.
6. Lin, B., & Ogunyemi, F. (1996). Implications of total quality management in federal services: The US experience. *International Journal of Public Sector Management, 9,* 4–11.
7. Munizu, M. (2013). Total quality management toward product quality performance: Case at food and beverage industry in Makassar. *Indonesia Journal of Business and Management, 9,* 55–61.
8. Terziovski, M. (2006). Quality management practices and their relationship with customer satisfaction and productivity improvement. *Management Research News, 29,* 414–424.
9. Ritchie, J., & Spencer, L. (1994). Qualitative data analysis for applied policy research. In A. Bryman & R. G. Burgess (Eds.), Analyzing qualitative data, pp. 173–194.
10. Douglas, T. J., & Judge, W. Q. (2001). Total quality management implementation and competitive advantage: The role of structural control and exploration. *Academy of Management Journal, 44,* 158–169.
11. Arawati, A. (2005). The structural linkages between TQM, product quality performance, and business performance: Preliminary empirical study in electronics companies. *Singapore Management Review, 27,* 87–105.
12. Li, J. H., Anderson, A. R., & Harrison, R. T. (2003). Total quality management principles and practices in China. *International Journal of Quality & Reliability Management., 20,* 1026–1050.
13. Yasin, M. M., Alavi, J., Kunt, M., & Zimmerer, T. W. (2004). TQM practices in service organizations: an exploratory study into the implementation, outcome and effectiveness. *Managing Service Quality, 14,* 377–389.

14. Antony, J. P., & Bhattacharyya, S. (2010). Measuring organizational performance and organizational excellence of SMEs—Part 2: An empirical study on SMEs in India. *Measuring Business Excellence., 14,* 42–52.
15. Sharma, M., Kodali, R. (2008). TQM implementation elements for manufacturing excellence.*The TQM Journal, 20,* 599–621.
16. Lee, S. M., Rho, B. H., & Lee, S. G. (2003). Impact of Malcolm Baldrige National Quality Award criteria on organizational quality performance. *International Journal of Production Research, 41,* 9.
17. Sila, I., & Ebrahimpour, M. (2002). An investigation of the total quality management survey based research published between 1989 and 2000: A literature review. *International Journal of Quality and Reliability Management, 19,* 902–970.
18. Ismail, S., & Ebrahimpour, M. (2002). An investigation of the total quality management survey-based research published between 1989 and 2000—a literature review. *International Journal of Quality & Reliability Management, 19,* 902–970.
19. Ahire, S. L., Landeros, R., & Goldhar, Y. D. (1996). Total quality management : A literature review and agenda for future research. *Production and Operations Management, 4,* 277–306.
20. Anderson, J. C. (1994). A theory of quality management underlying the Deming management method. *Academy of Management Review, 19,* 472–509.
21. Flynn, B. B., Schroeder, R. G., & Sakakibara, S. (1994). A framework for quality management research and an associated measurement instrument. *Journal of Operations Management, 11,* 339–366.
22. Saraph, J. V., Benson, P. G., & Schroeder, R. G. (1989). An instrument for measuring the critical factors of quality management. *Decision Sciences, 20,* 810–829.
23. Corbett, L. M., & Rastrick, K. N. (2000). Quality performance and organization culture: A New Zealand study. *International Journal of Quality & Reliability Management, 17,* 14–26.
24. Prabhu, V. B., & Robson, A. (2000). Impact of leadership and senior management commitment on business excellence: An empirical study in north east of England. *Total Quality Management, 11,* 399–409.
25. Anderson, M., & Sohal, A. S. (1999). A study of the relationship between quality management practices and performance in small businesses. *International Journal of Quality & Reliability Management, 16,* 859–877.
26. Sun, H. (2000). A comparison of quality management practices in Shanghai and Norwegian companies. *International Journal of Quality & Reliability Management, 17,* 636–650.
27. Hendricks, K. B., & Singhal, V. R. (2003). The effect of supply chain glitches on shareholder wealth. *Journal of Operations Management, Science Direct Elsevier, 21,* 501–552.
28. Choi, T. Y., & Eboch, K. (1998). The TQM paradox: Relations among TQM practices, plant performance, and customer satisfaction. *Journal of Operations Management, 17,* 59–75.
29. Forker, L. B., Vickery, S. K., & Droge, C. L. M. (1996). The contribution of quality to business performance. *International Journal of Operations & Production Management, 16,* 44–62.
30. Steiber, A., & Alänge, S. (2015). Organizational innovation: Verifying a comprehensive model for catalyzing organizational development and change. *Triple Helix, 2*(1), 1–28.
31. Sweis, R., Obeidat, B., & Kanaan, R. K. (2019). Reviewing the literature on total quality management and organizational performance. *Journal of Business & Management, 7,* 192–215.
32. Mahmood, S., & Ahmed, A. (2015). The role of organizational learning in understanding relationship between total quality management and organizational performance. *Pakistan Journal of Commerce and Social Sciences, 9,* 282–302.
33. Al-Dhaafri, H. S., Yusoff, R. Z., & Al-Swidi, A. K. (2013). The effect of total quality management, enterprise resource planning and the entrepreneurial orientation on the organizational performance: The mediating role of the organizational excellence---a proposed research framework. *International Journal of Business Administration.*
34. Bhadury, B., & Mandal, P. (1998). Adoption of quality management concepts amongst Indian manufacturers. *Productivity, 39.*
35. Motwani, J. G., Mahmoud, E., & Rice, G. (1994). Quality practices of Indian organizations: An empirical analysis. *International Journal of Quality & Reliability Management, 11,* 38–52.

36. Mohanty, R. P., & Lakhe, R. R. (2000). *Handbook of total quality management*. Mumbai: Jaico Publishing House.
37. Majumdar, N. (1999). *TPM INC*. New Delhi: Business Today.
38. Badiger, S. K., & Laxman, R. (2013). Total quality management and organisation development. *International Journal of Business and Management Invention, 2,* 34–37.

Analyzing the Effect of Electromagnetic Radiations' Risk Factors with the Intention on the Usage of Smart Phones and Mobile Services

Monika Sharma and Navita Mahajan

1 Introduction

Over the last few years, there has been incredible progress in the use of smartphones in India. According to the TRAI report [1], the telecom industries in India is the second-largest market across the globe [1]. The progress of the telecom market is due to the introduction of new radio technologies, mobile value-added services, 3G, 4G, and mobile networks [2]. This growth has led to increasing concerns by stakeholders about the harmful effects of the Electromagnetic Field with regards to towers and mobile phones. The concern raised by [3] reports related to the possible health impacts from contact to electromagnetic radiation (EMF) has been taken seriously by the World Health Organization. Cancer, changes in behavior, memory loss, and related adverse health effect are a few of the major health hazards due to overexposure to EMF [4]. In this context, smartphone radiations are one of the biggest contributors. Typically, smartphones work within the radiofrequency range 30 kHz to 300 GHz of the spectrum. The smartphone depends on radiofrequency communication between handsets and telecom towers. There have, however, been concerns raised by the stakeholders about possible health effects of the mobile phone radiations from mobile towers and handsets. Such concerns, if left unaddressed will affect the development and progress of the mobile services in the country [5]. This leads to the motivation behind the present paper.

Numerous research papers have mentioned the possible health problems of EMF. However, the impact of people perceiving risks associated with mobile phones has not been delved upon in the literature. It is an important aspect to study how people's knowledge differs from experts in this aspect to understand the implications of the results of this study on the sustainability of the mobile phone business. In the present

M. Sharma (✉) · N. Mahajan
AIBS, Amity International Business School, Noida, India
e-mail: msharma28@amity.edu; monikasharma2011@gmail.com

© The Author(s), under exclusive license to Springer Nature Singapore Pte Ltd. 2021   245
P. K. Kapur et al. (eds.), *Advances in Interdisciplinary Research*
in Engineering and Business Management, Asset Analytics,
https://doi.org/10.1007/978-981-16-0037-1_19

study, the observations of laypeople towards the consequences of mobile phone radiations are examined. The main concentration of the present study is to measure the level of the perceived hazards and attitudes towards regulating authorities and the EMF associated with mobile phones.

2 Literature Survey

The use of mobile phones has become ubiquitous to modern society over the past decade, and the increased demand of mobile phones also raised concerns about possible health effects from the exposures of the electromagnetic radiations which emits from the mobile phones [5]. A review of literature investigates some aspects of the negative perception of the user towards mobile phone radiations in developing country like India.

WHO established the International Electromagnetic Fields (EMF) has done a study to evaluate health and environmental impacts of radiations exposure? The WHO concerned area is the following.

"How the general public observes health hazards, and how regulatory authorities, telecom industries, and scientists can more effectively solve these problems, is essential to our overall understanding of all facts of the electromagnetic issue".

In developing countries, smartphones have become a part of modern life style. After all, the towers and smartphones radiations have been associated in some research work with possible DNA problems [6] and some other effects.

On the other side, a variety of other health symptoms like headache, blood pressure fluctuation, generally denoted as hypersensitivity were reported in many studies of smartphone users [7]. In 1987, Slovic did psychometric analysis which was focused on the identification of factors that determine public awareness of different hazards [8]. A study [9] has observed the Indian users' perception towards mobile value-added services and pointed out that they have become a necessary part of mobile services.

Progress in Telecom sector is also differentiated by Average Revenue per User (ARPU). ARPU from value-added services has also come down due to high competition and decline in cost and also concerns about the Electromagnetic Field with regards to towers and mobile phones. The users' mainly look for reliable, highly secure, timely, good quality, cost-effective, and context-aware mobile value-added services, and expect a good quality level of services, and a trustable service provider while choosing particular network operator. In 2003, research work mentioned the factors that affect trust in the regulatory authorities responsible for regulating rules for smartphone usage [10]. The outcome of the survey indicated that lack of trust, less reliability, and brand value also affect trust. A study [11] displayed that people are associated with unknown health effects with smartphones. There are multiple factors that have contributed to the fast growth of mobile connections in India. These include favorable demographics, good regulatory environment, changes in consumer behavior, low cost, and low handset prices.

Numerous studies have been done on the different aspects of the telecom industries, but not many studies concentrated on the users' intention to use the mobile value-added services and smart phones [8]. The objective of the study is to better understand the users' response to the mobile phone's radiations. The study examines how much smartphone users valuate EMF risk from smartphone devices. More specifically, the influence of the risk on the choice of mobile services is explored in evaluation with the impact of other factors, which are regarded as the most important factors in users' decision-making. In this study, we explore the users' point for mobile phone radiations impact on the acceptance/use of mobile services and smart phones which can be considered an emerging risk.

Identification of the factors:

Health Impacts: the use of mobile phone has created some health effects like short-term health problems including headaches, and tingling in the head increased with the more mobile phone use. Some studies have given some concern to health problems which occur due to Electromagnetic radiations (EMF) from the towers and mobile phones [12]. Repacholi [13] has also raised this issue that mobile phone radiations could affect the peoples' health. So there have been very limited studies that have observed thoroughly at such users.

Environmental Impact: The emission of mobile phone affects reproductive system which is an environmental health hazard, and one of the most general complaints made about the use of mobile phones after health concerns is the quantity of noise pollution they produce: on public transport and other public places.

Social Media Impact: The UK media mainly focused on the use of mobile phone handsets and has characterized it by headlines such as, "Will your phone kill you?" and "My mobile gave me cancer". There are important differences in the way in which the issue has developed country by country, the factors identified have international relevance and the principles required to address public concerns are broadly transferable. Such types of issues have produced a negative perception in society related to adverse outcomes, no matter whether the risk is real or perceived.

Average Revenue per User (ARPU): Average Revenue per User (ARPU) is one of the major contributing factors in the growth of the telecom sector. ARPU from value-added services has also come down due to increased high competition, decline in call rates, and also concerns about the Electromagnetic Field with regards to towers and mobile phones. As ARPUs decline, the challenge for service providers would be to retain customers, create alternative revenue streams, and develop a basis for service differentiation [1].

Users' intention to purchase: The mobile phone radiation factors affect users' intention to use the mobile services while introducing new services like 3G, 4G, and other radio technologies.

Brand image of provider: The consumer's perception towards service providers is an important aspect. The users are interested to buy services from those service provider's that have a reliable image in the market. The users prefer to purchase the services of such provider's who have provided reliable services in spite of their high tariffs. We recommend those provider's to friend circles for services, who have excelled at their services. People also discuss their bad experiences with relatives and friends.

Government policy: Government policy is a major contributor to the external environment. The rules and regulations have a high influence on the performance of telecom industries. According to the specialists and higher authorities, policies for tariffs, auctions, and reliable services have a major influence on the future campaigns and approaches of the telecom sector.

Quality of Telecom services: The service quality was a strategic aspect mentioned by most of the participants. Call clarity, network connection quality, and high data speed are the important points of consumers while measuring the performance of a service provider. Some participants were concerned that network connectivity should be offered in remote villages; call connection and the network quality are major concerns towards service quality. The Internet connectivity, 3G, and 4G network speed are also major contributors to the convenient use of smartphone services.

3 Methodology

The growing pressure of stakeholders regarding health and environmental problems caused by mobile phone radiations (EMF) is forcing the concerned authorities to redesign their operational strategies. In developing economy such as India, most companies focused on the economic viability and neglect the health consequence of the SMF radiations. It can be seen that the literature has already delved upon the possible health consequences of mobile phone radiations. However, very few research studies have observed the users' perception towards risks associated with mobile phone and services. Thereby, the present study attempts to provide a framework to assess the major factors that influence the consumer perspective towards the use of mobile phone services. These metrics are identified through literature review and detailed discussions with the DMs to shortlist eight factors. DEMATEL methodology is utilized to segment these factors into cause and effect relationships. Therefore, the research methodology evaluates various metrics to identify the key factors which affect the other metrics, and therefore focusing on them will enhance the remaining factors. Following are the major steps of integrated fuzzy DEMATEL approach:

DEMATEL Approach

Step 1: Identifying the decision body and short listing the factors: In this step, a formal decision-making body is formed, who will assist in identifying the goal of the study followed by short listing the essential factors.

Step 2: Identifying the scale: The scale for evaluation of the factors is identified in this step.

Step 3: Determining the assessments of decision-making body. In this step, the relationship between short-listed metrics was determined to generate initial direct matrix A.

Step 4: Calculating normalized direct-relation matrix: A normal direct matrix D is obtained using the following formulas (1) and (2)

$$D = [d_{ij}]_{\max} = \frac{A}{s} \tag{1}$$

$$s = \max \left(\max_{1 \le i \le n} \sum_{j=1}^{n} a_{ij}, \max_{1 \le j \le n} \sum_{j=1}^{n} a_{ij} \right) \tag{2}$$

Step 5: Calculating total relation matrix: The total-relation matrix T is obtained using the formula (3), in which I denote the identity matrix.

$$T = [r_i]_{n \times 1} = \lim_{m \to \infty} (D + D^2 + D^3 \ldots + D^\infty) = D(I - D)^{-1} \tag{3}$$

Step 6: Analysis of the relationship to identify the cause and effect group. In this step, the criteria are filtered to understand the dispatcher (cause) and receiver (effect) group. Ri and Cj represent the row-sum and column-sum of the T. Ri represents the total effect generated by ith criteria on other criteria while Cj represents the total effect received by jth criteria. Thus, (Ri + Ci) indicates the level of significance of ith criteria while (Ri − Ci) indicates the resultant effect of ith criteria. This dataset (Ri + Ci, Ri − Ci) are mapped in the form of casual diagram (Table 1).

Following Step 1, in the next step the scale for evaluation was identified. The scale for evaluation of the factors is identified in this step. The following scores 0, 1, 2, 3, and 4 representing 'No influence', 'Very Low influence', 'Low influence', 'High influence' and 'Very high influence', respectively, are used. In step 3, the relationship between short listed metrics was determined to generate initial direct matrix A as given in Table 2.

See Table 3.

Table 1 Contextual relationship with factors

Factor name	Description	Contextual relationship
C1	Media influence	Smart phone radiations factors will impact the use of the services
C2	Health hazards	
C3	Environmental concerns	
C4	Users' intention to purchase	
C5	Government policy	
C6	Quality of Telecom services	
C7	Average Revenue Per User (ARPU)	
C8	Service provider's image	

Table 2 Intial direct matrix

	C1	C2	C3	C4	C5	C6	C7	C8
C1	0	2.33	2.66	3.66	3	0.33	3	2.33
C2	0.33	0	0.33	3.3	3.66	1.3	2	3
C3	0.33	2.66	0	2	3	1.6	1.33	2.33
C4	2.66	3	0.33	0	2.33	3	2.33	2.66
C5	3.33	3	2	3	0	1.66	2	1.66
C6	2.33	2	2.33	3	2.33	0	3.33	3.33
C7	1.33	3	2.66	4	2.33	2.3	0	3.33
C8	1.33	0	1.66	1.6	0.66	2	2.66	0

Table 3 Total relation matrix

	C1	C2	C3	C4	C5	C6	C7	C8
C1	0	0.11	0.12	0.17	0.14	0.015	0.14	0.11
C2	0.015	0	0.015	0.16	0.17	0.064	0.096	0.14
C3	0.015	0.12	0	0.096	0.14	0.080	0.064	0.11
C4	0.12	0.14	0.015	0	0.11	0.14	0.11	0.12
C5	0.16	0.14	0.096	0.14	0	0.080	0.096	0.080
C6	0.11	0.096	0.11	0.14	0.11	0	0.16	0.16
C7	0.064	0.14	0.12	0.19	0.11	0.11	0	0.16
C8	0.064	0	0.080	0.080	0.031	0.096	0.12	0

4 Result and Discussion

All the statistical results are provided in Table 4.

The study reveals that customers' perception towards EMF radiations can affect their behavior. The emphasis on the safety of EMF radiation can attract some types

Table 4 Result analysis

	ROW total (R)	Colum total (C)	R + C	R − C
C1	4.1443	4.11212	8.25642	0.03218
C2	0.6021	4.11212	4.93956	0.6021
C3	4.13666	4.11212	7.39536	0.87796
C4	3.21133	4.11212	8.7404	-2.31774
C5	3.97905	4.11212	10.02239	6.04334
C6	8.81336	4.11212	13.00784	4.61888
C7	4.49369	4.11212	9.77601	-0.78863
C8	4.47284	4.11212	9.4254	-0.47972

of customers who are aware of electromagnetic fields. In particular, smart phone manufacturers and service providers can strategically use the EMF security aspect in their marketing plans. The results of our research study suggest that a huge number of customers intend to change their brands in accordance with the security aspects.

5   Conclusion and Future Scope

The purpose of this paper is to utilize DEMATEL method in order to find the causal relationship among the factors. The study revealed that customer satisfaction has a positive impact on continuance intention. Consequently, satisfied users are likely to continue using the same smartphone and its services. The findings of the current research have several theoretical and practical implications for smartphone manufacturers, service providers, and academic researchers. Service providers and decision-makers must have a clear understanding of the drivers of smartphone use. In addition, potential users, women, and younger people are adversely affected by the perceived health consequences related to smartphones. Service providers and legislators must take this finding into account and design strategies to eliminate this negative perception of radiation from smartphones. These guidelines should be adequately communicated to the general populations of India to counter the misconception towards electromagnetic radiation of technology.

References

1. Annual Report's 2010–11, 2011–12 and 2012–13, Telecommunication Regulatory Authority of India. Accessed on 15th to 31st March 2014 from https://www.trai.gov.in.
2. National Telecom Policy. (2012). https://www.dot.gov.in/telecom-polices/national-telecom-policy-2012.
3. Brusick, D., Albertini, L., McRee, D., Peterson, D., Williams, G., Hanawalt, P., & Preston, J. (1998). Genotoxicity of radiofrequency radiation. DNA/Genotox expert panel". *Environmental and Molecular Mutagenesis, 32,* 1–16.
4. Lai, H., & Singh, N. P. (1995). Acute low intensity microwave exposure increases DNA singlestrand breaks in rat brain cells . *Bioelectromagnetics, 16,* 207–210.
5. Background Paper. (2014). *Second International Conference on Health & Safety Aspects of Mobile Telecommunications,* 17 Nov 2014. New Delhi.
6. Bernhardt, J. H., Matthes, R., Repacholi, M. H. (Eds). (1997). Non-thermal effects of RF electromagnetic fields. International Commission on Non-Ionizing Radiation Protection (ICNIRP), vol. 3. Munich, Germany, pp. 1–244.
7. Hocking, B. (1998). Preliminary report: Symptoms associated with mobile phone use. *Occupational Medicine, 48,* 357–360.
8. Slovic, P. (1987). Perception of risk. *Science, 236,* 280–285.
9. Singh, S., Sharma, G. D., & Mahendru, M. (2011). Role of value added services in shaping Indian telecom industry
10. Shine. (2013). Telecom industry set for growth in 2013. https://info.shine.com/Industry-Information/Telecom/183.aspx.

11. Sjoberg, L. (2002). Attitudes toward technology and risk: Going beyond what is immediately given. *Policy Sciences, 35,* 379–400.
12. Ahlbom, A., Cardis, E., Green, A., Linet, M., Savitz, D., & Swerdlow, A. (2001). Review of the epidemiologic literature on EMF and health. *Environmental Health Perspectives, 109,* 911–993.
13. Repacholi, H. M. (2001). Health risks from the use of mobile phones. *Toxicology Letters, 120,* 323–331.

Crime Forecasting Using Time Series Analysis

Neetu Faujdar, Yashita Verma, Yogesh Singh Rathore, and P. K. Rohatgi

1 Introduction

Enforcement of laws is one of the most important jobs in the current world and has been for many hundreds of years since the dawn of human civilization. Crimes have been occurring since the beginning of recorded history and we as a civilization have been recording and storing such data. This data is valuable in finding any rises and falls in crime rates and finding what kind of crime is being committed the most. The amount of data being collected has aggressively accelerated in the past few decades and data of all kinds is now available for analysis. With the invention of computers and the incredible rate at which technology has improved and advanced in the last five decades, collection and storage of data has become very easy and affordable. Improvements in data storage technology have been helping companies, corporations, and governments to store incredibly large amounts of data. Previously, analyzing this amount of data was far too expensive for most corporations and companies aside from the largest such Google. But with advancements in data distribution technology and the invention of algorithms such as MapReduce by Google and its implementations such as Hadoop which was funded by Yahoo, analyzing massive amounts of data has become easy and affordable [1, 2]. The field of data mining and analysis has emerged from this new age of information technology and has quickly become one of the most important assets for corporations. Corporations around the world are looking for a data analysts and data scientists to dig for novel patterns and knowledge in the large data banks they collect [3]. Forecasting the future has been one of the important applications of data mining and analysis. There are many other applications in the fields of healthcare, education, fraud detection, etc. [4–7]. Using statistical data analysis techniques from the fields of data mining, machine

N. Faujdar (✉) · Y. Verma · Y. S. Rathore · P. K. Rohatgi
GLA University, Mathura, India
e-mail: neetu.faujdar@gmail.com

© The Author(s), under exclusive license to Springer Nature Singapore Pte Ltd. 2021
P. K. Kapur et al. (eds.), *Advances in Interdisciplinary Research in Engineering and Business Management*, Asset Analytics,
https://doi.org/10.1007/978-981-16-0037-1_20

learning, and predictive modeling to forecast the future is called predictive analysis [8]. Applications of predictive analysis include fraudulent claim detection, weather forecasting, marketing, customer behavior predictions, etc. Time Series Analysis is one of the major techniques that has been developed to forecast the future. In this paper, time series analysis is used to predict future crime rate of specific crimes. The United Nations Office of Drugs and Crime dataset is used which contains data for each specific type of crime from 2003 onwards for all countries.

2   Related Work

Criminal activities are a massive burden to society and have existed since the beginning of civilization. Analyzing trends in the crime can prove to be an incredible tool to help curb criminal activities and improve law enforcement. Police and law other law enforcement authorities benefit greatly from the development of new preventative measures. A study in 2005 suggested that the most effective and important aspect to the global decreasing trend in crime is the development and use of new preventative measures [9]. Many studies have shown that the global rate of common crimes has been steadily decreasing for the past few decades. In recent times, with the advancement of computer science and data analysis, crime data is being analyzed to find patterns and knowledge that may help the police [10].

Nine sorts of crimes were categorized into four divisions in paper. Rossmo's equation was used in order to communicate the concept of predictive policing and sentiment access. The resultant weights and corresponding graph is calculated for every division and deeply inspected in order to find the occurrence which is most likely to happen. By analyzing the various graphs and understanding the behavioral designs of former crimes, this strategy has proved to decrease the number of crimes occurring at present [11].

The paper uses the auto-regressive model strategy to predict the crime flow in the downtown or urban regions of the country. The architecture and execution of the model are effectively depicted in the paper. It can easily determine and predict the figure of various crimes that may occur in rolling time horizons [12].

Generally, one of the enormous challenges faced while using crime data is witnessing incomplete datasets. Most of the time, the data is missing, scattered, or flawed. The dearth of appropriate guidelines and standards across countries to record crimes causes a delay in the data mining process. About 80% of the time is occupied in only collection and pre-processing of data [13].

Various crime blockage models, namely, Spatiotemporal Model, Point Model, Spatial Ellipse Model, and Kernel density estimation Models were deeply studied and analyzed in paper 4. The analysis showed the kernel density estimation model works the best in comparison to other models [14].

Paper provides an algorithm which in turn helps to analyze the crime rich areas, corresponding crime mediums, and a spatiotemporal crime prediction model. This strategy is implemented on live and real data and displays excellent precision in its performance. It is established on auto-regressive models and spatial inspection to determine the areas of enhanced risk in downtown or urban regions [15].

The impact of data mining and analysis has been useful in helping not only the local law enforcement but also the government authorities such as the Federal Bureau of Investigation (FBI). The field of crime forecasting is one of the methods that has come up from research in crime data analysis. Forecasting crime can prove to be one of the most effective techniques in crime prevention. If law enforcement authorities know when and what kind of crime is on the rise or fall, they can aptly prepare for it. Forecasting crime rates is another relatively less used application of predictive analysis. Forecasting crime rates can help federal authorities and governments make accurate decisions in budgeting and planning. Accurate forecasts are an important factor already in the fields of weather prediction and natural disaster prediction. Prediction of rising or falling crime rates can allow these authorities in preplanning for these events and help curb overall crime rates in their jurisdictions.

3 Data Mining

The field of data mining has been experiencing great improvements and gaining importance in the information age. Data mining is part of the Knowledge Discovery in Databases (KDD) process and involves finding new patterns and hidden information or knowledge in datasets. Such hidden information usually manifests in the form of patterns, trends, and groups in the data. Data mining has many subfields and several applications. Some subfields of data mining are Graph Data Mining, Web Data Mining, and Link Mining. Social network analysis is also considered to be a type of data mining. Applications of data mining include fraud detection, healthcare applications, airport network analysis, market basket analysis, crime analysis, biochemistry applications, etc. [11, 12].

The main types of tasks in data mining are as follows:

1. **Classification**: Classification is a machine learning technique in which the computer is trained on a labeled dataset. A labeled dataset has set values for a certain target attribute and this target attribute is called class label. The computer is then given an unlabeled test dataset and must predict the class label. It is a supervised learning process which is used to predict values based on previously made generalizations [13].
2. **Clustering**: Grouping of similar variable based on some criteria is called clustering. Clustering is an unsupervised machine learning process and it is used for exploratory data mining and analysis [14].
3. **Association Rule Mining**: Association rule mining is a technique used to find strong rules that are frequently found in a dataset. Rules exist between variables and attributes and more frequently found rules can help find useful patterns. AR Mining is often called market basket analysis because it is often used to find what items are frequently purchased together.

4.   **Anomaly Detection**: This technique is used to try and find observation which does not match a predefined pattern. Observations which do not match the pattern are often considered to be actionable or critical. Anomaly detection has a wide variety of uses such as detecting fraud, safety systems fault detection, and even medical problems [15].
5.   **Regression Analysis**: Regression involves using mathematical concepts of regression to find the correlation and covariance between two variables.

3.1   Time Series Analysis

A time series is an ordered series of data points over a certain period of time. Data analysis when done over a period of time to analyze a variable is accomplished using time series analysis. It is widely used technique commonly used in the fields of finance, economics, and retail [6]. Weather forecasting and earthquake prediction also employ time series analysis to predict the relevant variables. The main use of time series analysis is making short-term predictions or forecasts in a time series.

A time series consists of four main parts:

1.   **Trend**: Movement in a time series that is long term. A trend can usually be estimated using other methods but time series analysis allows in-depth and accurate predictions of trend. The trend component may also be needed to be removed to predict variables in a manner in which variables are not affected by a long-term trends.
2.   **Cycle/Periodic Fluctuations**: Periodic rises and falls in a time series often appear over time. Regularities which are not bound by seasonal aspects are considered here.
3.   **Seasonality**: A major component of a time series is seasonal fluctuations in a time series. Seasonal fluctuations appear when variables regularly appear during certain parts of the year repeatedly throughout the time series.
4.   **Random**: The random component of a time series is a mixture of noise and certain other properties which may be able to help predict future values. There are two main types of models suggested for time series analysis: the auto-regressive models and the moving average models. In 1976, George Box and Gwilym Jenkins developed a methodology for time series analysis in which they combined the AR and MA models to form the ARMA TSA. A generalization of ARMA called the Auto-regressive Integrated Moving Average (ARIMA) model is used in this paper.

The ARIMA model has the following mathematical equation:

$$(1 - \sum_{i=1}^{p} (1 - L)^d X_t = \delta + (1 + \sum_{i=1}^{q} \theta_i L^i)\varepsilon_i \tag{1}$$

Differencing may be done on a time series to make it a stationary time series. For analysis purposes, ARIMA models require time series to be stationary [16]. An implementation of ARIMA in R programming language is used for forecasting in this paper. The ARIMA (p,d,q) model also requires the use of autocorrelation and partial autocorrelation to find the value of p and q.

4 Methodology

The following methodology is followed:

1. **Data Collection**: The dataset used in this paper was collected from the United Nations Office of Drugs and Crime site. Yearly data for all different types of crime is included in the dataset for all countries in the world.
2. **Data Cleansing**: The dataset contains a lot of missing values and incomplete data. UNODC dataset does not contain a consistent set of data for each type of crime and these missing values must be removed.
3. **Time Series Building**: The data points for the time series must be collected and converted into the correct format for analysis. In R, this format is the time series object. The data for U.S.A. is converted into a numeric vector and then transformed into a time series object.
4. **ARIMA analysis:** The ARIMA model is used to conduct analysis of the time series and forecast crime rates in the future. The Autocorrelation Function (ACF) and Partial Autocorrelation Function (PACF) are used to find the relevant values for a ARIMA (p, d, q) model.

4.1 Experimental Setup

Dataset: The data for this project was collected from the United Nations Office of Drugs and Crime (UNODC) data portal. The portal contains an interface which can be used to query the UNODC database and also download the queries. For this project, all the data related to crimes were collected in Comma Separated Values (CSV) file format. The dataset contains data of several different crimes such as Assault, Robbery, Sexual Crimes, Burglary, Theft, and Vehicle Theft and these have been separated into distinct CSV files. They are all extracted as separate files from the portal. The dataset also contains many columns three of which are names for geographical regions with distinctions for sub-regions and countries or territories. The columns under the "Count" header display data for the absolute number of instances of a specific crime in the region or country from the year 2003 till 2015. Under the "Rate per 100,000" header, the columns represent per capita crime which is a much better indicator of the severity of occurrences of the crime. Each CSV file for the data for many countries of the world. The data is often incomplete for most African and Asian nations but is complete for Europe and the Americas.

Tool Used: The main tool used in this paper is the R programming language. The Rstudio IDE is used and several packages from the CRAN package collection are used. Mainly, the forecast package is used for ARIMA modeling and time series analysis.

5   Result Analysis

Case Study 1: The Assault dataset from the UNODC data repository is retrieved. The U.S.A. data is retrieved from the dataset. It is converted into a numeric vector for the purposes of analysis using the as.numeric() function in R.

 The ACF and PACF functions can be used to find values p and q for the ARIMA (p, d, q) model. The acf() and pacf() functions are used to find the value of p and q for the Assault time series. The value of p is found to be 3 and the value of q is found to be 0. Thus, the ARIMA (3,0,0) model is used for TSA. The ARIMA model analysis shows us a net decrease in future assault crime rates in the United States. This model shows the forecast for the next 5 years from the last data available. This is the forecast for the years 2016–2020. A 10-year forecast is also done showing the predicted assault rate from 2016 to 2025. The dark blue region represents the 85% accuracy interval and the light blue regions represent the 95% accuracy interval. These intervals can be used to conclude maximum possible change in the time series.

Case Study 2: The theft data is retrieved from the UNODC data repository. It contains the numbers from 2003 to 2015. The same process is used on this dataset to find the correct ARIMA model. Using ACF and PACF, we find that the best model for the theft time series is the ARIMA (2,0,0) model. The theft rate is expected to see a

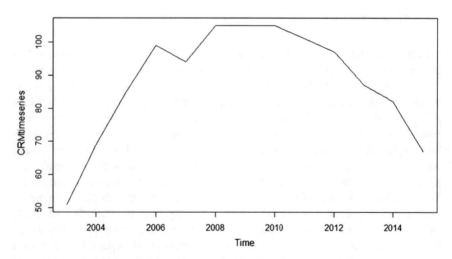

Fig. 1 Time Series for USASSAULT RATES from 2003 to 2015

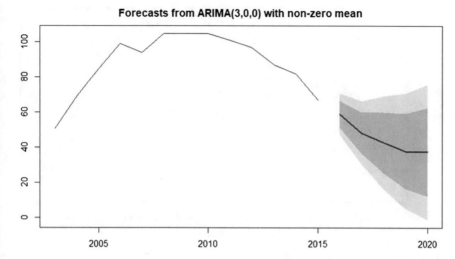

Fig. 2 Assault forecast for 2016–2020

historic decrease around 2017 or 2018 but is then expected to rise. The dark blue region shows the 85% accuracy interval and light blue region shows the 95% accuracy interval.

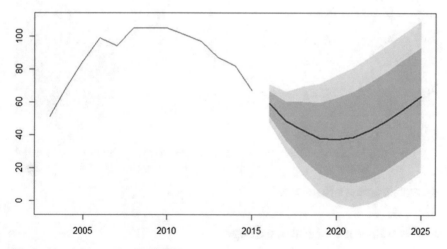

Fig. 3 Assault forecast for 2016–2025

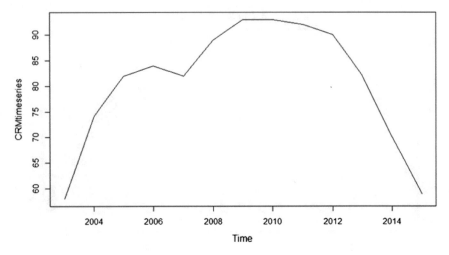

Fig. 4 Time series for us theft rate from 2003 TO 2015

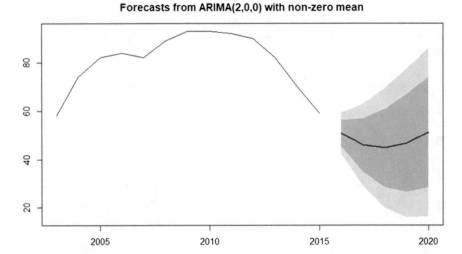

Fig. 5 Theft forecast for 2016–2020

6 Conclusion and Future Scope

From the time series analysis of the United States crime data found in the UNODC dataset, we have forecasted that the rate of assault in the U.S.A. is going to decrease over the period of 2015–2020. The rate of assault will start increasing after 2020 and is predicted to rise until at least 2025. The rate of theft in the U.S.A. is also predicted to reduce over this period of time. The forecasts show a consistent decrease in theft rates until 2019 after which the rate of theft will increase. This information is valuable

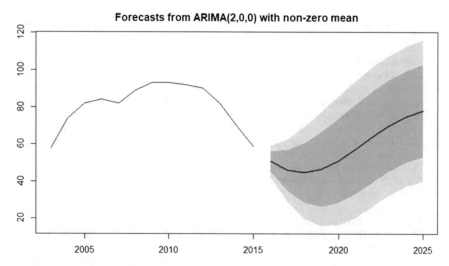

Fig. 6 Theft forecast for 2016–2025

in planning and budgeting for U.S.A. law enforcement agencies. The methodology used in this paper can be used to predict the rate of many different types of crimes in many countries provided complete and consistent data is available.

Contributions can be made either to the data which is incomplete in many places or to website design or through addition of visualizations. The many different ways of visualizing geo-spatial and spatial data can be explored and a variety of visualizations can be made through the visualization techniques explored. There are many candidate countries for time series analysis which can also be analyzed and general crime trends can be identified in those countries.

References

1. White, T. (2012). *Hadoop: The definitive guide.* O'Reilly Media, Inc.
2. Dean, J., & Ghemawat, S. (2008). MapReduce: Simplified data processing on large clusters. *Journal of Communications of the ACM, 51*(1), 107–113.
3. Han, J., Pei, J., & Kamber, M. (2011). Data mining: Concepts and techniques. Elsevier.
4. Koh, H. C., & Tan, G. (2011). Data mining applications in healthcare. *Journal of Healthcare Information Management, 19*(2).
5. Baker, R. S. J. D. (2008). Data mining for education. *International Encyclopedia of Education, 7*(3), 112–118.
6. Chan, P. K., Fan, W., Prodromidis, A. L., & Stolfo, S. J. (1999). Distributed data mining in credit card fraud detection. *IEEE Intelligent Systems and Their Applications, 14*(6), 67–74.
7. Nyce, C., & CPCU, A. (2007). Predictive analytics white paper American Institute for CPCU. *Insurance Institute of America,* 9–10.
8. Van Dijk, J. J. M., Manchin, R., Van Kesteren, J. N., & Hideg, G. (2005). The burden of crime in the EU: A comparative analysis of the European Survey of Crime and Safety. EU ICS.

 9. McCue, C. (2014). *Data mining and predictive analysis: Intelligence gathering and crime analysis*. Butterworth-Heinemann.
10. Joshi, A., Bansal, A., Sabitha, A. S., & Choudhury, T. (2017). An efficient way to find frequent patterns using graph mining and network analysis techniques on United States airports network. In *Smart computing and informatics* (pp. 301–316). Springer.
11. Ahmad, F., Syal, S., & Tinna, M. (2018). Criminal policing using Rossmo's equation by applying local crime sentiment. In S. Satapathy, V. Bhateja, K. Raju, & B. Janakiramaiah (Eds.), *Data engineering and intelligent computing* (pp. 627–637). Singapore: Springer.
12. Cesario, E., Catlett, C., & Talia, D. (2016). Forecasting crimes using autoregressive models. In *IEEE 14th International Conference on Dependable, Autonomic and Secure Computing, 14th International Conference on Pervasive Intelligence and Computing, 2nd International Conference on Big Data Intelligence and Computing and Cyber Science and Technology Cogress (DASC/PiCom/DataCom/CyberSciTech)* (pp. 795–802).
13. Garg, R., Malik, A., & Raj, G. (2018). A comprehensive analysis for crime prediction in smart city using R programming. In *8th International Conference on Cloud Computing, Data Science Engineering (Confluence)* (pp. 14–15).
14. Barreras, F., Daz, C., Riascos, A., & Riber, M. (2016). Comparison of different crime prediction models in Bogot. IOS Andes.
15. Catlett, C., Cesario, E., Talia, D., & Vinci, A. (2018). A data-driven approach for spatio-temporal crime predictions in smart cities. In *IEEE International Conference on Smart Computing (SMARTCOMP)* (pp. 17–24).
16. Piatetsky-Shapiro, G. (1996). *Advances in knowledge discovery and data mining* (p. 21). Menlo Park: AAAI press, Vol.

Software Quality and Reliability Improvement in Open Environment

Chetna Choudhary, P. K. Kapur, Sunil K. Khatri, and Rana Majumdar

1 Introduction

The modernized civilization is irreversibly and practically reliant on information technology, and more specifically on software products. Software has been a key technology in general protection, management, business, industry, manufacturing, energy, and invades services of all kinds. Conversely, empirical evidence disagrees; science studies have established GNU software to be more accurate than proprietary software equivalent. Software malfunction can result in severe emanation of which few are common and devastating [1]. Reliable software problems have become common in nature, including how to measure code integrity with additional data as a required crisis and challenges. Previously, a great deal of effort was exhausted in methods for a variety of reliable software facets [2–10], which mainly incorporated basic concepts, expressions, inquiries, capacity and software process reliability assessment, etc. Founded on these studies, criteria such as functionality, coexistence, reliability, security, maintenance, and usability emerge as the most important quality properties committed to software fidelity. As a result, this work, considered software reliability evaluation (SRE) as a multi-criteria decision-making (MCDM) issue considering both quantitative and qualitative criteria's

C. Choudhary (✉) · P. K. Kapur · S. K. Khatri
Amity University Uttar Pradesh, Noida, India
e-mail: chetna.choudhary13@gmail.com

P. K. Kapur
e-mail: pkkapur1@gmail.com

S. K. Khatri
e-mail: sunilkkhatri@gmail.com

R. Majumdar
MSIT Techno India Group, Kolkata, India
e-mail: rana.majumdarwb@gamil.com

© The Author(s), under exclusive license to Springer Nature Singapore Pte Ltd. 2021
P. K. Kapur et al. (eds.), *Advances in Interdisciplinary Research in Engineering and Business Management*, Asset Analytics,
https://doi.org/10.1007/978-981-16-0037-1_21

The MCDM process is carried out appropriately with classical MCDM issues [11–13] where ratings and criteria weights are measured in crisp numbers. There are also other classes in the context of fuzzy multi-criteria decision-making issues [14] where the scores and weights of the parameters are calculated on the basis of complexity, uncertainty, and elusiveness and are typically expressed by linguistic variables and then put into dynamic numbers [15]. The rationale behind this study is to devise software reliability assessment as a MCDM model. The evaluation outcome supports the contributors to decide on a trustworthy software artifact from projected options. Among the methods offered by MCDM, analytic hierarchy process [16, 17], entropy theory [18], and order preference by resemblance are considered perfect techniques and have been extensively applied in numerous domains to pacts with multiple criteria assessment or selection problems [20–32]. Conversely, this perception is hardly ever used in the area of software reliability assessment. Therefore, unusual from the realistic applications expressed in the earlier sections, this study endorses amalgamation of weighting method to assess the software reliability, thus quality.

This work is organized as section II defines problem statement. In Sect. 3, the conceptual framework is projected. Section 4 illustrates the evaluation of the proposed method of software. The final section draws conclusions for potential research areas and makes suggestions.

2   Problem Statement

The projected model evaluation process consists of two phases. Considering that core factors contribute to a software system's reliability, SRE criteria are preferred in phase 1. Phase 2 determines the weights of the criteria and is made up of two segments. Segment one is used with AHP to achieve subjective weights, this process is widely used to articulate evaluators' findings in dealing with multi-criteria evaluation. Whereas Shannon entropy theory [18] calculates two objective weights in the segment, which is well matched to measure the correlated contrast intensities of elements to indicate the standard inherent data propagated to the DM [33]. Subsequently, a combination weighing method is used to resolve the comprehensive weights of criteria [34]. Finally, perpendicular projection distance is used to acquire the software reliability ranking that is decisive.

3   Conceptual Framework

Greenfield analysis is not a basic criteria test. Steffen et al. believe that accuracy, security, quality of service (performance, reliability, and accessibility), security, and privacy determine the credibility of software [6]. Tan et al. spoke about the crucial fidelity—including attribute-based properties such as flexibility, protection, usability, stability, portability, durability, and reliability [7]. As reported by ISO 9126-1, six main attributes of quality software are functionality, reliability, efficiency, maintenance, usability, and portability [38]. Fenton et al. provides an accurate and realistic technique in their work to identify criteria such as maintenance, reliability, ability to learn, and operability [39]. Zhao et al. believed that software trustworthiness should be assessed on the basis of five criteria availability, reliability, maintenance, security, and security [37]. In this study we identified readiness of software, robustness, testing as parameters with respect to effort, recourses, integrity as alternatives and denoted, respectively.

As shown in Fig. 1, the evaluation of the projected model is based on two phases. Considering that significant variables contribute to software willingness from a

Table 1 Identified parameters

Parameters	
Readiness of software	Modern software systems include many dissimilar applications, environments, processes, and tools, as well as wide range of users and uses. To fully appraise whether a system is ready for go-live or production, you must truthfully appreciate the production-readiness criteria
Robustness	Designed or evolved in such a way as to be defiant to total failure despite fractional damage. Resistant or impervious to failure in spite of user input or unanticipated situations
Testing with respect to effort, resources, integrity	Recent testing has encouraged port scanning to include testing software behavior as a critical aspect of system activities to ensure software integrity

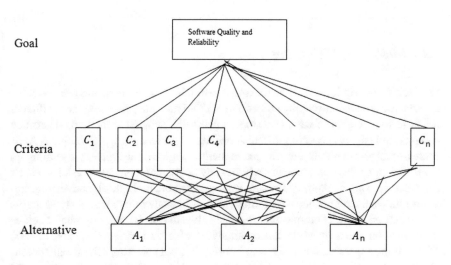

Fig. 1 Goal is achievable based on criteria and alternatives

Fig. 2 Evolution formation

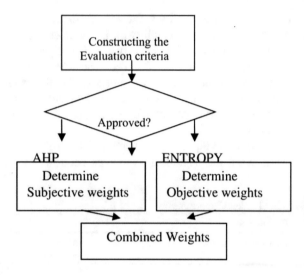

quality and reliability view and at this point SQR criteria are selected; interoperability, user-friendliness, safety. Phase 2 determines the criteria weights and consists of two parts. Part one gave subjective weights using AHP, [refer Tables 1, 2, 3, 4, 5, 6, 7, 8 and 9]. Part two identifies objective weights by using Shannon entropy theory [18], for representing the average intrinsic information refer (Tables 10, 11 and 12). Finally, the comprehensive weights of criteria are solved using a combination weighing method (refer Tables 13 and 14) (Fig. 2).

The main quality and reliability criteria for accurate evaluation result from all-inclusive investigation and conversation with various specialists from a variety of fields.

3.1 *Method of Weighting*

Precisely its own influencing role for software reliability assessment has been selected criteria in any decision-making process. In MCDM methods, the influence coefficient refers to the weights of each criterion. Universally, subjective methods determine weights exclusively according to decision-makers' preference, which includes AHP, professional test methods, etc. Objective methods agree on weights by operating on statistical equations automatically without any deliberation on the decision-maker's fondness, entropy method, optimizing variance method, multiple objective programming, etc. On the other hand, weighting methods like other methods also have merits and demerits. There is no question that the subjective weighting approach clarifies the evaluation primarily while the quantitative one is slow to do so. Objective method of weighting expressed the analysis in information, but sometimes the weight coefficients of a few indexes often conflict with the actual implications of these metrics, and it is more difficult to explain spontaneously than the subjective method of weighting [44]. Therefore, in this work, AHP and entropy are consistently promoted in order to achieve equilibrium of the advantages, deficiency of subjective and objective weights, and a combined weighting method.

3.2 Objective Form of Weighting Entropy

Entropy is a parameter used to measure the amount of functional information given in the theory of information. The higher the discrepancy of evaluating objects for the same identical criterion, the supposedly decreased entropy value implies that this criterion generates extra precious data and the weight of this criterion is expected to be quite big, and vice versa [32]. The theory of entropy is based on mathematical analysis and avoids bias of the evaluator, making it a process that is fairly unbiased. The entropy weighting procedure is consistent with the following steps:

(i) *Construction of Original data Matrix*

$$
X = A_n \begin{bmatrix} c_1 & c_2 & \cdots & c_n \\ x_{11} & x_{12} & \cdots & x_{1n} \\ x_{21} & x_{22} & \cdots & x_{2n} \\ \vdots & & \ddots & \vdots \\ x_{s1} & x_{s2} & \cdots & x_{sn} \end{bmatrix} \tag{1}
$$

where $A_i(i = 1 \ldots s)$ is the substitutes for the candidate alternatives, C_j $(j = 1 \ldots n)$ are evaluation criteria, and $X_{ij}(i = 1 \ldots s; j = 1 \ldots n)$ is a criteria-based alternative performance.

(ii) *Normalizing the Data Matrix*

Because each criterion has its own dimension, the conversion into a similar scale with different criteria varies from 0 to 1. The original data can therefore be interpreted as

$$
v_{ij} = \frac{x_{ij}}{\sum_{i=1}^{s} x_{ij}} \tag{2}
$$

The normalized matrix can therefore be displayed as $V = [v_{ij}]_{s \times n}$ V = [v_{{ij}}]_{{s\times n}}

(iii) *The entropy value for the jth criterion is calculated as*

$$
N_j = -k \sum_{i=1}^{s} v_{ij} \ln(v_{ij}) \tag{3}
$$

where k is constant and is given by $\frac{1}{\ln(n)}$. If all x_{ij}'s are same then $v_{ij} = 1/s$ and $N_j = 1$.

(iv) *Calculation of the entropy-based objective weight vector and the component as*

$$w_j^{(1)} = \frac{(1 - N_j)}{\sum_{k=1}^{n} (1 - N_k)}, \ j = 1, \ 2, \ \ldots n \tag{4}$$

where $w_j^{(1)}$ is the criterion's objective weights.

3.3 Subjective Weighting Method

Saaty [44] established AHP and discusses how a multi-criteria decision problem can determine the comparative significance of a set of criteria. In predictable AHP formulation, crisp real numbers characterize the rational thought. However, individual experiences may require uncertainty in many applications in the real world constantly. This level of subjectivity is acknowledged, and experts will be evaluated on the grounds of their perspective in some other methodology as well.

Assume m staff members are included in an assessment group. The method for determining the assessment criteria's subjective weights through AHP can be outlined as follows:

i. Create matrices for pairs between all requirements
ii. Aggregate the subjective judgment of m members.
iii. Calculate the subjective weights of each criterion.

$$w_j^{(2)} = \frac{x_{ij}}{\sum_{j=1}^{n} x_{ij}}; \ j = 1, \ 2, \ \ldots n \tag{5}$$

3.4 Combination Weighting Process

Weighting methods incorporate the objective weights with the subjective weights. The results of normalization acknowledging from Eq. (2) are a normalized matrix with dimensional unit-free information that represents the methodological differences of raw information with regard to all criteria and is apt for calculating weights centered on entropy hypothesis. In this analysis, the hybrid weighting technique is applied for achieving the synthesized weights to match the subjective weighting with the objective entropy weighting. Standardizing contributes to the standardized matrix and unit-free information that represents the difference in raw information from each criterion and is appropriate for entropy-based weights. On the other side, it is regarded that a number of evaluation strategies benefit criteria (i.e., the better the higher) and price criteria (i.e., the better the lesser). The matrix should be further standardized by the following formula in order to get rid of the irregularity arising from the variance of the criterion classification:

If C_j is a cost criterion:

$$U_{ij} = \frac{(\max v_{ij} - v_{ij})}{(\max v_{ij} - \min v_{ij})} \quad i = 1 \dots m, \ j = 1 \dots n \tag{6}$$

If C_j is a benefit criterion:

$$U_{ij} = \frac{(\max v_{ij} - \min v_{ij})}{(\max v_{ij} - \min v_{ij})} \quad i = 1 \dots m, \ j = 1 \dots n \tag{7}$$

Wang et al. suggested the generation equation for the combined weights vector and can be expressed as follows [4]:

$$W = \sum_{k=1}^{2} \lambda_k W^k \tag{8}$$

where $w^{(1)}$, $w^{(2)}$ represents the objective weights of entropy and subjective weights are indicated by AHP, respectively, where λ_k the linear combination coefficient is deducted on the basis of the optimization technique and λ_k is given as

$$\lambda_k = \frac{\exp\left\{-\left[1 + q \sum_{i=1}^{s} \sum_{j=1}^{n} W_j^{(k)} \frac{(1-u_{ij})}{(1-q)}\right]\right\}}{\sum_{k=1}^{2} \exp\left\{-\left[1 + q \sum_{i=1}^{s} \sum_{j=1}^{n} W_j^{(k)} \frac{(1-u_{ij})}{(1-q)}\right]\right\}} \tag{9}$$

where q is a balance-coefficient with range value as $0 < q < 1$ and $0 \le \lambda_k \le 1$, where $\sum_{k=1}^{2} \lambda_k = 1$. The combined weights vectors can therefore be acquired and expressed as $W = (w_1, w_2, w_3, \dots, w_n)$

The website usability score is calculated as follows:

$$\text{Usability} = \sum_{i=1}^{s} w_i x_i \tag{10}$$

where s is the quantity of criteria, w is the separate weights of the criteria, and x is simply the estimation of the criteria.

4   Implementation and Results

Our application is related to the selection of reliable parameters for enhancing software reliability and quality which is needed after the failure of software and changed people thinking that higher version is more reliable than previous version. So in this study three parameters are considered they are, namely, Readiness of Software, Robustness, testing with respect to effort, recourses, integrity are considered as alternative and denoted as, respectively.

Step 1: Build pair-wise matrices among all criteria

Weights can be determined by a pair-wise comparison within each pair of attributes, according to the AHP methodology. Experts or researchers were asked to create pair-wise comparisons to determine the relative weights. Then each comparison was converted into a numerical value. Based on Saaty's (1977) eigen-vector method, the pair-wise comparison information were arranged in a matrix format and summarized as

Step 2: Add m members' subjective judgment.

Step 3: Calculate each criterion's subjective weights.

To measure the expert's perception of a system's different attributes, it is incomparable to use an effective and structured questionnaire. The aim is to capture how a system expert/customer counts and weights that process attributes in a given context. The expert/user is chosen and asked to complete the survey using an appropriate

Table 2 Comparison matrix

Criteria	SWRFE	RSP	TIRE
SWRFE	1	0.5	0.2
RSP	2	1	3
TIRE	5	0.3333333	1
Sum	8	1.8333333	4.2

Table 3 Criteria weight

	SWRFE	RSP	TIRE	Sum	Avg	Weights
SWRFE	0.125	0.2727273	0.047619	0.445346	0.148449	14.84
RSP	0.25	0.5454545	0.714286	1.50974	0.503247	50.32
TIRE	0.625	0.1818182	0.238095	1.044913	0.348304	34.83
	1	1	1	3	1	99.99

(Probabilistic/Non-Probabilistic) sampling method and a well-defined sample size based on the vastness and reliability of the researcher. An expert is requested to rate each particular attribute on a preference scale based on his process experience with that particular attribute (10—Very Good 8—Great, 6—Neutral, 4—Bad & 2—Very Poor). We repeat the process for all three solutions to determine and pick the most suitable solution from the three shortlisted solutions.

Step 4: Based on each criterion, calculate CI.

The consistency ratio for each of the matrices and general incoherence for the hierarchy were calculated to regulate the outcome of the technique. The following equation expresses the deviations from consistency and the consistency measure is called the consistency index (CI).

$$CI = \frac{\lambda\mathrm{max} - n}{n - 1} \tag{11}$$

Table 4 Readiness of software

Criteria	Weights	Ratings				
		Very Good	Good	Neutral	Bad	Very Poor
		10	8	6	4	2
SWRFE	11.96	0.5	0.3	0.2	0	0
RSP	54.92	0.1	0.4	0.5	0	0
TIRE	33.12	0.3	0.3	0.2	0.2	0

Table 5 Robustness

Criteria	Weights	Ratings				
		Very good	Good	Neutral	Bad	Very poor
		10	8	6	4	2
SWRFE	11.96	0.1	0.3	0.5	0.1	0
RSP	54.92	0.4	0.4	0.1	0.1	0
TIRE	33.12	0.3	0.3	0.3	0.1	0

Table 6 Testing with respect to effort, resources, integrity

Criteria	Weights	Ratings				
		Very Good	Good	Neutral	Bad	Very Poor
		10	8	6	4	2
SWRFE	11.96	0	0.1	0.5	0.3	0.1
RSP	54.92	0.6	0.2	0.2	0	0
TIRE	33.12	0.4	0.4	0.2	0	0

Step 5: Constructing original data matrix

Taking the subjective weights of criteria from Tables [4] to [6] and using Eq. (5) the subjective weights calculated are shown in Table [8].

Step 6: Normalizing the data matrix

The combined weights are again normalized w.r.t to cost and benefit criteria as per Eqs. (6) and (7).

The combination weights are finally determined using the combination weighting technique using Eq. (9)

Table 7 CI calculated for given criteria

Criteria	SWRFE	RSP	TIRE	3rd root of product		Priority Vector	
SWRFE	1	0.333333333	0.2	0.066666667	0.405480133	0.110449084	
RSP	3	1	3	9	2.080083823	0.566595833	
TIRE	5	0.333333333	1	1.666666667	1.185631101	0.322955082	
Sum	9	1.666667	4.2		3.671195	1	
Sum*PV	0.994042	0.94433	1.356411				
					Lambda Max=		3.29478
Now Calculating CI= (Lambda Max-n)/n-1 =				0.019			

Table 8 Subjective values for three criteria

	SWRFE	RSP	TIRE
Software readiness	8.6	6.8	5.2
Robustness	6.8	8.2	8.8
Testing	7.4	7.6	8.4
Sum	22.8	22.6	22.4

Table 9 Normalized values

	SWRFE	RSP	TIRE
Software readiness	0.377192982	0.300884956	0.232142857
Robustness	0.298245614	0.362831858	0.392857143
Testing	0.324561404	0.336283186	0.375

Table 10 Combined weights

Weights	SWRFE	RSP	TIRE
Subjective weight	0.1196	0.5492	0.3312
Objective weight	0.769389	0.211519	0.019092
combined weight	0.428732	0.388551	0.182717

Step 7: Computing the jth criterion entropy calculation:

Using the Eqs. (3)–(4) the objective weights are calculated, and the outcome result is presented as in Table [11]

Step 8: Calculation of the entropy-based objective weight vector $w^{(1)}$ and calculation of the element as

Using the Eqs. (6)–(7) to calculate the utility values for criteria
  The use of Eq. (8) for the combination of objective and subjective weights is shown in Tables [9] and [11], respectively:

Step 9: To determine the final combination weights, use the combination weighting method. The optimal combination linear weighting method is $q = 0.5$

The proposed metric of usability given in Eq. (10) three parameters are used to test the usability. The weights determined using a composite method for different criteria are taken from Table [13] and the functionality for the three solutions is as follows:
  usability $= 0.428732*$ SWRFE $+ 0.388551*$ RSP $+ 0.182717*$ TIRE
  Using the metrics indicated in Eq. (10), checking functionality in terms of effort, money, credibility using the uniform value as shown in Eq. (6) and the combined weights as shown in Table [13] shall be calculated as
  usability (Testing with respect to effort, resources, integrity) $= 0.428732* 0 + 0.388551* 1 + 0.182717* 1 = 0.571268488$
  The weighted result is specified in Table 14. The quality and reliability of candidate software open software scales under each criterion can be considered as profit-related accomplishment in the decision matrix. In other words, the higher the reliability score of the ith alternative relative to the jth criterion, the more effective is the value of this alternative. The goal can therefore be determined and presented as Testing > Software Readiness > Robustness is the well-placed parameters to improve the reliability and quality of the software.

Table 11 Entropy base objective weights

	SWRFE	RSP	TIRE
Nj	0.979611292	0.99439477	0.999494074
1-Nj	0.020388708	0.00560523	0.000505926
w1(j)	0.769389152	0.211519189	0.019091659

Table 12 Utility values for criteria

	SWRFE	RSP	TIRE
u_{ij}	1	0	0.333333333
	0.294117647	0.545454545	0
	0	1	1

Table 13 Values for λ

λ obj	0.475741263
λ sub	0.5242587360

Table 14 Final outcome

Reliability and Quality derived data Analysis	0.48963724	Software readiness	48.96372401
	0.338034579	Robustness	33.80345787
	0.571268488	Testing with respect to effort, resources, integrity	57.12684877

5 Conclusion

Nowadays, the reliability of software, a holistic property surrounding a set of well-defined criteria, instigates increased interest in academia and industry. This paper sets out a novel approach for evaluating the reliability of software in combination with MCDM technique to investigate and construct a new evaluation scheme that is responsible for other options as stated in the literature. The predicted approach is based on the combination's weighting process. In the evaluation process, the vagueness of the data is taken into account. AHP method is used to solve the importance and prevent weight uniformity subjectivity; entropy weighting approach is used by managing valuable raw data knowledge to solve objective weights. The linear combination weighting approach is then used to synthesize the subjective and objective weights. The basic method and structure of this assessment design can also be useful in testing other software systems. In addition, focusing on static criteria as the basis for evaluation criteria and ignoring other vibrant SRE factors in the software's dissimilar life cycle phase should be critical for this study. The upcoming investigation will reflect on formulating a construction of dynamic criteria and developing an approach to evaluation in conjunction with other existing methods.

References

1. Michael, J. B., Voas, J. M., & Linger, R. C. (2004). 2nd National Software & Workshop series (NSS2) Report: Trustworthy Systems Workshop Report. Retrieved from http://www.cnsoft ware.org/nss2report/.
2. Becker, S., Boskovic, M. et al. (2006, November). Trustworthy software systems: a discussion of basic concepts and terminology. *ACM SIGSOFT Software Engineering Notes, 31*(6), 1–18.
3. Hasselbring, W., & Reussner, R. (2006, April). Toward trustworthy software systems. *IEEE Computer, 39*(4), 91–92.
4. Liu, K., Shan, Z. G., Wang, J., et al. (2008). Overview on major research plan of trustworthy software. *Bulletin of National Natural Science Foundation of China, 22*(3), 145–151. (in Chinese).

5. Parnas, D. L., Schouwen, J., & Kwan, S. P. (1990). Evaluation of safety critical software. *Communications of ACM, 33*(6), 636–648.
6. Steffen, B., Wilhelm, H., Alexandra, P. et al. (2006, November). Trustworthy software systems: A discussion of basic concepts and terminology. *ACM SIGSOFT Software Engineering Notes Archive, 31*(6), 1–18.
7. Tan, T., He, M., Yang, Y., Wang, Q., & Li, M. S. (2008). An analysis to understand software trustworthiness. The 9th International Conference for Young Computer Scientists, pp. 2366–2371.
8. Yuan, Y. Y., & Han, Q. (2010, December). Data mining based measurement method for software trustworthiness. Proceedings-2010 International Symposium on Intelligence Information Processing and Trusted Computing (IPTC 2010), pp. 293–296.
9. Zhang, H. P., Shu, F. D., Yang, Y., Wang, X., & Wang, Q. (2010). A fuzzy based method for evaluating the trustworthiness of software processes. Lecture Notes in Computer Science (including subseries Lecture Notes in Artificial Intelligence and Lecture Notes in Bioinformatics), vol. 6195, pp. 297–308.
10. Zhu, L., Aurum, A., Gorton, I., & Jeffery, R. (2005, December). Tradeoff and sensitivity analysis in software architecture evaluation using analytic hierarchy process. *Software Quality Journal, 13*(4), 357–375.
11. Wang, T. C., & Lee, H. D. (July 2009). Developing a fuzzy TOPSIS approach based on subjective weights and objective weights. *Expert Systems with Applications, 36*(5), 8980–8985.
12. Wang, Y. J., & Lee, H. S. (April 2007). Generalizing TOPSIS for fuzzy multiple criteria group decision-making. *Computers & Mathematics with Applications, 53*, 1762–1772.
13. Feng, C. M., & Wang, R. T. (July 2000). Performance evaluation for airlines including the consideration of financial ratios. *Journal of Air Transport Management, 6*(3), 133–142.
14. Wang, Y. J., Lee, H. S., & Lin, K. (2003). Fuzzy TOPSIS for multicriteria decision-making. *International Mathematical Journal, 3*, 367–379.
15. Bellman, R. E., & Zadeh, L. A. (1970, December). Decision-making in a fuzzy environment. *Management Science, 17*(4), 141–164.
16. Van Laarhoven, P. J. M., & Pedrcyz, W. (1983). A fuzzy extension. *Fuzzy Sets and Systems, vol. 11*, pp. 199–227.
17. Buckley, J. J. (1985, December). Fuzzy hierarchical analysis. *Fuzzy Sets and Systems, 17*(3), 233–247.
18. Shannon, C. E., & Weaver, W. (1947). *The mathematical theory of communication.* Urbana: The University of Illinois Press.
19. Hwang, C. L., & Yoon, K. (1981). *Multiple attribute decision making Methods and application.* Berlin: Springer.
20. Hsiao, S. W., & Chou, J. R. (2006, February). A Gestalt-like perceptual measure for home page design using a fuzzy entropy approach. *International Journal of Human-Computer Studies, 64,* 137–156.
21. İ.". (2009, January). Performance evaluation of Turkish cement firms with fuzzy analytic hierarchy process and TOPSIS methods. *Expert Systems with Applications, 36,* 702–715.
22. Fausto, C. (February 2010). Fuzzy TOPSIS approach for assessing thermal-energy storage in concentrated solar power (CSP) systems. *Applied Energy, 87*(2), 496–503.
23. Gao, H. S., Ran, J. X., & Sun, Y. Q. (2008). Risk evaluation of communication network of electric power based on improved FAHP. *Systems Engineering—Theory & Practice, 28*(3), 133–138 (In Chinese).
24. Hsieh, T. Y., Lu, S. T., & Tzeng, G. H. (2004, October). Fuzzy MCDM approach for planning and design tenders selection in public office buildings. *International Journal of Project Management, 22*(7), 573–584.
25. Kima, S. S., Yang, I. O., Yeo, M. S., & Kim, K. W. (2005, August). Development of a housing performance evaluation model for multi-family residential building in Korea. *Building and Environment, 40*(8), 1103–1116.
26. Morteza, P. A. (2010, September). Project selection for oil-fields development by using the AHP and fuzzy TOPSIS methods. *Expert Systems with Applications, 37*(9), 6218–6224.

27. Pochampally, K. K., Gupta, S. M., & Kamarthi, S. V. (2003, October). Evaluation of production facilities in a closed-loop supply chain: a fuzzy TOPSIS approach. Proceedings of the SPIE International Conference on Environmentally Conscious Manufacturing III, Providence, Rhode Island, pp. 125–138.
28. Tsao, C. T. (2003). Evaluating investment values of stocks using a fuzzy TOPSIS approach. *Journal of Information & Optimization Science, 24*(2), 373–396.
29. Tsaur, S. H., Chang, T. Y., & Yen, C. H. (2002, April). The evaluation of airline service quality by fuzzy MCDM. *Tourism Management, 23,* 107–115.
30. Wang, T. C., & Chang, T. H. (2007, November). Application of TOPSIS in evaluating initial training aircraft under a fuzzy environment. *Expert Systems with Applications, 33*(4), 870–880.
31. Wu, H. Y., Tzeng, G. H., & Chen, Y. H. (2009, August). A fuzzy MCDM approach for evaluating banking performance based on Balanced Scorecard. *Expert Systems with Applications, 36,* 10135–10147.
32. Zou, Z. H., Yun, Y., & Sun, J. N. (2006, October). Entropy method for determination of weight of evaluating in fuzzy synthetic evaluation for water quality assessment indicators. *Journal of Environmental Sciences, 18*(5), 1020–1023.
33. Zeleny, M. (1999). *Multiple criteria decision making.* New York: Springer.
34. Wang, Z. Y., Gu, H. F., Yi, X. X., & Zhang, S. R. (2003). A Method of determining the linear combination weights based on entropy. *Systems Engineering—Theory & Practice, 23*(3), 112–116. (in Chinese).
35. Zhao, X. X., Shi, Y., Liu, Y., & Zhang, L. L. (2010, June). An empirical study of the influence of Software Trustworthy Attributes to Software Trustworthiness. 2nd International Conference on Software Engineering and Data Mining (SEDM 2010), pp. 603–606.
36. ISO 9126-1. (2001). Software engineering "Product quality " Part 1: Quality model. ISO/IEC, Published standard.
37. Fenton, N. E., & fleeger, L. P. (2004). Software metrics: A rigorous and practical approach, 2nd edn. Beijing: China Machine Press (in Chinese).
38. Zadeh, L. A. (1965). Fuzzy sets. *Information and Control, 8,* 338–353.
39. Naghadehi, M. Z., Mikaeil, R., & Ataei, M. (2009, May). The application of fuzzy analytic hierarchy process (FAHP) approach to selection of optimum underground mining method for Jajarm Bauxite Mine, Iran. *Expert Systems with Applications, 36*(4), 8218–8226.
40. Chen, S. J., & Hwang, C. L. (1993). *Fuzzy multiple attribute decision making, methods and applications* (Vol. 375)., Lecture notes in economics and mathematical systems New York: Springer.
41. Zadeh, L. A. (1975). The concept of a linguistic variable and its application to approximate reasoning. *Information Sciences, 8,* 199–249(I), 301-357(II).
42. Wang, J. J., Zhang, C. F., Jing, Y. Y., & Zheng, G. Z. (2008). Using the fuzzy multi-criteria model to select the optimal cool storage system for air conditioning. *Energy and Buildings, 40*(1), 2059–2066.
43. Saaty, T. L. (1980). *The analytic hierarchy process.* New York: McGraw-Hill.
44. Chou, T. Y., Hsu, C. L., & Chen, M. C. (June 2008). A fuzzy multi-criteria decision model for international tourist location selection. *International Journal of Hospitality Management, 27*(2), 293–301.

Analysis of Systolic Blood Pressure via Machine Learning

Ankit Kumar Yadav, Rahul Saxena, Piyush Kumar Singh, Vimal Vibhu, and Biswa Mohan Sahoo

1 Introduction

The high pervasiveness and the social and financial outcomes of foundational blood vessel hypertension (SAH) portray it as a general medical issue in Brazil1. In 2006, 17 million.

Brazilians exhibited hypertension, which spoke to 35% of grown-up people more seasoned than 40 years2 [1].

One of the analytic criteria of hypertension is the nearness of BP esteems >140/90 mmHg, estimated in no less than two distinctive times4. In epidemiological examinations, estimations of high BP levels gauge the commonness of hypertension. Populace-based national examinations have been completed since the 90 s in all locales of the nation, aside from the northern district [2].

The present examination went for assessing the pervasiveness of the high BP among grown-ups living in the urban zone of the town of Lages, territory of Santa Catarina, southern Brazil. Age, blood cholesterol, and circulatory strain are natural event which are related to man and different creatures. This procedure is critical as it decides the condition of prosperity and well-being level of a man. Systolic circulatory

A. K. Yadav · R. Saxena (✉) · P. K. Singh · V. Vibhu · B. M. Sahoo
Amity University, Greater Noida, India
e-mail: ryansaxena03@gmail.com

A. K. Yadav
e-mail: Okankit1312@gmail.com

P. K. Singh
e-mail: rockpiyush16640@gmail.com

V. Vibhu
e-mail: drvimal@gn.amity.edu

B. M. Sahoo
e-mail: Bmsahoo@gn.amity.edu

strain measures the most extreme weight in the corridors amid the cardiovascular cycle, which happens when the heart contract or beat to pump blood [3].

Circulatory strain is influenced by drug, cardiovascular or urological disarranges, neurological conditions, and mental factors, for example, stress or outrage. A portion of the approaches to bring down the circulatory strain perusing is by expanding physical exercises, controlling liquor utilization, and expanding sustenance content in the eating regimen. One uncommon strategy for diminishing the pulse of man is subject to a few factors that imply there are a few factors that impact the level of circulatory strain perusing in the body such factors are Age and blood cholesterol [2].

Blood cholesterol is a greasy substance that happens normally in the body and which is essential for hormone creation, cell digestion, and other crucial processes. Having elevated cholesterol level in the blood can expand the danger of heart sicknesses and stroke. Age being the quantity of long stretches of presence from birth to the present time isn't dictated by any factor, it is a mandatory organic change that can't be controlled. Age joins development, advancement, and passing [3]. The more years you spend as a grown-up the less dynamic your body framework moves toward becoming and this is the place age is motivated something to do with circulatory strain identified with the pulse. Age achieves amassing and devaluation and with this more serious danger of therapeutic precariousness, so age realizes an adjustment in the circulatory strain of man as it is appeared in this paper.

2 Data and Methodology

This exploration is gone for seeing if Age and Blood cholesterol has any impact on the pulse. Information for the examination were gotten from JS Memorial Hospital.

Relapse examination was utilized in this investigation. We broke down information containing in excess of two straight out factors by utilizing numerous relapse methodology. PASW (18) was utilized in evaluating the parameters of the model [2]. Relapse models are utilized when the conduct of one variable is 'clarified' by changes in different factors. A factual model assesses the characteristic changeability of the information: this will emerge in view of estimating or observational mistakes or chance conditions.

3 Materials and Methods

3.1 Machine Learning

Machine adapting more often than not alludes to the adjustments in frameworks that perform assignments related with man-made consciousness (AI) [2]. Such assignments include acknowledgment, analysis, arranging, robot control, expectation, and so forth (Fig. 1).

4 Classical Programming Versus Machine Learning

In classical programming, the paradigm of symbolic man-made intelligence, people input guidelines (a program) and information to be handled by these tenets, and result answers. With machine learning, people input information just as the appropriate responses anticipated from the information, and result the standards [2]. These standards would then be able to be connected to new information to create unique answers. Machine learning framework is prep instead of expressly customized (Figs. 2 and 3).

5 Types of Machine Learning

In this segment, we will investigate the three sorts of machine learning: directed learning, unsupervised learning, and fortification learning (Fig. 4).

Supervised Learning:

The primary objective of managed learning is to take in a model from named preparing information that enables us to make expectations about inconspicuous

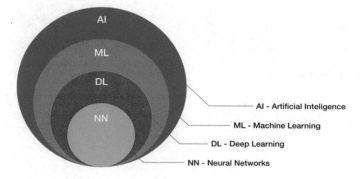

Fig. 1 Technologies

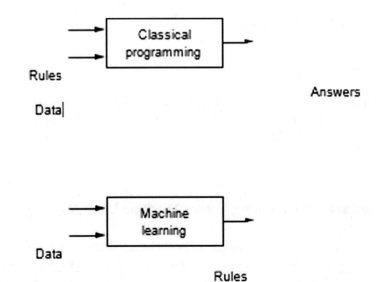

Fig. 2 Rules

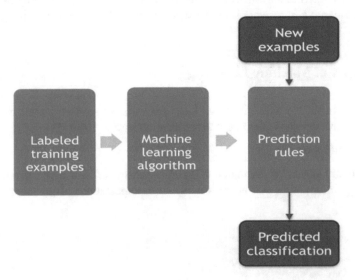

Fig. 3 Block diagram

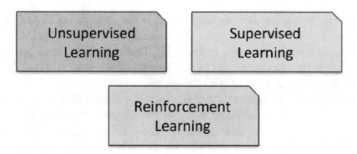

Fig. 4 Types of machine learning

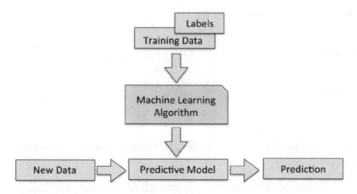

Fig. 5 Supervised learning

or future information [4]. Here, the term regulated alludes to a lot of tests where the ideal yield signals (names) are as of now known (Fig. 5).

6 Classification (Supervised Learning)

Order is the way toward finding or finding a model (work) which helps in isolating the information into numerous downright classes.

In characterization, the gathering participation of the issue is recognized, which implies the information is arranged under various names as indicated by a few parameters and afterward the marks are anticipated for the information [2].

The fundamental objective of arrangement is to anticipate the objective class (Yes/No).

On the off chance that the prepared model is for anticipating any of two target classes, it is known as parallel grouping. Considering the understudy profile to foresee whether the understudy will pass or fizzle [5]. Thinking about the client, exchange points of interest to foresee whether he will purchase the new item or not. These kind of issues will be tended with paired arrangement.

In the event that we need to anticipate increasingly the two target classes it is known as multi-order. Considering every single subject detail of an understudy to anticipate which subject the understudy will score more. Recognizing the question in a picture. These kinds of issues are known as multi-grouping issues.

7   Classification (Supervised)

Mathematically,

$$O = if(I)i + IE$$

Here,

F will be the relation between the marks and number of hours the student prepared for an exam.

I is the iINPUT (NUMBER of hours he prepared), O is the output (Marks the student scored in the exam).

E iwill ibe irandom ierror...

The ultimate goal of the supervised learning algorithm is to predict iO with the maximum accuracy for a given new input iI i [6]. iThere are several ways to implement supervised Ilearning.

8   Regression (Supervised Learning)

If the desired yield contains somewhere around one steady factors, by then the task is called backslide. Backslide is the path toward finding a model or limit with respect to perceiving the data into relentless honest to goodness characteristics rather than using classes. Numerically, with a backslide issue, one is endeavoring to find the limit estimation with the base misstep deviation [7]. In backslide, the data numeric dependence is foreseen to remember it. The essential target of backslide figurings is envisioning, the discrete or a return with regard. 2.7.1 Regression (Supervised Learning) Relapse models include the accompanying parameters and factors:

- The obscure parameters, meant as B, which may speak to a scalar or a vector.
- The free factors I.

- The subordinate variable O.
- A relapse demonstrate relates O to a component of I and B.
- $O = f\,(I, B)$

9 Technologies Used

Bivariate Linear Regression

The tip amount is dependent on the total bill amount so the tip amount is the dependent variable and the bill amount is the independent variable (Figs. 6 and 7).

Correlation

$$R = (n(\textstyle\sum xy) - (\sum x)(\sum y))/\mathbf{Sqrt}([n(\sum x^2) - (\sum x)^2][n(\sum y^2) - (\sum y)^2])$$

Elucidating insights and the centroid play a noteworthy come in relapse line so the main thing we need to do is locate the mean of every factor so our normal bill sum or a mean bill sum with 74 dollars for the tips it was $10 we'll take our mean of the bills of 74 dollars then we'll take the mean of the tips which was $10 and we'll really diagram a point there now is vital and it's known as the centroid and here's the reason it's imperative the best fit or the slightest squares relapse line will or should go through the centroid so whatever our relapse line happens to be it needs to experience the centroid which is included the mean of the X-variable and the mean of the Y-variable [8] (Fig. 8).

Total bill ($)	Tip amount ($)
34.00	5.00
108.00	17.00
64.00	11.00
88.00	8.00
99.00	14.00
51.00	5.00

Fig. 6 Total bill versus tip amount

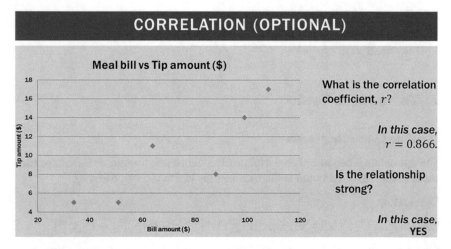

Fig. 7 Bill versus tip amount

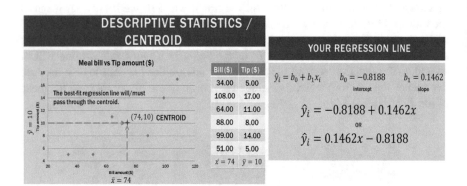

Fig. 8 Centroid

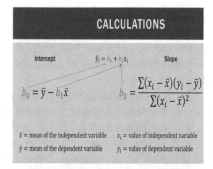

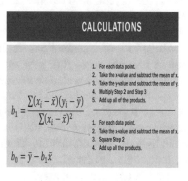

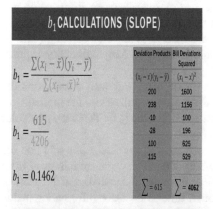

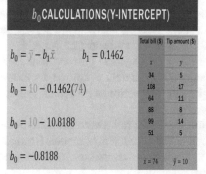

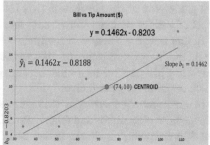

9.1 Technologies Used

Multivariate Regression

Estimations are utilized in the answer for information depiction and prompting. Inferential encounters are utilized to answer ask for regarding the information, to test theories (orchestrating the decision or invalid hypotheses), to make a degree of impact, regularly a degree of rates or hazards, to portray (affiliations), or to demonstrate affil-iations (fall away from the faith) inside the information and, in different particular cutoff points [7]. In the estimation framework, the sporadic slip isn't avoidable. One approach to manage address will be to figure p-values for a degree of conceivable parameter respects (counting the invalid) [9].

The degree of qualities for which the p-respect beats a predestined alpha dimen-sion (every now and again 0.05) is called confirmation interim. An interim estimation system will, in 95% of accentuations (indistinct examinations in all regards with the exception of unusual oversight), pass on limits that contain the true-blue parame-ters. Frequentist approaches choose surveys by utilizing probabilities of information (either p-qualities or probabilities) as degrees of closeness among information and theories, or as degrees of the relative help that information gives hypotheses [10]. Another reasoning, the Bayesian, utilizes information to enhance existing (earlier) assesses in light of new information. Genuine utilization of any rationality requires mindful understanding of statistics 1, 2.

The target in any data examination is to isolate from unrefined information the careful estimation. A choice to answer this demand is to utilize lose the faith exami-nation recalling a definitive target to display its relationship [11]. There are particular sorts of losing the faith examination.

The kind of the break faith show relies on the sort of the dissipating of Y; on the off chance that it is endless and around ordinary we utilize organize fall away from the faith represent; if dichotomous we utilize discovered apostatize; if Poisson or multinomial we use log-straight examination; if time-to-occasion information inside observing modified cases (survival-type) we use Cox descend into sin as a system for appearing. By demonstrating we try to anticipate the result (Y) in light of estimations of a game-plan of marker factors (Xi) [10]. These systems enable us to survey the effect of various segments (covariates and factors) in the equivalent model 3, 4.

In this article we center in facilitate lose the faith. Organize descend into sin is the methodology that estimates the coefficients of the straight condition, including something like one free factor that best imagines the estimation of the reliant variable which ought to be quantitative.

In most honest gatherings, a bend estimation approach produces contort estimation lose the faith estimations and related plots for a broad assortment of models (straight, logarithmic, pivot, quadratic, cubic, control, S-wind, discovered, exponential, and whatnot.). It is essential to plot the information with a specific extreme target to understand which model to use for each dependent variable [12]. On the off chance that the factors show up, obviously, to be related explicitly, a reasonable direct fall away from the faith model can be utilized yet for the situation that the components are not straightly related, information change may help. In the event that the change does not help then a progressively tangled model might be required. It is decidedly taught to see early a scatterplot as for your information; if the plot takes after a numerical point of confinement you see, fit the information to that sort of model [7]. For instance, if the information take after an exponential breaking point, an exponential model is to be utilized. Of course, on the off chance that it isn't clear which show best fits the information, an alternative is to attempt two or three models and select among them.

It is positively embraced to screen the information graphically (for example, by a scatterplot) to pick how the free and ward factors are related (direct, exponentially, etc.) 4–6.

The most sensible model could be a straight line, a higher degree polynomial, a logarithmic, or exponential. The techniques to locate a proper model join the forward framework in which we begin by enduring the specific clear model, for example, a straight line ($Y = a + bX$ or $Y = b0 + b1X$) [12] By then we locate the best check of the ordinary model. In the event that this model does not fit the information satisfying, by then we expect an increasingly cluttered model, for example, a second-degree polynomial ($Y = a + bX + cX2$) and so forth. In a retrogressive strategy we expect a trapped show, for example, a high degree polynomial, we fit the model and we attempt to disentangle it. The clearest case of direct lose the faith examination is that with one marker vary variable 6, 7.

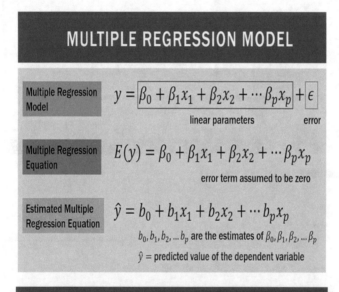

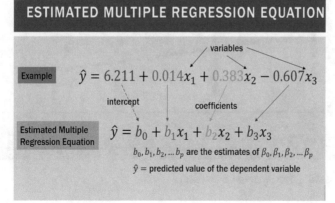

9.2 Results and Screenshots of Graph

See Figs. 9, 10, 11, 12, 13 and 14.

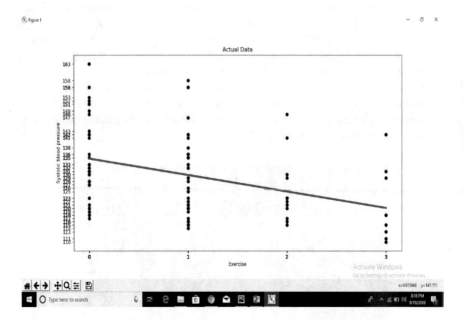

Fig. 9 Actual graph (systolic vs. exercise)

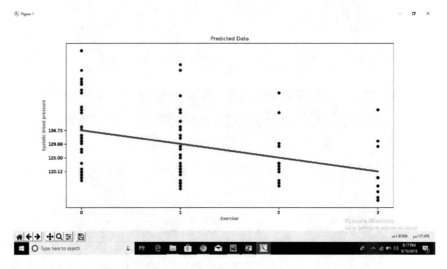

Fig. 10 Predicted graph (systolic vs. exercise)

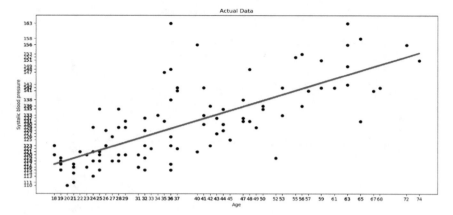

Fig. 11 Actual graph (systolic vs. age)

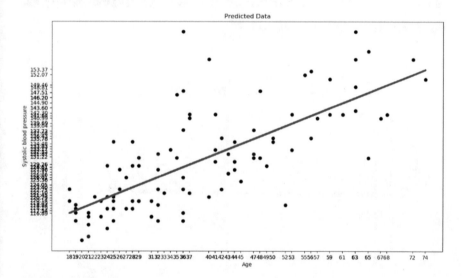

Fig. 12 Predicted graph (systolic vs. age)

10 Conclusion

In light of the results obtained we induce that there exists an important association between age and blood cholesterol on systolic circulatory strain. It could be seen from the examination that the circulatory strain of a man is related to his age and blood cholesterol. The association between the circulatory strain, blood cholesterol, and age using least square backslide procedure has been taken care of to a sensible degree [11]. Government at each level should hone their subjects on the need to decrease any sustenance that can manufacture the level of blood cholesterol in the body by

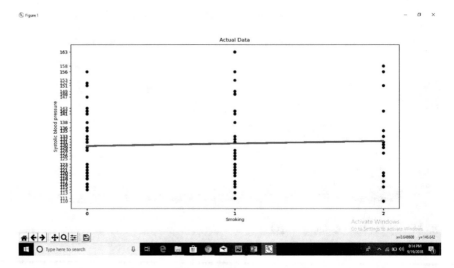

Fig. 13 Actual graph (systolic vs. smoking)

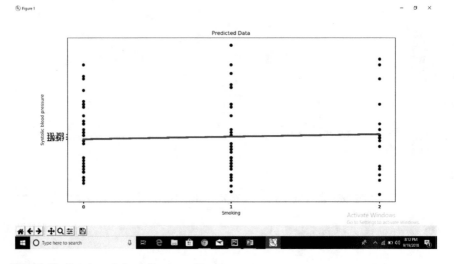

Fig. 14 Predicted graph (systolic vs. smoking)

methods for oily sustenance. It is doubtlessly knowing from this examination that the level of circulatory strain scrutinizing in the body can be influenced by age and blood cholesterol, thusly it is fitting for every individual beyond 18 years old to go for restorative enlistment at any rate once in 3 months. Along these lines, if there exist any changes in their body system, it can without quite a bit of a stretch be taken after and be supervised.

References

1. Kannel, W. B., et al. (1972). Role of blood pressure in the development of congestive heart failure: The Framingham study. *New England Journal of Medicine, 287*(16), 781–787.
2. Stevens, S. L., et al. (2016). Blood pressure variability and cardiovascular disease: systematic review and meta-analysis. *BMJ, 354*, i4098.
3. Chobanian, A. V., et al. (2003). The seventh report of the joint national committee on prevention, detection, evaluation, and treatment of high blood pressure: The JNC 7 report. *JAMA, 289*(19), 2560–2571.
4. Izzo, J., Levy, D., Black, H. R. (2000). Clinical warning explanation. Significance of systolic circulatory strain in more established Americans. *Hypertension, 35*, 1021–1024 [PubMed].
5. Jiao, X., Fang, X. (2002). Research on continuous measurement of blood pressure via characteristic parameters of pulse wave. *Journal of Biomedical Engineering, 2*.
6. Nathanson, B. H., & Higgins, T. L. (2008). An introduction to statistical methods used in binary outcome modeling. In *Seminars in cardiothoracic and vascular anesthesia* (Vol. 12, No. 3). Sage CA: Los Angeles, CA: SAGE Publications.
7. Martel, E., et al. (2013). Comparison of high-definition oscillometry—A non-invasive technology for arterial blood pressure measurement—With a direct invasive method using radio-telemetry in awake healthy cats. *Journal of Feline Medicine and Surgery, 15*(12), 1104–1113.
8. Yuansheng, L. (2007). New blood pressure measure method based on characteristic point. In *2007 8th International Conference on Electronic Measurement and Instruments*. IEEE.
9. Allen, J., & Murray, A. (1999). Modelling the relationship between peripheral blood pressure and blood volume pulses using linear and neural network system identification techniques. *Physiological Measurement, 20*(3), 287.
10. Golino, H. F., et al. (2014). Predicting increased blood pressure using machine learning. *Journal of Obesity, 2014*.
11. Pulido, M., Melin, P., & Prado-Arechiga, G. (2019). Blood pressure classification using the method of the modular neural networks. *International Journal of Hypertension, 2019*.
12. Lacson, R. C., et al. (2018). Use of machine-learning algorithms to determine features of systolic blood pressure variability that predict poor outcomes in hypertensive patients. *Clinical Kidney Journal, 12*(2), 206–212.

User Growth-Based Reliability Assessment of OSS During the Operational Phase Considering FRF and Imperfect Debugging

Vibha Verma, Sameer Anand, and Anu G. Aggarwal

1 Introduction

Software's role in smooth functioning and productivity enhancement of almost every sector cannot be denied. The advancements in computing capabilities of devices have proved to be one of the significant reasons for introduction of software to time-critical systems. OSS has brought major breakthrough in development trend for large scale software systems. For such software the prototype is developed and is evolved with voluntary contribution across the globe. The development pattern followed by OSS doesn't rigidly comply with standards described under traditional software engineering lifecycle. With enormous rise in software demand, IT industry has become highly competitive. The quality is the most critical characteristic that ensures sustainability of software in market and reliability is an important metrics to quantify system's quality. In past decades numerous researches have been carried out in the direction of reliability modelling and assessment [1].

V. Verma (✉) · A. G. Aggarwal
Department of Operational Research, University of Delhi, Delhi, India
e-mail: vibhaverma.du.aor@gmail.com

A. G. Aggarwal
e-mail: anuagg17@gmail.com

S. Anand
Department of Computer Science, Shaheed Sukhdev College of Business Studies, University of Delhi, Delhi, India
e-mail: sananddu@gmail.com

© The Author(s), under exclusive license to Springer Nature Singapore Pte Ltd. 2021   293
P. K. Kapur et al. (eds.), *Advances in Interdisciplinary Research in Engineering and Business Management*, Asset Analytics,
https://doi.org/10.1007/978-981-16-0037-1_23

1.1 The Concept of User Growth in OSS

Open Source Software (OSS) varies characteristically from closed source software. The development lifecycle of OSS is described in Fig. 1. Features such as irregular volunteer participation, lack of hierarchical management, no delivery schedules, and lack of rigid SDLC principles makes it different from closed source systems. OSS development is initiated with an idea or concept and a working prototype is worked out by few developers or small team. The core system with limited functionality is then released over the internet along with source code for further enhancement and refinement with global volunteer participation. The most distinguishing feature over closed source software is that this is released with very little functionality testing and therefore it is the user population who adopt such systems become the deciding factor in reliability growth phenomenon for such software. Figure 1 shows that user reports bug in bug tracking system after the release and thereby explicate the significance of user growth-based reliability modelling.

Li, Li, Xie and Ng [2] discussed the volunteer participation in testing of software after release and suggested that an initial rise is observed in adopter population due to attractiveness of software which reaches a peak and then slowly declines with loss of attractiveness of software. Based on this argument they assumed fault detection rate to follow a hump-shaped curve. Alhazmi and Malaiya [3] related user growth with vulnerability detection phenomenon and presented three phases of usage environment changes in terms of population. The phases discussed are learning phase, where people like hackers work on to learn or understand the system followed by an increasing trend in user population attracted towards new challenges in system. After

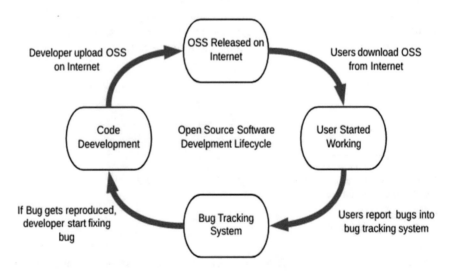

Fig. 1 Process of OSS development

reaching a peak, termed as linear phase, a saturation phase is observed which marks the decline in interest of users in software due to introduction of newer systems.

1.2 Modelling User Growth

Bass [4] has given the mathematical formulation for the phenomenon of Diffusion of Innovation given by Rogers [5]. This theory explains the user growth concept and Bass model described adoption process in terms of innovators and imitators. Those who adopt a product on their own without seeking any reviews regarding the product from others are termed as innovators. People who take up the product based on review/ratings given by others are defined as Imitators. The expression due to Bass is given in Eq. (1) describes both the processes. The model assumed the number of innovators to be proportional to the remaining people in population and p is called the coefficient of innovation. The imitators are assumed to be proportional to the remaining population by a factor of people already adopted and the coefficient of proportionality is termed as Imitation coefficient and is denoted by q. This model explained user growth to take hump-shape curve.

$$\frac{dN(t)}{dt} = \left(p + q\frac{N(t)}{\bar{N}}\right)\left(\bar{N} - N(t)\right) \tag{1}$$

where $N(t)$ is the number of users, $\bar{N}$ is the number of potential users. Solving Eq. (1) using $N(0) = 0$ gives S-shaped cumulative adopter distribution (Eq. 2)

$$N(t) = \bar{N}\left(\frac{1 - e^{-(p+q)t}}{1 + \frac{q}{p}e^{-(p+q)t}}\right) \tag{2}$$

As per Fig. 1, after the release over Internet, users download it and the process of debugging gets escalated.

1.3 The Process of Debugging

The OSS keeps on evolving due to the contributions from volunteers across the world. The software is developed by core team with few members and released over Internet. Later, with time core developers and volunteers form a community. The OSS community generally follows a layered architecture which describes the users into various categories like core members, active developers, peripheral developers, bug fixer, bug reporter, passive users, etc. from core to periphery. The users are assigned participation levels which may change with time and contributions. People start working on software and as and when they find some glitch in software workflow, they report it as failure and the fault is identified. After fault identification developers try to remove the fault. This process is called debugging and it is further categorized as perfect debugging and imperfect debugging.

After the identification of a fault, the process of rectification starts and the phenomenon of correction in which no further error generation takes place are called perfect debugging in contrast to the imperfect debugging which refers to the phenomenon of generation of addition faults while rectifying the previous ones. With the growth in number of users, the community for OSS grows and the process of bug reporting and rectification gets accelerated which correspondingly increases error generation. Therefore fault content function can be assumed to act as function of usage factor of OSS. In this study, we assume the rate of introduction of new faults to be based on user growth as a function of faults that have been already removed from the software and is expressed through Eq. (3).

$$a(N) = a + \alpha m(N) \tag{3}$$

where a is the initial number of faults present in the system, α is the rate of error generation, and $m(N)$ is the Mean Value Function (MVF), i.e., expected number of faults removed when number of users is $N(t)$.

The relation of net fault removed to faults experienced can be associated with user growth phenomenon. It has been explained in the following section.

1.4   User Growth-Based FRF

Fault Reduction Factor (FRF) is proportion of total number of faults rectified to the total failures experienced [6]. It denotes the time difference between fault encounter and its correction. It has been observed in realistic situations that after a fault being reported, it takes some time to correct the reported fault. Also many times the development team is not able to remove all the acknowledged faults due to various environmental conditions.

This may be due to the complexity associated with the code due to its size. It is also affected by the time it takes to detect the actual cause of the failure and debug it. There the notion of FRF comes into existence. In context of open source systems, this ratio can be expressed in terms of user growth and assumed to follow hump-shaped curve with the adopter population. With the rise in user population FRF is expected to increase till it reaches a peak value and then slowly decreases with fall in user population.

OSS has proved to be a very cost-effective solution for sectors like education, healthcare, agriculture, etc. and a huge shift towards these has been observed in recent past. As a particular OSS rolls out and become popular, reliability assessment becomes extremely important so as to avoid any kind of potential loss to the adopting customers. This necessitates reliability assessment and prediction for such systems.

In this paper we have developed a user growth-based SRGM incorporating FRF and Imperfect debugging. We have considered that the number of faults is generated with a constant rate. The rest of the paper is structured as section two discusses the brief literature survey followed by model development in section three. In the next section the proposed model is validated on two real-life fault dataset of Gnome and Firefox. Finally conclusions are drawn in the last section.

2 Literature Survey

Reliability assessment is fundamentally a five-step process as depicted by Fig. 2 and has been a very sought-after area of research in past few decades and numerous SRGMs have been proposed to assess the reliability of software systems. An SRGM expresses the software failure process mathematically with certain parameters and factors. Such mathematical representation is based on few assumptions and one of the very popular assumption is that software fault process follows non-homogeneous Poisson distribution with Mean MVF, $m(t)$, that categorizes models into a class known as non-homogenous Poisson process-based SRGMs.

Some of the very initial models in this category were Goel and Okumoto model [7], Delayed S-shaped model [8, 9], Inflection S-shaped model [10], etc. Numerous studies have been carried out for assessing the reliability of OSS systems. Tamura and Yamada [11] investigated reliability assessment combining it with neural networks. Some of the latest SRGMs proposed to assess reliability growth of software systems

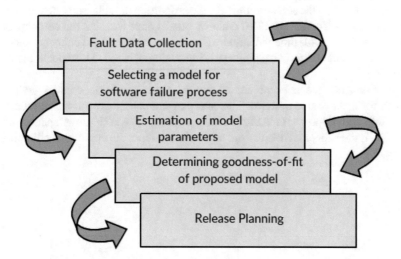

Fig. 2 Reliability modelling process

incorporating practical phenomenon are Chatterjee and Shukla [12], Li Q and Pham [13], Zhu and Pham [14].

In [11], stochastic differential equation-based modelling is proposed for OSS and further discussed its application in optimal version update problem. Li et al. [2] presented another model to predict optimal version update time for OSS. Yang et al. [15] included process of fault detection and correction and performed reliability modelling for multi-release OSS. Many other studies tried to model reliability for OSS. Rahmani et al. [16, 17] and Zhou and Davis [18] studied OSS reliability model by experiments and also performed the comparative analysis.

Kapur and Younes [19] presented the SRGM framework under imperfect debugging. Pham, Nordmann, and Zhang [20] modelled the reliability growth of system corresponding to S-Shaped fault detection rate. Kapur, Pham, Anand, and Yadav [21] proposed SRGMs incorporating error generation under imperfect debugging. Lin [22] studied the effects of imperfect debugging phenomenon on faults removed by means of simulation-based model. Jain, Manjula, and Gulati [23] also considered the imperfect debugging phenomenon in their proposed SRGM along with FRF and multiple change points.

Various other researches have incorporated FRF into reliability assessment and enhancement in estimation capabilities of models has been experienced. Observing the relationship between faults to failures Musa [6] introduced the concept of FRF and included it as a constant in proposed SRGM. Later, considering the environmental influence over FRF, studies started considering it as time-variable function. Chatterjee and Shukla [12] discussed a Weibull-type FRF for software debugged under imperfect conditions. Anand, Verma, and Aggarwal [24] proposed a FRF-based 2-D SRGM under imperfect debugging for software released in multiple versions. Hsu, Huang, and Chang [25] discussed three patterns in FRF, i.e., constant, increasing, and decreasing. It was suggested that due to the human learning process, testers over time get acquainted with the software and thus the proportion of faults corrected to failures may start rising (Increasing FRF) whereas other belief was that the complexity of bugs encountered with time increase and because of this the time delay between the detection of fault and its correction starts increasing, which leads to declining trend in FRF.

We extended this concept and expressed these trends in context with user growth for open source systems. This study explores user growth-based reliability modelling and proposes three SRGM incorporating three FRF growth patterns, i.e., constant, increasing, and decreasing, respectively, under the conditions of imperfect debugging.

3 Proposed Model

The rest of the notations used for mathematical model are given as.

t: Calendar time.
$m, m (N)$:   Expected number of faults removed when number of users are $N(t)$
$N, N (t)$:   Cumulative number of users in $(0, t]$
b: Fault removal rate.
$a (N)$:   User growth-based fault content function.
$r (N)$:   User growth-based Fault detection rate.
$B(N)$:   User growth-based Fault reduction factor.

The proposed model follows underlying assumptions:

1. Software failure process follows NHPP.
2. The number of failures occurring after software execution is proportionate to the number residual number of faults.
3. Any instance of unexpected behaviour is considered as a failure and the responsible fault is identified. The possibility of error generation while fixing of a previous fault is incorporated.
4. Fault removal is defined in terms of number of users using the software.
5. Users are represented by the diffusion model [4].

The failure process is expressed as follows:

$$\frac{dm}{dt} = \frac{dm}{dN}\frac{dN}{dt} \tag{4}$$

which is combination of two components.

First Component $\frac{dm}{dN}$ describes the detection rate in terms of users. The differential equation to describe the failure process is given by

$$\frac{dm}{dN} = r(N)(a(N) - m) \tag{5}$$

$$r(N) = rB(N) \tag{6}$$

$a(N)$ is given by expression in Eq. (3). The three cases of FRF are discussed in Table 1.

Second Component The second component $\frac{dN}{dt}$ denotes the concept of adoption population growth for OSS described by Bass Model [4]. Bass expressed potential customers based on innovators and imitators as given in Eq. (1). The solution of this equation for initial condition $N(0) = 0$ is given by Eq. (2).

Model-1: Considering case1 expression for constant FRF given in Table 2, and substituting Eq. (2), (3) (5), (6) in Eq. (4) and applying chain rule, we obtain the expression for Mean Value Function (MVF) as

Table 1 FRF expressions and interpretations

Case	Interpretation	Expression
Case1 (Constant FRF)	FRF is assumed to be a constant	$B(N) = B$
Case2 (Increasing FRF)	With time, the process of human learning takes place which results in rising FRF	$B(N) = 1-(1-B_0)e^{-kN(t)}$
Case3 (Decreasing FRF)	The increase in complexity of faults or the impact of environmental factors lead to decreasing FRF	$B(N) = B_0\, e^{-kN(t)}$

Table 2 Dataset information

Dataset	Calendar Time (Weeks)	Detected Bugs
DS-1 (Gnome 2.0)	24	85
DS-2 (Firefox 3.0)	53	65

$$m(t) = \frac{a}{\alpha - 1}\left(1 - e^{rB(\alpha-1)N(t)}\right) \tag{7}$$

Model-2: Considering case2 expression for increasing FRF given in Table 2, and substituting Eq. (2), (3), (5), (6) in Eq. (4) and applying chain rule, we obtain the expression for MVF as

$$m(t) = \frac{a}{1-\alpha}\left(1 - e^{-r(1-\alpha)\left(N(t)+\frac{\left(1-e^{-kN(t)}\right)(B_0-1)}{k}\right)}\right) \tag{8}$$

Model-3: Considering case3 expression for decreasing FRF given in Table 2, and substituting Eq. (2), (3), (5), (6) in Eq. (4) and applying chain rule, we obtain the expression for MVF as

$$m(t) = \frac{a}{1-\alpha}\left(1 - e^{-r\frac{(1-\alpha)B_0(1-e^{-kN(t)})}{k}}\right) \tag{9}$$

4 Model Validation

In this section we validate our model on two real-life fault datasets of Gnome 2.0 [1] and Firefox 3.0 [15]. The numerical example helps to illustrate the procedure. Both the datasets are obtained from Literature. Table 2 gives the information about datasets used for empirically judging the model capabilities.

Table 3 Estimated parameters

DS-1	a	B	r	α	k	N	p	q
Model-1	95.781	0.021	0.228	0.448	-	215.514	0.003	0.056
Model-2	87.567	0.160	0.092	0.638	0.012	181.23	0.022	0.113
Model-3	98.229	0.111	0.156	0.755	0.078	115.543	0.103	0.212
DS-2	a	B	r	α	k	N	p	q
Model-1	73.347	0.110	0.081	0.260	-	248.069	0.141	0.282
Model-2	65.496	0.122	0.037	0.388	0.087	220.996	0.231	0.462
Model-3	65.015	0.117	0.065	0.389	0.015	222.978	0.243	0.486

4.1 Parameter Estimation

Least square estimation technique is used for estimating parameters of the model using statistical tool SPSS. The estimated parameters of each model for both the datasets are shown in Table 3.

4.2 Goodness-of-Fit Evaluation

It is extremely important to determine the model performance after getting the estimation results. Goodness-of-fit criteria are appropriate indicators of prediction potential of model. In this study, we have used four fit indicators (MSE, PRR, PP, and R^2). The values obtained for comparison criteria discussed above are given in Table 4 and corresponding goodness-of-fit curves are shown in Fig. 3.

Table 4 Goodness of fit comparison

DS-1	R^2	MSE	PRR	PP
Model-1	0.992	2.331	0.039	0.041
Model-2	0.997	2.204	0.106	0.091
Model-3	0.998	2.166	0.065	0.073
DS-2	R^2	MSE	PRR	PP
Model-1	0.990	8.631	14.766	12.551
Model-2	0.994	5.789	9.541	8.601
Model-3	0.993	6.352	9.893	9.684

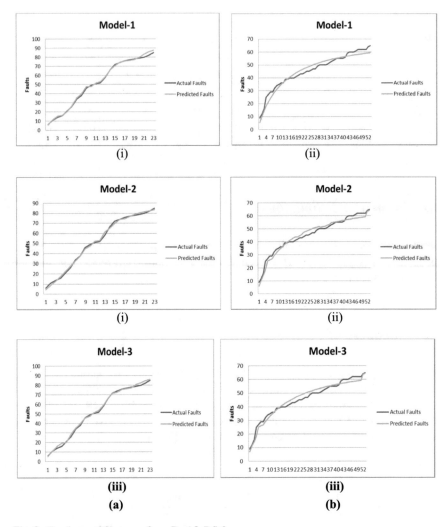

Fig. 3 Goodness of fit curves for **a** Ds-1 **b** DS-2

5 Discussions on Results

From the estimation results presented in Table 4, it can be concluded that user based software reliability growth models proposed in this study are competent predictors of data and thus are appropriate for reliability assessment. The values indicate that Model-2 and Model-3 modelled incorporating increasing and decreasing FRF trends are fitting the fault process better than Model-1 which portrays constant FRF and this is consistent to realistic situation of impact of environmental factors, human learning on FRF. Further it can be seen that value of p is less than value of q which signifies the role of spoken communication in the adoption of software by the users.

6 Conclusions

This study presented three NHPP-based SRGMs incorporating imperfect debugging situation and fault reduction factor. It is assumed that additional faults added are a function of already fixed faults. User growth-based fault content function and FRF have been combined with proposed SRGMs. Based on three trends of FRF observed in literature three SRGMs namely Model-1, Model-2, and Model-3 are proposed. Models are evaluated over fault dataset of GNOME 2.0 and Firefox software. The prediction abilities are concluded based on goodness-of-fit values calculated. The study can be extended in future for inclusion of change point or multiple change points in SRGM.

References

1. 1Kapur, P., Pham, H., Gupta, A., & Jha, P. (2011). *Software reliability assessment with OR applications*. Springer.
2. Li, X., Li, Y. F., Xie, M., & Ng, S. H. (2011). Reliability analysis and optimal version-updating for open source software. *Information and Software Technology, 53*(9), 929–936.
3. Alhazmi, O. H., & Malaiya, Y. K. (2008). Application of vulnerability discovery models to major operating systems. *IEEE Transactions on Reliability, 57*(1), 14–22.
4. Bass, F. M. (1969). A new product growth for model consumer durables. *Management Science, 15*(5), 215–227.
5. Rogers, E. M., Diffusion, O. I. (1962). *New York: The Free Press of Glencoe.*
6. Musa, J. D. (1975). A theory of software reliability and its application. *IEEE Transactions on Software Engineering, 3,* 312–327.
7. Goel, A. L., & Okumoto, K. (1979). Time-dependent error-detection rate model for software reliability and other performance measures. *IEEE Transactions on Reliability, 28*(3), 206–211.
8. Yamada, S., Ohba, M., & Osaki, S. (1984). S-shaped software reliability growth models and their applications. *IEEE Transactions on Reliability, 33*(4), 289–292.
9. Pham, H., & Zhang, X. (1997). An NHPP software reliability model and its comparison. *International Journal of Reliability, Quality and Safety Engineering, 4*(3), 269–282.
10. Ohba, M. (1984). Software reliability analysis models. *IBM Journal of Research and Development, 28*(4), 428–443.
11. Tamura, Y., & Yamada, S. (2008). A component-oriented reliability assessment method for open source software. *International Journal of Reliability, Quality and Safety Engineering., 15*(1), 33–53.
12. Chatterjee, S., & Shukla, A. (2016). Modeling and analysis of software fault detection and correction process through weibull-type fault reduction factor, change point and imperfect debugging. *Arabian Journal for Science and Engineering, 41*(12), 5009–5025.
13. Li, Q., & Pham, H. (2017). A testing-coverage software reliability model considering fault removal efficiency and error generation. *PLoS ONE, 12*(7), e0181524.
14. Zhu, M., & Pham, H. (2017). *A multi-release software reliability modeling for open source software incorporating dependent fault detection process.* Annals of Operations Research, pp. 1–18.
15. Yang, J., Liu, Y., Xie, M., & Zhao, M. (2016). Modeling and analysis of reliability of multi-release open source software incorporating both fault detection and correction processes. *Journal of Systems and Software, 115,* 102–110.
16. Rahmani, C., Siy, H., & Azadmanesh, A. (2009). *An experimental analysis of open source software reliability.* Department of Defense/Air Force Office of Scientific Research.

17. Rahmani, C., Azadmanesh, A. H., & Najjar, L. (2010). A comparative analysis of open source software reliability. *JSW, 5*(12), 1384–1394.
18. Zhou, Y., & Davis, J. (2005). *Open source software reliability model: an empirical approach.* In *ACM SIGSOFT Software Engineering Notes.* ACM.
19. Kapur, P., & Younes, S. (1996). Modelling an imperfect debugging phenomenon in software reliability. *Microelectronics Reliability, 36*(5), 645–650.
20. Pham, H., Nordmann, L., & Zhang, Z. (1999). A general imperfect-software-debugging model with S-shaped fault-detection rate. *IEEE Transactions on Reliability, 48*(2), 169–175.
21. Kapur, P., Pham, H., Anand, S., & Yadav, K. (2011). A unified approach for developing software reliability growth models in the presence of imperfect debugging and error generation. *IEEE Transactions on Reliability, 60*(1), 331–340.
22. Lin, C. T. (2011). Analyzing the effect of imperfect debugging on software fault detection and correction processes via a simulation framework. *Mathematical and Computer Modelling, 54*(11–12), 3046–3064.
23. 23Jain, M., Manjula, T., & Gulati, T. (2012). *Software reliability growth model (SRGM) with imperfect debugging, fault reduction factor and multiple change-point.* In *Proceedings of the International Conference on Soft Computing for Problem Solving (SocProS 2011) December 20–22, 2011.* Springer.
24. Anand, S., Verma, V., & Aggarwal, A. G. (2018). *2-Dimensional Multi-Release Software Reliability Modelling considering Fault Reduction Factor under imperfect debugging. Ingenieria Solidaria, 14.*
25. Hsu, C.-J., Huang, C.-Y., & Chang, J.-R. (2011). Enhancing software reliability modeling and prediction through the introduction of time-variable fault reduction factor. *Applied Mathematical Modelling, 35*(1), 506–521.

Unreliable Server Queue with Balking, Optional Service, Bernoulli Feedback and Vacation Under Randomized Policy

Madhu Jain, Sandeep Kaur, and Mayank Singh

1 Introduction

Queueing modeling and performance analysis have applicability in various congestion situations arising in communication network, manufacturing industry, computer operating system, etc. The reliable server queueing models under varied assumptions can be found in queueing literature. But due to real-life application of queueing model with server breakdown, some researchers have contributed their study towards unreliable server queue also [2, 4, 6]. Due to server breakdown, the output of queueing system may be affected and thus the efficiency of the system is reduced. In order to deal with such situations, queueing system must be supported with repair facility so that delays in service due to server failure can be reduced.

A vacationing server is also a realistic feature observed in many day-to-day queueing situations. The server vacation is a very important economic policy for improving the efficiency and output of system. The server can also utilize the vacation time for secondary jobs such as maintenance purpose. Queueing model with randomized vacation policy (RVP) is studied by various researchers in different contexts [5, 9]. Queueing models with essential and optional service has real-life application in banking, clinical labs, vehicle repair system, etc. Singh et al. [8] analyzed $M^X/G/1$ queueing system with the provision of some realistic features

M. Jain · M. Singh
Department of Mathematics, Indian Institute of Technology Roorkee, Roorkee 247667, India
e-mail: madhufma@iitr.ac.in

M. Singh
e-mail: singh.mayank0617@gmail.com

S. Kaur (✉)
Department of Mathematics, Guru Nanak Dev University, Amristar 143005, India
e-mail: sandeepsaini49@gmail.com

© The Author(s), under exclusive license to Springer Nature Singapore Pte Ltd. 2021   305
P. K. Kapur et al. (eds.), *Advances in Interdisciplinary Research in Engineering and Business Management*, Asset Analytics,
https://doi.org/10.1007/978-981-16-0037-1_24

including optional service, state-dependent arrival rate, and randomized policy with at most m vacations.

In several queueing scenarios, it can be noticed that the customers, who find the incomplete service or unsatisfied with its service, can repeat their service as feedback customer. Moreover, the phenomenon of balking behavior of the customers can also arise in particular when the service interruption occurs due to failure or vacation of the server. Balking can be observed in many congestion situations, namely, emergency department of hospital, where serious patients may balk from the hospital due to unavailability of the doctor, failure of the equipment being used for treatment or heavy crowd of the patients. An unreliable server queue including the concepts of bulk input, general service time, balking, Bernoulli feedback was investigated by Singh et al. [7] using supplementary variable technique (SVT) and maximum entropy principle (MEP). By getting motivation from the real-life applications of queueing models in various fields, our present investigation is devoted to the study of non-Markovian bulk arrival queueing model by considering the features (i) server breakdown, (ii) Bernoulli feedback, (iii) essential/optional service, (iv) randomized vacation policy, (v) balking, (vi) delay repair/repair. The remaining contents of the paper are described in different sections. The brief description of our model is presented in the form of assumptions and notations in Sect. 2. The construction of Chapman Kolomogorov equations with corresponding boundary conditions using supplementary variables is described in Sect. 3. In Sect. 4, queue length distributions in terms of PGFs are derived. In Sect. 5, we present the various performance indices, viz., server status probabilities, mean queue size, reliability indices, and cost function. Numerical illustration is provided in Sect. 6. The conclusion along with salient features of model studied is given in Sect. 7.

2 Model Description

Consider a non-Markovian model with server breakdown and randomized vacation policy. The brief description of various assumptions and notations considered, while developing the model and framing the governing equations are described as given in the following sub-sections.

2.1 Assumptions

The bulk input queue is analyzed by incorporating the features of general distributed service time, feedback, balking, vacation, optional service, and server breakdown. Following are the basic assumptions made for formulating the reference model.

(i)   **Arrival process and balking**: The arrival of units is in group of size (X) following Poisson process with rate λ and $c_j = $ Prob.$[X = j]$. The arriving unit may balk if server is in busy, delayed for repair/under repair and vacation state and joins the system with respective joining probability of each state given by b, b_1 and b_2.

(ii) **Service process:** The unit joining the system requires first essential phase (FEP) service. After getting (FEP) service, the arriving unit may go for second optional phase (SOP) service also, where r_i is the probability with which units can demand for i^{th} ($1 \leq i \leq m$) (SOP) service or can depart the system with probability $r_0 = 1 - \sum_{i=1}^{m} r_i$.

(iii) **Server vacation under randomized policy**: After serving all the accumulated units, when the server becomes free, it can go for the vacation under RVP with at most l vacations. After returning from the vacation, the server will instantly start executing the service if some units are waiting in the queue. In case if no unit is present, the server may remain in a dormant state with probability p or may depart for one more vacation with complementary probability. The server continues this sequence up to l vacations and after returning from lth vacation, the server stays idle if no more units in the system are left for service or otherwise immediately serves the unit from the queue according to FCFS rule.

(iv) **Bernoulli feedback:** If a unit finds its FEP/SOP service unsatisfactory, he can rejoin the queue as feedback unit with probability θ ($0 \leq \theta \leq 1$) for getting FEP/SOP service again or departs from the system with probability $\bar{\theta} = (1 - \theta)$.

(v) **Repair process:** It is considered that the server is failure prone as such random breakdown of it may occur in providing FEP/SOP service. The inter-failure time of the server is exponentially distributed with rate α_i ($0 \leq i \leq m$) while fails during ith phase service. After failure, the server is recovered for service by repair and delay in repair may occur before the actual repair starts.

(vi) The FEP/SOP services, delay to repair and repair processes, vacation times are i.i.d. and general distributed.

The notations used in developing the models are as follows ($i = 0$ for FEP service and $1 \leq i \leq m$ for SOP service):

Notations:	
X	Random variable denoting the batch size
c, $c_{(2)}$	1st and 2nd factorial moments of batch size (X)
N_q	Queue size together with unit in service
$B_i(x)$	CDF (Cumulative distribution function) of FEP/SOP service time
β_i, $\beta_i^{(2)}$	First two moments of $B_i(x)$
$D_i(y), R_i(y)$	CDF of delay in repair time and repair time, respectively, while the server is failed in FEP/SOP service
γ_i, $\gamma_i^{(2)}$	First and second moments of $D_i(y)$

(continued)

(continued)

Notations:	
$g_i, g_i^{(2)}$	First and second moments of $R_i(y)$
$V(x)$	CDF of vacation time
$v, v^{(2)}$	First two moments of $V(x)$
P_0	Probability that there is no customer and the server is idle
$P_{n,i}(x)dx$	Probability of n units in the system when the server is rendering FEP/SOP service and the elapsed service time lies in $(x, x + dx]$
$P_{n,i}^D(x, y)dy$	Probability of n units in the system when the server fails in FEP/SOP service and waiting for repair with elapsed delay time lies in $(y, y + dy]$
$P_{n,i}^R(x, y)dy$	Probability of n units in system when the server fails in FEP/SOP service and under repair; the elapsed repair time lies in $(y, y + dy]$
$P_{n,j}^V(x)dx$	Probability of n units in system when the server is on jth vacation and the elapsed vacation time lies in $(x, x + dx]$
$\widetilde{F}(u)$	Laplace Stieltjes transform (LST) of CDF $F(u)$
$f(u)du$	The hazard rate function of distribution function $F(u)$, where $F(.)$ can take values $B_i(.), V(.), D_i(.), R_i(.)$ with corresponding values of $f(.)$ as $\mu_i(.), v(.), \eta_i(.), \xi_i(.)$

The probability generating functions (PGFs) used for mathematical analysis are as follows:

$$P_i(x, z) = \sum_{n=1}^{\infty} z^n P_{n,i}(x); \ P_i^D(x, y, z) = \sum_{n=1}^{\infty} z^n P_{n,i}^D(x, y); \ P_i^R(x, y, z)$$

$$= \sum_{n=1}^{\infty} z^n P_{n,i}^R(x, y);$$

$$P_j^V(x, z) = \sum_{n=0}^{\infty} z^n P_{n,j}^V(x); \text{ for } 0 \le i \le m \text{ and } 1 \le j \le l.$$

Also above PGFs hold for $x = 0$ and $y = 0$.

2.2 Governing Equations

In accordance with Cox [3], the governing equations under steady-state conditions are obtained for ($i = 0$ for FEP service and $1 \le i \le m$ for SOP service):

$$\lambda P_0 = \int_0^\infty P_{0,l}^V(x)v(x)dx + p\sum_{j=1}^{l-1}\int_0^\infty P_{0,j}^V(x)v(x)dx \qquad (1)$$

$$\frac{d}{dx}P_{n,i}(x) + [\lambda b + \alpha_i + \mu_i(x)]P_{n,i}(x) = \lambda b\sum_{k=1}^n c_k P_{n-k,i}(x) + \int_0^\infty \xi_i(y)P_{n,i}^R(x,y)dy;$$
$$x, y > 0, n \geq 1 \qquad (2)$$

$$\frac{d}{dy}P_{n,i}^D(x,y) + [\lambda b_1 + \eta_i(y)]P_{n,i}^D(x,y) = \lambda b_1\sum_{k=1}^n c_k P_{n-k,i}^D(x,y); x, y > 0, n \geq 1 \qquad (3)$$

$$\frac{d}{dy}P_{n,i}^R(x,y) + [\lambda b_1 + \xi_i(y)]P_{n,i}^R(x,y) = \lambda b_1\sum_{k=1}^n c_k P_{n-k,i}^R(x,y); x, y > 0, n \geq 1 \qquad (4)$$

$$\frac{d}{dx}P_{n,j}^V(x) + (\lambda b_2 + v(x))P_{n,j}^V(x) = \lambda b_2(1-\delta_{n,0})\sum_{k=1}^{n-1} c_k P_{n-k,j}^V(x), x > 0, \qquad (5)$$
$$n \geq 0, 1 \leq j \leq l.$$

The boundary conditions (BCs) at $x = 0$ are

$$P_{n,0}(0) = \lambda c_n P_0 + \sum_{j=1}^l \int_0^\infty P_{n,j}^V(x)v(x)dx$$

$$+\bar{\theta}\left(r_0\int_0^\infty P_{n+1,0}(x)\mu_0(x)dx + \sum_{i=1}^m\int_0^\infty P_{n+1,i}(x)\mu_i(x)dx\right)$$

$$+\theta\left(r_0\int_0^\infty P_{n,0}(x)\mu_0(x)dx + \sum_{i=1}^m\int_0^\infty P_{n,i}(x)\mu_i(x)dx\right), n \geq 1 \qquad (6)$$

$$P_{n,i}(0) = r_i\int_0^\infty \mu_0(x)P_{n,0}(x)dx, 1 \leq i \leq m, n \geq 1 \qquad (7)$$

$$P_{n,1}^V(0) = \delta_{n,0}\left[\bar{\theta}\left(r_0\int_0^\infty P_{n+1,0}(x)\mu_0(x)dx + \sum_{i=1}^m\int_0^\infty P_{n+1,i}(x)\mu_i(x)dx\right)\right.$$

$$\left.+\theta\left(r_0\int_0^\infty P_{n,0}(x)\mu_0(x)dx + \sum_{i=1}^m\int_0^\infty P_{n,i}(x)\mu_i(x)dx\right)\right], n \geq 0 \qquad (8)$$

$$P_{n,j}^V(0) = \delta_{n,0}\bar{p}\int_0^\infty v(x)P_{n,j-1}^V(x)dx, 2 \leq j \leq l, n \geq 0. \qquad (9)$$

The BCs at $y = 0$ are

$$P_{n,i}^D(x, 0) = \alpha_i P_{n,i}(x), \quad n \geq 1, 0 \leq i \leq m \tag{10}$$

$$P_{n,i}^R(x, 0) = \int_0^\infty \eta_i(y) P_{n,i}^D(x, y)dy, \quad n \geq 1, 0 \leq i \leq m. \tag{11}$$

Also

$$P_0 + \sum_{i=0}^m \sum_{n=1}^\infty \left[\int_0^\infty P_{n,i}(x)dx + \int_0^\infty \int_0^\infty P_{n,i}^D(x, y)dxdy + \int_0^\infty \int_0^\infty P_{n,i}^R(x, y)dxdy \right]$$

$$+ \sum_{j=1}^l \sum_{n=0}^\infty \int_0^\infty P_{n,j}^V(y)dy = 1. \tag{12}$$

3 Queue Length Distribution

In this segment, we describe the queue length distribution in terms of PGFs in the following theorems obtained on solving Eqs. (1)–(12).

Theorem 1 The joint PGFs of the queue length distribution are given as

$$P_0(x, z) = U(z) \exp\{-\tau_0(z)x\} \overline{B_0(x)}(\bar{\theta} - \chi_2)(\sigma S(z))^{-1} \tag{13}$$

$$P_i(x, z) = r_i U(z) \exp\{-\tau_i(z)x\} \widetilde{B}_0(\tau_0(z)) \overline{B_i(x)}(\bar{\theta} - \chi_2)(\sigma S(z))^{-1}, 1 \leq i \leq m \tag{14}$$

$$P_i^D(x, y, z) = \alpha_i P_i(x, z) \exp\{-\phi_3(z)y\} \overline{D_i(y)}, \quad 0 \leq i \leq m \tag{15}$$

$$P_i^R(x, y, z) = \alpha_i P_i(x, z) \widetilde{D}_i(\phi_3(z)) \exp\{-\phi_3(z)y\} \overline{R_i(y)}, \quad 0 \leq i \leq m \tag{16}$$

$$P_j^V(x, z) = \lambda(\bar{\theta} - \chi_2)(\bar{p}\chi_0)^{j-1}(1 - \bar{p}\chi_0) \exp(-\phi_4(z)x) \overline{V(x)}\sigma^{-1}, 1 \leq j \leq l \tag{17}$$

where

$$\overline{F(u)} = 1 - F(u), \overline{\widetilde{F}(u)} = 1 - \widetilde{F}(u)$$

$$\phi_1(z) = \lambda \overline{X(z)}, \phi_2(z) = \lambda b \overline{X(z)}, \phi_3(z) = \lambda b_1 \overline{X(z)}, \phi_4(z) = \lambda b_2 \overline{X(z)},$$

$$\tau_i(z) = \phi_2(z) + \alpha_i\{1 - \tilde{D}_i(\phi_3(z))\tilde{R}_i(\phi_3(z))\}, \quad 0 \le i \le m$$

$$U(z) = z\{\chi_1\phi_1(z) + \lambda \overline{\tilde{V}(\phi_4(z))}(1 - (\bar{p}\chi_0)^l)\}$$

$$S(z) = (\bar{\theta} + \theta z)\tilde{B}_0(\tau_0(z))[r_0 + \sum_{i=1}^{m} r_i \tilde{B}_i(\tau_i(z)) - z$$

$$\chi_0 = \tilde{V}(\lambda b_2), \quad \chi_1 = \chi_0\{(\bar{p}\chi_0)^{l-1}(1 - \bar{p}\chi_0) + p(1 - (\bar{p}\chi_0)^{l-1})\}$$

$$\chi_2 = \lambda c[\beta_0(b + h_0 b_1) + \sum_{i=1}^{m} r_i \beta_i(b + h_i b_1)],$$

$$\chi_3 = \lambda c[\beta_0(1 + h_0) + \sum_{i=1}^{m} r_i \beta_i(1 + h_i)]$$

$$h_i = \alpha_i(\gamma_i + g_i); 0 \le i \le m,$$
$$\sigma = \chi_1(\bar{\theta} - \chi_2 + \chi_3) + \lambda v(1 - (\bar{p}\chi_0)^l)(\bar{\theta} - \chi_2 + b_2\chi_3).$$

Proof On multiplying Eqs. (1)–(5) by appropriate powers of z and using boundary Conditions (6)–(12), we obtain the required results (cf. [8])Using the following equation

.

Theorem 2 The marginal PGFs of the queue length distribution are

$$P_0(z) = U(z)\overline{\tilde{B}_0(\tau_0(z))}(\bar{\theta} - \chi_2)(\sigma S(z)\tau_0(z))^{-1} \tag{18}$$

$$P_i(z) = r_i U(z)\tilde{B}_0(\tau_0(z))\overline{\tilde{B}_i(\tau_i(z))}(\bar{\theta} - \chi_2)(\sigma S(z)\tau_i(z))^{-1} \quad 1 \le i \le m \tag{19}$$

$$P_i^D(z) = \alpha_i P_i(z)\overline{\tilde{D}_i(\phi_3(z))}(\phi_3(z))^{-1}, \quad 0 \le i \le m \tag{20}$$

$$P_i^R(z) = \alpha_i P_i(z)\tilde{D}_i(\phi_3(z))\overline{\tilde{R}_i(\phi_3(z))}(\phi_3(z))^{-1}, \quad 0 \le i \le m \tag{21}$$

$$P_j^V(z) = \lambda(\bar{\theta} - \chi_2)(\bar{p}\chi_0)^{j-1}(1 - \bar{p}\chi_0)\overline{\tilde{V}(\phi_4(z))}(\sigma \phi_4(z))^{-1}, \quad 1 \le j \le l \tag{22}$$

Proof Utilizing the expression $\int_0^\infty e^{-su}(1 - M(u))du = (1 - \tilde{M}(s))/s$, for $u = x$ or $u = y$, we can easily obtain required Eqs. (18)–(22) by Integrating (13)–(17) with appropriate variable.

Theorem 3 The PGF for stationary queue length distribution at random epoch is

$$\psi(z) = (\bar{\theta} - \chi_2)\sigma^{-1}\left[U(z)(S(z))^{-1}\{M_0(z) + \tilde{B}_0(\tau_0(z)) \sum_{i=1}^m r_i M_i(z)\} \right.$$
$$\left. + \lambda(1 - (\bar{p}\chi_0)^l)\overline{\tilde{V}(\phi_4(z))}(\phi_4(z))^{-1} \right] \tag{23}$$

where

$$M_i(z) = (1 - \tilde{B}_i(\tau_i(z)))[\tau_i(z)]^{-1}[1 + \alpha_i\{1 - \tilde{D}_i(\varphi_3(z))\tilde{R}_i(\phi_3(z))\}(\phi_3(z))^{-1}],$$
$$0 \leq i \leq m$$

Proof Using the following equation

$$\psi(z) = P_0 + \sum_{i=0}^m \left(P_i(z) + P_i^D(z) + P_i^R(z)\right) + \sum_{j=1}^l P_j^V(z),$$

we can obtain the Eq. (23).

4 Performance Characteristics

In this section, various performance characteristics are formulated using PGFs of the queue size distributions for the different system states obtained in previous section. The different states of server being in idle, busy, delay in repair, repair, and vacation states are denoted by I, B_i, D_i, R_i, and V_j.

4.1 Long Run Probabilities

Theorem 4 Long run probabilities for different server's states are obtained as

(a)   The server is in idle state

$$P(I) = (\bar{\theta} - \chi_2)\chi_1\sigma^{-1} \tag{24}$$

(b)   The server is busy in FEP and SOP service

$$P(B_0) = \lambda c \beta_0 [\chi_1 + \lambda b_2 v (1 - (\bar{p}\chi_0)^l)] \sigma^{-1} \tag{25}$$

$$P(B_i) = \lambda c r_i \beta_i [\chi_1 + \lambda b_2 v (1 - (\bar{p}\chi_0)^l)] \sigma^{-1}, \, 1 \le i \le m \tag{26}$$

(c)   The server is waiting for repair in case of failure during FEP and SOP service

$$P(D_0) = \alpha_0 \lambda c \beta_0 \gamma_0 [\chi_1 + \lambda b_2 v (1 - (\bar{p}\chi_0)^l)] \sigma^{-1} \tag{27}$$

$$P(D_i) = \alpha_i \lambda c r_i \beta_i \gamma_i [\chi_1 + \lambda b_2 v (1 - (\bar{p}\chi_0)^l)] \sigma^{-1}, 1 \le i \le m \tag{28}$$

(d)   The server is under repair in case of failure during FEP and SOP service

$$P(R_0) = \alpha_0 \lambda c \beta_0 g_0 [\chi_1 + \lambda b_2 v (1 - (\bar{p}\chi_0)^l)] \sigma^{-1} \tag{29}$$

$$P(R_i) = \alpha_i \lambda c r_i \beta_i g_i [\chi_1 + \lambda b_2 v (1 - (\bar{p}\chi_0)^l)] \sigma^{-1}, 1 \le i \le m \tag{30}$$

(e)   The server is on j^{th} vacation

$$P(V_j) = \lambda(\bar{\theta} - \chi_2)(\bar{p}\chi_0)^{j-1}(1 - \bar{p}\chi_0)v(\sigma)^{-1}, 1 \le j \le l. \tag{31}$$

It is noted that

$$P(B) = P(B_0) + \sum_{i=1}^{m} P(B_i),$$

$$P(D) = P(D_0) + \sum_{i=1}^{m} P(D_i),$$

$$P(R) = P(R_0) + \sum_{i=1}^{m} P(R_i),$$

$$P(V) = \sum_{j=1}^{l} P(V_j).$$

Also, effective arrival rate is $\lambda_e = \lambda\bar{\theta}[\chi_1 + \lambda b_2 v(1 - (\bar{p}\chi_0)^l)]\sigma^{-1}$.

Proof The results (25)–(30) can be obtained by substituting $z = 1$ in Eqs. (18)–(22) and Eq. (24) is obtained by using $P(I) = 1 - \sum_{i=0}^{m}\{P(B_i) + P(D_i) + P(R_i)\} + \sum_{j=1}^{l} P(V_j)$.

The effective arrival rate is obtained by using the following equation.

$$\lambda_e = \lambda P_0 + \sum_{i=0}^{m}\{\lambda b P(B_i) + \lambda b_1 P(D_i) + \lambda b_1 P(R_i)\} + \sum_{j=1}^{l} \lambda b_2 P(V_j).$$

4.2 Mean Queue Length

Theorem 5 Mean queue length L_q is given by

$$L_q = \sigma^{-1}\Big[\lambda^2 b_2 c v^{(2)}(2)^{-1}(1 - (\bar{p}\chi_0)^l)(\bar{\theta} - \chi_2) + \lambda c\{\chi_1 + \lambda b_2 v(1 - (\bar{p}\chi_0)^l)\}\chi_4$$
$$+\{\beta_0(1 + h_0) + \sum_{i=1}^{m} r_i\beta_i(1 + h_i)\}\chi_5\Big]. \tag{32}$$

where

$$\chi_4 = \lambda c(2)^{-1}\Big\{\alpha_0\beta_0 b_1(\gamma_0^{(2)} + g_0^{(2)} + 2\gamma_0 g_0) + \beta_0^{(2)}(b + b_1 h_0)(1 + h_0)\Big\}$$
$$+ \beta_0\lambda c(b + b_1 h_0) \sum_{i=1}^{m} r_i\beta_i(1 + h_i) + \sum_{i=1}^{m} r_i\lambda c(2)^{-1}$$
$$\times \Big\{\alpha_i\beta_i b_1(\gamma_i^{(2)} + g_i^{(2)} + 2\gamma_i g_{i0}) + \beta_i^{(2)}(b + b_1 h_i)(1 + h_i)\Big\}$$

$$\chi_5 = \lambda c\{\chi_1 + \lambda b_2 v(1 - (\bar{p}\chi_0)^l)\} + (2(\bar{\theta} - \chi_2))^{-1}\Big[\lambda(\bar{\theta} - \chi_2)\{\chi_1 c_{(2)}$$
$$+(1 - (\bar{p}\chi_0)^l)(v^{(2)}(\lambda b_2 c)^2 + v\lambda b_2 c_{(2)})\} + \lambda c\{\chi_1 + \lambda b_2 v(1 - (\bar{p}\chi_0)^l)\}\chi_6\Big]$$

$$\chi_6 = 2\theta\chi_2 + \beta_0^{(2)}(\lambda c(b + b_1 h_0))^2 - \beta_0\tau_0''(1) + 2\beta_0(\lambda c)^2(b + b_1 h_0) \sum_{i=1}^{m} r_i(b + b_1 h_i)$$
$$+ \sum_{i=1}^{m} r_i\{\beta_i^{(2)}(\lambda c(b + b_1 h_i))^2 - \beta_i\tau_i''(1)\}$$

$$\tau_i''(1) = -[\lambda bc_{(2)} + \alpha_i\{(\gamma_i^{(2)} + g_i^{(2)} + 2\gamma_i g_i)(\lambda b_1 c)^2 + (\gamma_i + g_i)\lambda b_1 c_{(2)}\}], \ 0 \le i \le m.$$

Proof The Eq. (32) can be obtained by differentiating Eq. (23) w.r.t. z and taking limit $z \to 1$ i.e. $L_q = \left(\frac{d\psi(z)}{dz} \right)_{z=1}$.

Remark Mean waiting time is given by $W_q = L_q (\lambda_e c)^{-1}$.

4.3 Reliability Indices

Some reliability metrics are obtained using PGFs as follows:

(i)   The availability of the server is

$$
\begin{aligned}
A_v &= P_0 + \lim_{z \to 1} \left[\sum_{i=0}^{m} P_i(z) \right] \\
&= \left[\chi_1(\bar{\theta} - \chi_2) + \lambda c \{ \chi_1 + \lambda b_2 v (1 - (\bar{p}\chi_0)^l) \} \left\{ \beta_0 + \sum_{i=1}^{m} r_i \beta_i \right\} \right] \sigma^{-1}. \quad (33)
\end{aligned}
$$

(ii)   The steady-state failure frequency is

$$
\begin{aligned}
F_f &= \lim_{z \to 1} \left[\sum_{i=0}^{m} \alpha_i P_i(z) \right] \\
&= \lambda c \{ \chi_1 + \lambda b_2 v (1 - (\bar{p}\chi_0)^l) \} \left\{ \alpha_0 \beta_0 + \sum_{i=1}^{m} r_i \alpha_i \beta_i \right\} \sigma^{-1}. \quad (34)
\end{aligned}
$$

4.4 Cost Function

The expected cost function associated with different cost components is expressed as follows:

$$
TC = C_h L_q + C_S [E(C)]^{-1} + \sum_{i=0}^{m} C_{B_i} P(B_i) + \sum_{i=0}^{m} C_{D_i} P(D_i) + \sum_{i=0}^{m} C_{R_i} P(R_i)
$$

$$
+ \sum_{j=1}^{l} C_{V_j} P(V_j). \quad (35)
$$

where the expected length of cycle is obtained by Singh et al. [8].

$$E(C) = \sigma[\lambda_e c(\bar{\theta} - \chi_2)\chi_1]^{-1}.$$

The different cost components per unit time used in Eq. (35) are

C_h	Holding for keeping one unit in the system
C_S	start-up cost
$C_{B_0}(C_{B_i})$	cost associated with the server when he is rendering FEP/SOP service
$C_{D_0}(C_{D_i})$	cost associated with the server when he is waiting for repair on failure while rendering FEP/SOP service
$C_{R_0}(C_{R_i})$	cost associated with the server when he is under repair on failure while rendering FEP/SOP service
C_{V_j}	cost associated with the server when he is on jth ($1 \leq j \leq l$) randomized vacation

5 Special Cases

The results for some particular cases are established by fixing suitable parameters of our present model as follows:

Case 1 If $b = b_1 = b_2 = 1$, $r_i = 0$, $\theta = 0$, $\alpha_i = 0$, the Eq. (23) reduced to

$$\psi(z) = \frac{(1 - \chi_2')\tilde{B}_0(\varphi_1(z))(z - 1)\left(((\tilde{V}(\varphi_1(z)) - 1)(1 - (\bar{p}\chi_0')^l) + \chi_0'\chi_1'(X(z) - 1)\right)}{\{(1 - (\bar{p}\chi_0')^l)\lambda v + \chi_0'\chi_1'\}(z - \tilde{B}_0(\varphi_1(z)))(X(z) - 1)}.$$

(36)

where $\chi_0' = \tilde{V}(\lambda)$, $\chi_1' = [(\bar{p}\chi_0')^{l-1}(1 - \bar{p}\chi_0') + p(1 - (\bar{p}\chi_0')^{l-1})]$, $\chi_2' = \lambda c \beta_0$.
The result corresponds to Ke et al. [5].

Case 2 By setting $b = b_1 = b_2 = 1$, $r_i = 0$, $\theta = 0$, $\alpha_i = 0$, $p = 0$, $l = 1$ in Eq. (23), we have

$$\psi(z) = \frac{(1 - \chi_2')\tilde{B}_0(\phi_1(z))(z - 1)\left((\tilde{V}(\phi_1(z)) - 1) - \chi_0'\overline{X(z)}\right)}{(\tilde{B}_0(\phi_1(z)) - z)\{\lambda v + \beta_0\}\overline{X(z)}}.$$

(37)

The above result (37) matches with the result derived in Choudhury [1].

6 Numerical Results

This section presents numerical simulation for analyzing the effect of varied descriptors on some specific indices. For numerical illustration, we take the distribution functions for batch arrival as geometric distributed. The service time follows $k-$ Erlangian distribution with parameter μ_i. Moreover, each of the delays to repair/repair and vacation follows exponential distribution with parameters d_i, f_i and ω, respectively. For the computation purpose, we assume $f_i = 3\mu_i$, $d_i = 6\mu_i$.

Following are the default set of parameters to examine the behavior of system:

$$c = 2, m = 2, l = 2, k = 2, \theta = .02, p = .5, \lambda = 1.3, \mu_0 = \mu = 2.5, \mu_1 = 2\mu_0,$$
$$\mu_2 = 3\mu_0, b = .6, b_1 = .3, b_2 = .35, \alpha_0 = \alpha = 0.1, \alpha_1 = 2\alpha_0, \alpha_2 = 3\alpha_0, \omega = 2.$$

6.1 Steady-State Probabilities

Tables 1–2 display the response of various steady-state probabilities by varying θ and μ. It is observed that the probabilities $P(I)$, $P(V)$ decrease (increases) while $P(B)$, $P(D)$, $P(R)$ increases (decreases) with increment in values of θ (μ).

Table 1 The steady-state probabilities on varying Θ

Θ	$P(I)$	$P(B)$	$P(D)$	$P(R)$	$P(V)$
0.01	0.07840	0.80841	0.00539	0.01078	0.09702
0.02	0.07486	0.81616	0.00544	0.01088	0.09265
0.03	0.07126	0.82407	0.00549	0.01099	0.08819
0.04	0.06759	0.83212	0.00555	0.01109	0.08365
0.05	0.06384	0.84034	0.00560	0.01120	0.07901

Table 2 The steady-state probabilities on varying μ

μ	$P(I)$	$P(B)$	$P(D)$	$P(R)$	$P(V)$
2.1	0.00824	0.95873	0.00761	0.01522	0.01020
2.2	0.02702	0.91867	0.00696	0.01392	0.03343
2.3	0.04426	0.88179	0.00639	0.01278	0.05478
2.4	0.06016	0.84772	0.00589	0.01177	0.07445
2.5	0.07486	0.81616	0.00544	0.01088	0.09265

6.2 Mean Queue Size and Mean Waiting Time

The effect of λ and μ on L_q and W_q for varying some specific parameters θ, α, m, l and number of service phases can be observed in Tables 3, 4, 5. Figures 1–2 show the impact of λ and μ on L_q for different values of θ. It can be noticed from figures that L_q increases (decreases) with increasing values of λ (μ) on changing θ. Also, if θ increases, L_q goes on increasing

Table 3 L_q and W_q on varying λ for different values of (Θ, l)

	λ	$\Theta = .02$		$\Theta = .04$		$\Theta = .06$	
		L_q	W_q	L_q	W_q	L_q	W_q
$l = 1$	1.1	6.84	4.80	7.30	5.15	7.83	5.55
	1.2	9.17	6.03	9.98	6.59	10.95	7.26
	1.3	13.14	8.13	14.78	9.18	16.91	10.56
	1.4	21.46	12.54	25.99	15.26	33.02	19.48
	1.5	50.05	27.75	82.28	45.83	233.89	130.88
$l = 10$	1.1	6.60	5.04	7.06	5.39	7.59	5.80
	1.2	8.90	6.28	9.71	6.84	10.67	7.52
	1.3	12.85	8.39	14.49	9.45	16.61	10.83
	1.4	21.14	12.82	25.67	15.54	32.69	19.76
	1.5	49.70	28.03	81.93	46.11	233.53	131.17

Table 4 L_q and W_q on varying λ for different values of (m, α)

	λ	$m = 2$		$m = 1$		$m = 0$	
		L_q	W_q	L_q	W_q	L_q	W_q
$\alpha = 0.01$	1.1	6.70	4.94	6.29	4.63	3.76	2.72
	1.2	9.01	6.17	8.31	5.69	4.5	3.03
	1.3	12.96	8.28	11.61	7.41	5.44	3.43
	1.4	21.26	12.71	17.96	10.72	6.68	3.95
	1.5	49.84	27.92	35.14	19.67	8.39	4.67
$\alpha = 0.05$	1.1	6.92	5.15	6.49	4.82	3.85	2.80
	1.2	9.39	6.49	8.65	5.96	4.62	3.13
	1.3	13.71	8.84	12.22	7.87	5.60	3.56
	1.4	23.23	14.01	19.36	11.66	6.9	4.11
	1.5	61.40	34.72	40.61	22.95	8.73	4.89

Table 5 L_q and W_q on varying μ for different service phases

μ	M/M/1		M/E$_2$/1		M/D/1	
	Lq	Wq	Lq	Wq	Lq	Wq
2.1	141.38	91.56	133.51	86.47	125.12	81.03
2.2	41.6	26.84	39.40	25.42	37.06	23.91
2.3	24.53	15.77	23.29	14.97	21.97	14.13
2.4	17.46	11.19	16.62	10.65	15.72	10.07
2.5	13.59	8.68	12.96	8.28	12.30	7.86

Fig. 1 Lq versus λ for different values of Θ

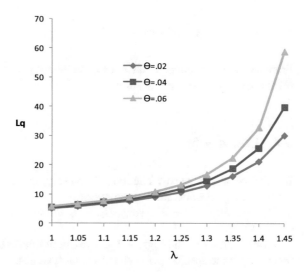

Fig. 2 Lq versus μ for different values of Θ

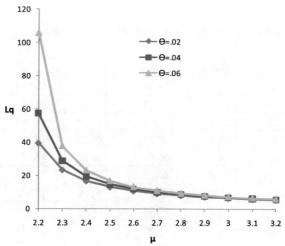

Table 6 A_v and F_f on varying α and l

α	$l = 1$		$l = 5$		$l = 10$	
	A_v	F_f	A_v	F_f	A_v	F_f
0.01	0.92686	0.01092	0.88555	0.01061	0.88446	0.01061
0.03	0.92428	0.03270	0.88347	0.03179	0.88239	0.03176
0.05	0.92171	0.05439	0.88140	0.05289	0.88034	0.05285
0.07	0.91916	0.07601	0.87934	0.07393	0.87829	0.07387
0.09	0.91661	0.09753	0.87729	0.0949	0.87625	0.09483

6.3 Reliability Indices

From Table 6, it is noticed that $A_v(F_f)$ decreases (increases) with the increment in α for different number of vacations.

6.4 Cost Analysis

In order to analyze the cost function (TC) on varying specific parameters, we consider the following set of default cost components:
$C_h = \$2/\text{day}$, $C_s = \$150/\text{day}$, $C_{b_i} = \$20/\text{unit}$, $C_{d_i} = \$25/\text{day}$, $C_{r_i} = \$30/\text{unit}$, $C_{V_j} = \$20/\text{day}$

From Tables 7–8 and Figs. 3–4, convexity of total cost is observed on varying μ and fixing some specific parameters to default values.

Table 7 TC on varying (μ, l) for different service phases

μ	$l = 1$			$l = 10$		
	$M/M/1$	$M/E_2/1$	$M/D/1$	$M/M/1$	$M/E_2/1$	$M/D/1$
2.3	80.86	78.37	75.7	77.24	74.78	72.16
2.4	70.94	69.24	67.43	66.03	64.36	62.59
2.5	67.22	65.95	64.59	60.98	59.75	58.43
2.6	**66.15**	**65.15**	**64.08**	58.58	57.61	56.58
2.7	66.42	65.60	64.73	**57.5**	56.72	55.88
2.8	67.41	66.73	66.00	57.17	**56.51**	**55.81**
2.9	68.83	68.24	67.62	57.26	56.71	56.11
3	70.49	69.98	69.44	57.62	57.14	56.62
3.1	72.28	71.84	71.36	58.14	57.72	57.27
3.2	74.15	73.76	73.33	58.75	58.38	57.98

Table 8 TC on varying (μ, l) for different values of Θ

μ	$m = 1$			$m = 2$		
	$\Theta = .02$	$\Theta = .04$	$\Theta = .06$	$\Theta = .02$	$\Theta = .04$	$\Theta = .06$
2.3	70.18	75.96	85.26	76.19	85.48	101.58
2.4	63.74	66.23	70.03	66.26	70.04	75.92
2.5	61.10	62.18	63.89	62.14	63.84	66.46
2.6	**60.21**	60.56	61.27	60.50	61.19	62.37
2.7	60.24	**60.18**	60.32	**60.10**	60.24	60.67
2.8	60.79	60.46	**60.27**	60.37	**60.19**	**60.17**
2.9	61.62	61.13	60.72	61.04	60.64	60.35
3	62.63	62.03	61.48	61.94	61.40	60.93
3.1	63.74	63.06	62.41	62.97	62.33	61.74
3.2	64.9	64.16	63.45	64.08	63.37	62.70

Fig. 3 *TC* on varying (μ, Θ)

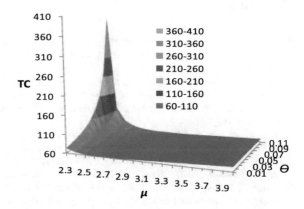

Fig. 4 *TC* on varying (μ, α)

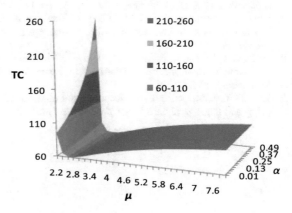

7 Conclusion

The present model investigates the bulk arrival non-Markovian queueing model with randomized vacation policy including several realistic features such as balking, feedback, and optional service. Various performance characteristics, viz., server status probabilities, mean queue size, reliability indices, and cost function are derived in explicit form and numerical tractability is validated by taking suitable illustration. Our model has applications to reduce delay in service facilitated in manufacturing industry, call center, health care system, etc. Furthermore, the results of our model can be used by the system designers and model developers to analyze various queueing systems for which our model is well fitted. The present research can be further generalized by incorporating working vacation policy, multi-phase repair, etc.

References

1. Choudhury, G. (2002). A batch arrival queue with a vacation time under single vacation policy. *Computers & Operations Research, 29,* 1941–1955.
2. Choudhury, G., & Tadj, L. (2009). An M/G/1 queue with two phases of service subject to the server breakdown and delayed repair. *Applied Mathematical Modelling, 33,* 2699–2709.
3. Cox, D. R. (1955). The analysis of non-markovian stochastic processes by the inclusion of supplementary variables. *Mathematical Proceedings of the Cambridge Philosophical Society, 51,* 433–441.
4. Jain, M., & Jain, A. (2014). Batch arrival priority queueing model with second optional service and server breakdown. *International Journal of Operational Research, 11,* 112–130.
5. Ke, J. C., Bin, Huang K., & Pearn, W. L. (2010). The randomized vacation policy for a batch arrival queue. *Applied Mathematical Modelling, 34,* 1524–1538.
6. Singh, C. J., Kaur, S. (2019). M^X/G/1 queue with optional service and server breakdowns BT—Performance prediction and analytics of fuzzy, reliability and queuing models : Theory and applications. In K. Deep, M. Jain, S. Salhi (Eds.), pp. 177–189. Singapore: Springer Singapore
7. Singh, C. J., Kaur, S., & Jain, M. (2017). Waiting time of bulk arrival unreliable queue with balking and Bernoulli feedback using maximum entropy principle. *Journal of Statistical Theory and Practice, 11,* 41–62.
8. Singh, C. J., Kumar, B., & Kaur, S. (2018). M^X/G/1 state dependent arrival queue with optional service and vacation under randomised policy. *International Journal of Industrial and Systems Engineering, 29,* 252–272.
9. Yang, D. Y., & Ke, J. C. (2014). Cost optimization of a repairable M/G/1 queue with a randomized policy and single vacation. *Applied Mathematical Modelling, 38,* 5113–5125.

Time-Based Mobile Application Adoption Model: A Firm's Perspective

Abhishek Tandon, Himanshu Sharma, and Anu G. Aggarwal

1 Introduction

With the technological advancements, the usage of mobile application for providing information and connecting with customers and traders has become a common practice. Organizations that still do not own an application may face some disadvantages in the future prospects of the firm's growth [1]. Assessing the dimensions that influenced the application adoption decision of the firms at different points of time can provide greater insights for other organizations that have not yet adopted the technology or are rechecking their decisions [2]. Previous studies have mainly focused on characteristics that affect adoption intention [3, 4] however, few have touched upon the importance of time point of adoptions. A grass-root theory that describes the time-based adoption of innovation is the diffusion of innovation (DOI) theory [5]. The theory argues that over time the social system gains idea and vision about the technology and ultimately succeeds in taking the adoption decision. DOI considers five elements, namely, relative advantage, compatibility, observability, trialability, and complexity. A time-based adoption theory is considered apt as not all the components of social system adopt the innovation at same time. Generally, it has been observed that those who perceive the technology to be efficient and effective show an earlier sign of usage in comparison to late adopters. Therefore, based on the above arguments DOI suggested that it is possible to classify system into five adopter categories, namely, innovators, early adopters, early majority, late majority, and laggards.

A. Tandon (✉)
Shaheed Sukhdev College of Business Studies, University of Delhi, Delhi, India
e-mail: abhishektandon@sscbsdu.ac.in

H. Sharma · A. G. Aggarwal
Department of Operational Research, University of Delhi, Delhi, India

© The Author(s), under exclusive license to Springer Nature Singapore Pte Ltd. 2021   323
P. K. Kapur et al. (eds.), *Advances in Interdisciplinary Research
in Engineering and Business Management*, Asset Analytics,
https://doi.org/10.1007/978-981-16-0037-1_25

Here, we consider the social system to be the collection of organizations that strive in industry together where mobile application plays a key role in the business operations. As knowledge and usage of technology diffuse at a faster rate regardless of industry's population, place, or offerings, therefore, system is aptly defined instead of considering niche groups [2]. Based on the m-commerce application adoption patters, we classify the firms into five adopter categories under six determinants, namely, relative advantage, organizational compatibility, technological incompatibility, top management support, complexity, and external pressure. The motivation for considering these constructs lies in the technology–organization–environment (TOE) theory [6], an extension of DOI in organizational framework. TOE describes the entire innovation process starting from its development and extending to its adoption and implementation by the firms. It considers adoption of innovation under three broad parameters: technological, organizational, and environmental. Therefore, the objective of study is to determine if there is a significant relationship between these determinants and the timing of mobile application by a firm. The remaining paper organization is as follows: Sect. 2 discusses the TOE framework along with the hypotheses development; Sect. 3 presents the adopted methodology; data analysis results are provided in Sect. 4; Sect. 5 presents discussions and implications related to the model; lastly conclusion and future scope is provided in Sect. 6.

2 Hypotheses Development

The technological adoption takes place over time and it is argued that firms which find it beneficial in their operation of business firm so as to gain competitive advantage in the industry. With time, the technology becomes popular and less complex to understand, thus later adopters adapt it so that they are not left detached from the industry [7]. Therefore, for digital products, a time-based theory may be apt for studying technological adoptions. One of the pioneering studies in this domain is the diffusion of innovation (DOI). According to this theory, it is assumed that not all of the social system adopts the technology at the same time and the process also depends on their relative innovativeness. Theory argues that the organizations can be classified into five categories on the basis of technology innovation adoption in comparison to other organizations, namely innovators, early adopters, early majority, late majority, and laggards.

The technology–organization–environment (TOE) framework is an organizational level theory to explain the information system (IS) adoption process of firms at three levels, i.e., technological, organizational, and environmental [6]. The underground concept of TOE is the adaptation and operation of the technological innovations by the representatives of a firm. The framework is considered to be consistent with the diffusion of innovation (DOI) theory proposed by M Rogers [5]. It is believed that both the theories portray same sentiments but the ultimate users show differentiation. The "organizational context" of TOE can be considered parallel to the "internal characteristics of organizational structure" of DOI; the "external characteristics of organizational structure" of DOI may be considered as a counterpart of TOE's "environmental context"; and finally the "technological characteristics of innovation" can be contrasted with the "technological context" of TOE framework [8]. The technological concept incorporates all the know-how relevant for a firm which are already in use by the firm or available in the market but not currently adopted [9]. The organizational concept covers the features and resources corresponding to a firm consisting of inter-employee linking structures, intra-firm communication process, firm size, and volume of slack resources. Organizational structure plays an important part in identifying the importance of adoption process within a firm [10]. Environmental concept includes the industrial structure, technology service providers, and adjusting atmosphere. On evaluating industrial structures it has been observed that intense competition stimulates adoption process and leading firms can stimulate others to promote innovation [11].

2.1 Perceived Benefits (PB) and Adoption Time

A firm will opt for technological adoption if it provides greater profits in comparison to the present ones. The firm will look forward for an innovation that solves its problems or help to extend its business boundaries [12]. Researchers have also denoted PB as relative advantage. This includes cost reduction, better cash flow, higher efficiency, and efficient service [13]. It may also incorporate increased ability to compete, extending its reach, better customer relationship management, and enhanced operational efficiency. It can be argued that earlier users expected that adoption will profit them more in comparison to late adopters. Thus, we posit that.

H1: Earlier adopters place greater importance on perceived benefits of having a mobile application than do later adopters.

2.2 Organizational Compatibility (OC) and Adoption Time

A firm will opt for adopting an innovation if is considered suitable with firm's ethics, morals, business practices, and current IS arrangement. M-commerce application affects firm's adoption decision by modifying present operations and obtain profits [14]. It has been argued that there exists direct association between organizational compatibility (OC) and IS technology adoption. Therefore, the speed with which a firm establishes a mobile application is influenced by its impact on cultural norms and business code of ethics [15]. Thus, we posit that.

H2: The mobile application is perceived as more organizationally compatible by earlier than by later adopters.

2.3 Technical Incompatibility (TI) and Adoption Time

The adoption of an innovation can be inhibited by the presence of incompatibility with a firm's present hardware, software, networking, or telecommunication architecture. Therefore, the firm's adoption decision depends on the extent to which the innovation can be incorporated into the present industrial framework [16]. Practitioners observe that IS mangers do not involve whole staff in application development process as many of them don't have a technological vision for the firm. Also, they argue that the application is incompatible with their current technologies [17]. Therefore, a firm that feels that adopting the new technology can be integrated into the firm's infrastructure and adopt easily. Thus, we posit that.

H3: The mobile application is perceived to be more technically incompatible by later adopters than by earlier ones.

2.4 Complexity (COM) and Adoption Time

It includes the extent to which a firm feels that it will be difficult to adopt an innovation. Introducing new technological innovations in the organizations might be difficult for the employees to adopt as it may require new skills and improvement in their existing business practices [18]. Incorporating technological innovations means to merge the telecommunications and advanced network application with the present ones. Previous studies based on IS systems have emphasized the detrimental effect of technological complexity over its adoption by firms [19]. On the basis of technological experiences of the firm and their employee's technical skills, establishing a mobile application may be a difficult job. It has been argued that an organization is less probable to adopt a technological innovation if intense skill transformations are required by its fellows. Thus, we posit that.

H4: Earlier adopters perceive mobile application to be less complex than later adopters.

2.5 Top Management Support (TMS) and Adoption Time

Top management support is highly desirable for an organization to adopt an innovation. This is particularly true for technological innovations like m-commerce as it requires a transformation in the present organizational strategies, business ethics, and their relationship with other trading partners such as retailers and wholesalers [20]. Adopting a technological innovation through higher management underlines their commitment to support the innovation at each level of organizational structure. It has been argued that though technological adoption takes place at grassroots level but without the consent of the top management this practice will halt [21]. Thus, we posit that.

H5: Earlier adopters have greater top management support for a mobile application than do later adopters.

2.6 External Pressure (EP) and Adoption Time

This covers the pressure on the firm to adopt an innovation that comes from the customers or internal and external environment, in order to improve its efficiency. Therefore, the major sources of environmental pressures in technological innovation adoption are to attain competitive position in the market as well as those coming from customers, traders, suppliers, or wholesalers [22]. It has been believed that these environmental pressures will strengthen the competitive position of the firm as well as achieving superior performance. Researchers suggested that by adopting innovation, firms will be able to leverage improved strategies in accordance with the technological advancements that will differentiate them from the competitors and instill in them a strategic change in comparison to other industries [23]. Delivering high level of technological services and better communication with traders, results in enhanced level of satisfaction received by suppliers and customers. Thus, we posit that.

H6: Earlier adopters can better handle external pressure for a mobile application than do later adopters.

3 Methodology

The paper focus on studying the key determinants of adoption time of a mobile application using six independent variables namely the PB, OC, TI, COM, TMS, and EP. A pictorial representation of the proposed conceptual model is provided in Fig. 1. The model developed here involved previous works and suggestions of IT experts from industry and academia. Then the model was empirically tested with the help of a suitable research instrument, i.e., using survey method. The measuring scale used in the survey was adopted from previously published materials and their statements along with the source are provided in Appendix. To validate the hypothesized model, a self-administered questionnaire was developed consisting of 20 constructs. To get accurate response and reduce ambiguity, pilot study was conducted on 60 respondents which mainly consisted of academicians and management experts dealing with m-commerce applications. The pilot group correctly filled the questionnaire and suggested few changes in the questionnaire language. The final questionnaire with closed-ended questions on 5-point likert scale was prepared after incorporating the suggested corrections. Around 510 questionnaires were e-mailed, of which 230

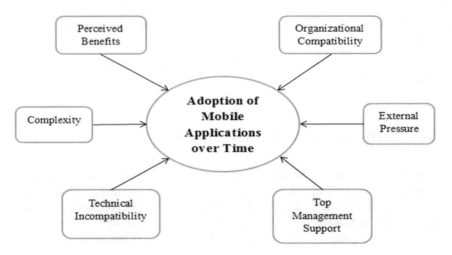

Fig. 1 Mobile application adoption model

responses were registered. These responses were filtered and 204 useful responses were retained for analysis. The respondents consisted of the employees in the IT department of various small and medium enterprises (SMEs) operational in India, guiding managerial decisions to establish a mobile commerce application for the firm's usage. The survey was carried from October 15, 2018 to December 30, 2018.

4 Data Analysis

4.1 Demographic Profile of the Respondents

Table 1 represents the profile of the respondents. On the basis of the time since which the organization had an operational mobile application, the respondents can be classified into five categories, as proposed in DOI. Firms which possess an operational application for more than 8 years were regarded as "innovators"; for five to eight years as "early adopters"; three to five years as "early majority"; one to three years as "late majority", and lastly less than one year or thinking of creating an app are termed as "laggards". The basic assumption underlying the categorization of firms is that they all belong to the same social system, regardless of industry.

Table 1 Demographic characteristics

Respondent's demographic profile	Variables	Usable responses
Job designation	System engineer	90
	Chief information officer	40
	Chief technical officer	50
	Manager	24
Highest level of academic qualification	Diploma/Advance Diploma	30
	Graduate Degree	70
	Postgraduate Degree	90
	Other	14
Monthly salary (in Rs)	Less than 40,000	20
	40,000 to 50,000	70
	50,000 to 60,000	94
	More than 60,000	20
Time of having operational mobile application (in years)	Currently developing/less than 1	36
	1 to 3	30
	3 to 5	70
	5 to 8	50
	More than 8	18

4.2 Validity and Reliability of Measures

The reliability of the model was supported by applying exploratory factor analysis (EFA) over 20 attributes in total. For EFA, principal component analysis with varimax rotation and eigenvalue greater than 1 were considered and attributes with factor loadings above 0.5 were retained for further analysis. Kaiser-Mayer-Olkin (KMO) value of 0.781 and valid Bartlett's Sphericity Test suggest the usefulness of factor analysis. All constructs obtained had Cronbach alpha value above 7 and composite reliability (CR) value above 7, which substantiated construct reliability. Confirmations of discriminant validity of the constructs were observed with AVE (average variance extracted) values above 0.5.

4.3 Hypothesis Testing Results

To test the proposed hypotheses, multivariate analysis of variance (MANOVA) was applied using SPSS software. The Wilks' lambda value for the overall model had an F-value of 5.085 with p-value less than 5% level of significance. This indicates that mean of the adopter categories contains significant differences at 5% level of significance. Moreover, all the six independent variables were found to be significant

Table 2 Tukey's honestly significant difference test results

Category	PB	OC	TI	COM	TMS	EP
Innovators	4.00	3.40	2.50	3.33	2.00	3.83
Early adopters	3.62	3.25	3.09	3.28	3.10	3.52
Early majority	3.14	3.25	3.19	3.10	3.21	3.24
Late majority	2.80	3.25	3.26	3.04	3.39	3.07
Laggards	2.40	3.14	4.33	2.89	3.50	3.00

at 5% level of significance under the evaluated F-values (PB $= 2.172$; OC $= 2.593$; TI $= 3.914$; COM $= 4.090$; TMS $= 6.543$; and EP $= 3.127$). To get greater insights into the results, Tukey's honestly significant difference test was used to further assess how each adopter category impacts the factors that determine mobile application adoption over time. Table 2 provides the results.

5 Discussions and Implications

The acceptance of hypothesis H1 confirms that perceived benefits have positive association with the m-commerce adoption time. Further insights using Tukey's test show that PB is more important for innovators as compared laggards and late adopters. This is consistent with the results provided by [12, 13]. Organizations that were the first to adopt mobile application in their firm believed that its usage will benefit their existing business operation and simplify their information sharing systems. However, those firms that were late did not expect any immediate business benefits. They decided to adopt the application only due to industrial influence and to stay connected within the industry. The positive relationship between organizational compatibility and adoption timing can be observed by the support of hypothesis H2. Tukey's test shows that OC is greater for innovators as compared to laggards and late adopters. This result is consistent with those of others [14, 15]. Therefore, organizations that consider the application to be consistent with their existing culture and internal business are highly likely to adopt it early. This may be attributed to the fact that they might already be using electronic commerce systems in their organization.

The support for hypothesis H3 verifies that technical incompatibility affects the application adoption timing. Further, the Tukey's test outlines that the effect is more for laggards as compared to the innovators and early adopters. This is consistent with previous researches [16, 17]. This may be due to the fact that firms that were the first to adopt the application perceive its development and implementation to be consistent with its existing IS infrastructure. Complexity is found to have a significant impact over timing of adopting an application, by support of hypothesis H4. Tukey's test validates that this impact is more intense for laggards in comparison to innovators and early adopters. This is consistent with previous studies [18, 19]. Such behavior may be observed as many of the firms that adopted the application earlier were risk takers and did not have enough knowledge regarding the technology. But, later adopters might have noticed the difficultly in fitting the technology into the skill structure of present employees and also in its technological framework.

Acceptance of hypothesis H5 validates the significant positive relationship between top management support and the timing of application adoption. The effect is more for innovators and early adopters as compared to laggards and late adopters. Previous works show similar results [20, 21]. This may be because most of the firms that were quick to respond to adoption decision had strong support from the top management. Those management members who have already used an application for their personal use might show more intent to adopt the innovation. The support for hypothesis H6 verifies the impact of external pressure over the adoption timing. The impact is observed to be more intense for innovators as compared to laggards and late adopters. This is consistent with previous studies [22, 23]. Those firms that already had an operational electronic commerce system might be the first to adopt mobile application as they encountered the expectations of suppliers, traders, and customers. Also, in order to gain competitive edge over its competitors, firms are ready to adopt the innovations as early as possible. Late adopters, on the contrary show intent to just hang-around in the industry.

6 Conclusion, Limitation, and Future Scope

The study aimed to examine the factors that impact the time of adopting a mobile commerce application with respect to a firm. The factors namely perceived benefits, organizational compatibility, technical incompatibility, complexity, top management support, and external pressure, were adopted from technology–organization–environment (TOE) framework. A hypothesized model was formed assuming how early as well as late adopters of mobile application were affected by these exogenous variables. To test the hypotheses, multivariate analysis of variance (MANOVA) was applied on a sample of 204 employees administering the IT department of their respective firms. The analysis validated the significance of all the hypotheses as well as the overall model proposed.

Research findings indicated that earlier adopters put more emphasis on perceived benefits in comparison to late adopters. Earlier adopters feel that the application is more compatible with their current organizational culture and practices. They also garnered mobile application to be consistent with the organization's existing technological infrastructure. Also, earlier adopters laid stress on greater support from the top management rather than the late adopters. As earlier adopters were the first to implement the technological changes without having much knowledge about the innovation, they faced less complexity in comparison to late adopters. Also, earlier adopters face the consumer's and trader's expectation of implementing a mobile application in their business operations. Even though the paper is well documented and is consistent with previous works, it incurs some limitations. The survey respondents considered representatives of firms operating in only two metropolitan cities of India, and facing geographic limitations. Future studies may incorporate other cities in order to generalize the results. Present model may be further extended to electronic commerce in future.

References

1. Chong, A.Y.-L., & Bai, R. (2014). Predicting open IOS adoption in SMEs: An integrated SEM-neural network approach. *Expert Systems with Applications, 41*(1), 221–229.
2. Shaikh, A. A., & Karjaluoto, H. (2015). Mobile banking adoption: A literature review. *Telematics and Informatics, 32*(1), 129–142.
3. Mtebe, J., & Raisamo, R. (2014). Investigating students' behavioural intention to adopt and use mobile learning in higher education in East Africa. *International Journal of Education and Development using ICT, 10*(3).
4. Kang, S. (2014). Factors influencing intention of mobile application use. *International Journal of Mobile Communications, 12*(4), 360–379.
5. M Rogers, E. (1983). Diffusion of innovations. The Free Press.
6. Tornatzky, L. G., Fleischer, M., & Chakrabarti, A. (2013). The processes of technological innovation. Issues in organization and management series. *Lexington Books, 10,* 2013. Retrieved June 1990 from https://www.amazon.com/Processes-Technological-Innovation-Organization/Management/dp/0669203483.
7. Rogers, E. M. (2004). A prospective and retrospective look at the diffusion model. *Journal of Health Communication, 9*(S1), 13–19.
8. Gangwar, H., Date, H., & Ramaswamy, R. (2015). Understanding determinants of cloud computing adoption using an integrated TAM-TOE model. *Journal of Enterprise Information Management, 28*(1), 107–130.
9. Picoto, W. N., Bélanger, F., & Palma-dos-Reis, A. (2014). A technology–organisation–environment (TOE)-based m-business value instrument. *International Journal of Mobile Communications, 12*(1), 78–101.
10. San-Martín, S., Jiménez, N., & López-Catalán, B. (2016). The firms benefits of mobile CRM from the relationship marketing approach and the TOE model. *Spanish Journal of Marketing-ESIC, 20*(1), 18–29.
11. Ghezzi, A., Rangone, A., & Balocco, R. (2013). Technology diffusion theory revisited: A regulation, environment, strategy, technology model for technology activation analysis of mobile ICT. *Technology Analysis & Strategic Management, 25*(10), 1223–1249.
12. Jeong, B. K., & Yoon, T. E. (2013). An empirical investigation on consumer acceptance of mobile banking services. *Business and Management Research, 2*(1), 31.
13. Al-Ajam, A. S., & Nor, K. M. (2013). Internet banking adoption: Integrating technology acceptance model and trust. *European Journal of Business and Management, 5*(3), 207–215.
14. Son, H., Lee, S., & Kim, C. (2015). What drives the adoption of building information modeling in design organizations? An empirical investigation of the antecedents affecting architects' behavioral intentions. *Automation in Construction, 49,* 92–99.
15. Kaur Kapoor, K., Dwivedi, Y. K., & Williams, M. D. (2014). Innovation adoption attributes: A review and synthesis of research findings. *European Journal of Innovation Management, 17*(3), 327–348.
16. Wang, Y.-S., et al. (2016). Factors affecting hotels' adoption of mobile reservation systems: A technology-organization-environment framework. *Tourism Management, 53,* 163–172.
17. Dahnil, M. I., et al. (2014). Factors influencing SMEs adoption of social media marketing. *Procedia-Social and Behavioral Sciences, 148,* 119–126.
18. Gitau, L., & Nzuki, D. (2014). Analysis of determinants of m-commerce adoption by online consumers. *International Journal of Business, Humanities and Technology, 4*(3), 88–94.

19. Joubert, J., & Van Belle, J. (2013). The role of trust and risk in mobile commerce adoption within South Africa. *International Journal of Business, Humanities and Technology, 3*(2), 27–38.
20. Chong, A.Y.-L., et al. (2015). Predicting RFID adoption in healthcare supply chain from the perspectives of users. *International Journal of Production Economics, 159,* 66–75.
21. Alrawabdeh, W. (2014). Environmental factors affecting mobile commerce adoption-an exploratory study on the Telecommunication firms in Jordan. *International Journal of Business and Social Science, 5*(8).
22. Ali, A. A., & Kamran, A. (2017). Barriers in adopting m-banking system in universities. In: Proceedings of the tenth international conference on management science and engineering management. Springer.
23. Dey, A., Vijayaraman, B., & Choi, J. H. (2016). RFID in US hospitals: An exploratory investigation of technology adoption. *Management Research Review, 39*(4), 399–424.

Analysis of Clustering-Based Stock Market Prediction

Neetu Faujdar, Karan Gupta, Rishabh Kumar Singh, and P. K. Rohatgi

1 Introduction

The money invested in the stock market is staggering; there are about 60 major stock indices in the world with a total value of nearly dollar 69 trillion [1]. In Fig. 1, we can see the top 16 stock exchanges in terms of money invested and their geographical positions. This amount of money invested also brings in a huge amount of data which could be worked upon to find trends for the coming future [2]. We consider several things like a group of things, e.g., historical data stock, stock quote, stock volume, etc. Clustering can be used in deciding which groups these things shall belong to [3]. Financial data is considered as complex data to forecast and predict. Fluctuation in any stock's price can be due to the number of reasons, such as increase in crude prices, decrease in currency monetary value, tremendous amount of selling of stock, bad public news related to the company, etc. [4]. As the data related to the stock market is huge, most work done has been on forming a unified efficient algorithm to predict prices. Pint-size work has been completed on demonstrating suitable methods for mining the give datasets [5]. Applying clustering on stocks to identify their classes and then using machine learning algorithms according to each class would give a better understanding of data and hence we get predict stock prices with better accuracy [6]. In this paper, we have laid down the stress on clustering of different stocks. We will talk about different clustering techniques that could be used and then finally show the results of the analysis done by k-means clustering. The data is taken from indices all over the world and about 270 stock data is monitored and using clustering we are able to divide the dataset into categorical groups or classes [7].

N. Faujdar (✉) · K. Gupta · R. K. Singh · P. K. Rohatgi
GLA University, Mathura, India
e-mail: neetu.faujdar@gmail.com

© The Author(s), under exclusive license to Springer Nature Singapore Pte Ltd. 2021   335
P. K. Kapur et al. (eds.), *Advances in Interdisciplinary Research in Engineering and Business Management*, Asset Analytics,
https://doi.org/10.1007/978-981-16-0037-1_26

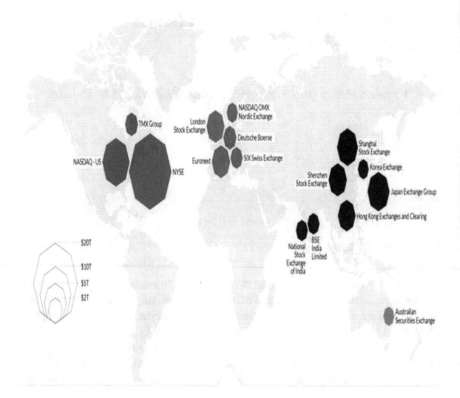

Fig. 1 Top 16 stock indices of the world

2   Related Work

Gopi Krishna Suvanam et al. [8] conveyed that the effect of fudging of structured things on exchange transacted equity things.

Romero et al. [9] stressed the need of classifying data and studying of pre-diction constraints for data analysis.

Abhishek Gupta [10] in his paper, Clustering Classification-Based Prediction of Stock Market Future Prediction, applied clustering classification approach for forecasting the stock market like practicing clustering algorithms like k-means and decision tree.

Argiddi and Apte [11] put through a section-based approach which was very favorable for obtaining nearly association instructions in the stock market. It could be applied for retelling in stock trading agendas. Similarly, they showed a segment-based tactic to deal with projecting association instructions and suggesting the clients. In addition, they used association instructions mined from stock exchange info to do upcoming pattern forecast.

Wang et al. [12] projected and showed that the KMSVM algorithm can build the reaction time of the classifiers venture up by fading the collective number of vectors while keeping up a suitable resemblance with SVM.

Nanda et al. [13] used k-Means, SOM (self-observing maps), and Fuzzy C-Means clustering algorithms to categorize Indian stocks in diverse clusters and then practice the consequences to make portfolios.

Kim and Ahn [14] smeared a genetic algorithm-based k-Means clustering algorithm to advance recommender schemes for online shopping market.

Wu et al. [15] used a method in which they pooled ANN with back circulation neural network and data mining practices (data sampling, data translation, and so on) and showed a highly accurate for predicting the Shanghai Stock Exchange Composite Index.

Nayak et al. [16] designed a neuro-genetic fusion system and equated the performances of numerous models with BSE data, and it advocated that FLANN-GA [17] model did well in several cases. There is a great number of machine learning algorithms developed to advance testing accurateness. SVM, a machine learning algorithm, was proposed by Boser et al. [18]. SVM analyses the data by its supervised learning base. It builds the risk function by forming an empirical error and a regularized term [19].

3 Clustering Methods

Clustering is an occupation in which we appoint certain gatherings or classes to specific questions to such an extent that the items inside a similar gathering or class are more comparable than those in the other recognized gathering or class [20]. In clustering, there is a need to and similarity measures between the objects which would lead to and some sort of pattern formation. This knowledge discovery found from pattern formation can be used to train data to and suitable results. Clustering is viewed as the most imperative unsupervised learning issue; it figures out how to discover a structure in the social affair of unlabeled data. As no class check is accessible in the readiness data, grouping is used to division enlightening accumulations into classes by creating names for them using the standard of expanding the intra-class closeness and constraining bury class likeness. Cluster analysis is the customized recognizing confirmation of gatherings of comparable questions or examples, on the chance that we mean an arrangement of information by 'm,' which is especially similar to a couple of different arrangements of information, we may instinctually tend to cluster 'm' and these arrangements of information into a characteristic group. By expanding bury gather closeness and constraining intra amasses similitude, different groups would frame on the perception space. We would then be able to easily see and appoint to the groups' proper name or highlight portrayal. Clustering algorithms can be categorized as.

- Exclusive

- Overlapping
- Hierarchical
- Probabilistic.

The list below provides the knowledge of four most commonly used techniques for clustering:

- K-means
- Fuzzy C-means
- Hierarchical
- Gaussians Mixture.

We will discuss fuzzy c-means in brief and then about k-means in detail. Fuzzy c-means: This algorithm is a membership assigned to each and every data point which is in accordance with each cluster center judging by the separation between cluster's center and its data point. If records are closer to center of the cluster, then the membership of cluster would be moving more toward that specific center of the cluster. It is quite obvious that the total of each membership data point should be one. After every repetition cluster centers and membership are informed accordingly using the formula:

$$\mu_{ij} = 1 / \sum_{k=1}^{c} (d_{ij}/d_{ik})^{2/m-1} \tag{1}$$

Here, 'c' is the number of total cluster centers, 'm' is the index of fuzziness, where m belongs to $[1, \infty]$ and d_{ij} is Euclidean distance.

$$V_j = \sum_{i=1}^{n} ((\mu_{ij})^{m_{x_i}}) / \sum_{i=1}^{n} (\mu_{ij})^{m}, \forall j = 1, 2, ...c \tag{2}$$

'v_j' represents the jth cluster center and 'μ_{ij}' represents the membership of ith data to jth cluster center. Main objective of fuzzy c-means algorithm is to minimize:

$$J(U, V) = \sum_{i=1}^{n} \sum_{j=1}^{c} (\mu_{ij})^{m} ||x_i - v_j||^2 \tag{3}$$

Here, $||x_i - v_j||$ is Euclidean distance between the cluster's center. The 2, 3, and 4 steps are repeated, where 'k' is the repetition step and β is the end criterion between $[0, 1]$.

K-means Clustering: It is a well-known unsubstantiated learning algorithm which is used to resolve clustering complications. It is one of the simplest methods to provide partitioning for automatic classification. The key objective is to state 'k,' i.e., total number of centroid, each for one cluster. The cluster centroid at the beginning must be chosen far from each other so that there is less conflict in assigning an object to a cluster. In the next step, we associate every nearest centroid to the dataset of the object or point. After all points have been taken into consideration and assigned a centroid, our first step is done and this gives us a stage of early grouping.

Algorithm 1 Fuzzy C-means

INPUT: X is equal to × 1, × 2, × 3, ... xn be dataset points and V is equal to v1, v2, v3 ..., vc be dataset of centers
OUTPUT: C-means clusters.
c is selected as clusters center at random
Fuzzy membership μ_{ij} is calculated using Eq. (1)
Fuzzy centers are found vj using Eq. (2).
The value of U(k), U(k + 1) is updated Until least value of J is found or ||U(k + 1) − U(k)|| < β.

At this stage, the recalculation of k new centroids for every center of the clusters must be done, with the help from the results of the earlier step. Presently, these new K centroids, a new tie should be made among the newest centroid and the same dataset points. We may observe that a loop is formed, and the K centroid will change their positions after every step until no additional variations are made (i.e., centroid don't move anymore) [21]. The objective function of k means is known by

$$J = \sum_{j=1}^{k} \sum_{i=1}^{n} ||(x_i)^j - c_j||^2 \tag{4}$$

Here, ||xi − cj|| is the Euclidean distance (xi)j is the data point, and cj indicates the distance of n data points from their respective clusters center.

Algorithm 2 K-means Clustering

Select k cluster at random
Distance between $(xi)j$ and cluster's center is calculated
The data point $(xi)j$ is assigned to cluster's center whose distance from cluster's center is the least of all.
Calculate the cluster's new center again using: n
$Ci = (1/n) * Pk j = 1(xi)j$
Distance between each xi and new obtained cluster's center is recalculated.
Stop if there is no data point was reassigned; otherwise, repeat the process from step 3.

Its working is shown in Fig. 2, where two clusters and their centroid relocation have been seen after each iteration. With fewer sample data, it is not very effective. Sometimes, it doesn't give the suitable result after a low number of iterations that is why running K means multiple times is advisable as it gives a better result. K-means is a very simple algorithm to use and it can be used in various domain problems. K-means uses three algorithms and they all have somewhat unique goals and so their results differ. These three algorithms are as follows: The Hartigan & Wong algorithm, the MacQueen algorithm, and the Lloyd algorithm. There is no best algorithm, they all are suitable for different kinds of datasets.

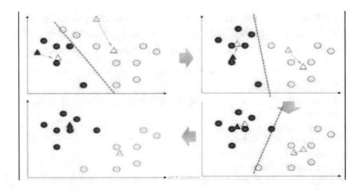

Fig. 2 Shifting of centroid and change in clusters

Fig. 3 K-means clustering
flowchart

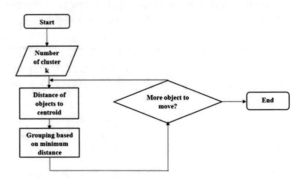

Column1	Open	High	Low	Close	Volume	Adjusted
1	325.640003	348.599995	324.840012	339.319996	2710383900	43.962146
2	341.299988	364.899994	337.720013	353.210007	2323987400	45.76173
3	355.470005	361.669987	326.259991	348.510006	2826614700	45.1528
4	351.110012	355.129997	320.160011	350.12999	2313457300	45.362684
5	349.739994	351.830002	329.42001	347.82999	1728015100	45.064697
6	348.869987	352.130009	310.500004	335.670006	2315962600	43.489255
7	335.950012	404.499992	334.200012	390.479992	2663486700	50.590412
8	397.779999	399.500011	353.019989	384.830009	4035649800	49.858403
9	385.819992	422.860012	366.480003	381.319996	2994362000	49.403647
10	380.369987	426.699993	354.23999	404.779999	3285356900	52.443115
11	397.410004	407.999992	363.320011	382.199989	2240925400	49.517659
12	382.539997	409.090004	377.679993	405.000004	1576633100	52.471619
13	409.399998	458.240021	408.999989	456.47998	1714963600	59.141341
14	458.410011	547.610023	453.980003	542.440025	2842138600	70.278286
15	548.169983	621.450005	516.219986	599.550011	3896084500	77.67743
16	601.830009	644	554.999992	583.979988	3899747600	75.660185
17	584.900009	596.759987	522.180016	577.730019	2776586400	74.850442

Fig. 4 Monthly data of Apple Inc. stock

Kindly assure that the Contact Volume Editor is given the name and email address of the contact author for your paper. The Contact Volume Editor uses these details to compile a list for our production department at SPS in India. Once the files have been worked upon, SPS sends a copy of the final pdf of each paper to its contact author. The contact author is asked to check through the final pdf to make sure that no errors have crept in during the transfer or preparation of the files. This should not be seen as an opportunity to update or copyedit the papers, which is not possible due to time constraints. Only errors introduced during the preparation of the files will be corrected.

This round of checking takes place about two weeks after the files have been sent to the Editorial by the Contact Volume Editor, i.e. roughly seven weeks before the start of the conference for conference proceedings, or seven weeks before the volume leaves the printer's, for post-proceedings. If SPS does not receive a reply from a particular contact author, within the timeframe given, then it is presumed that the author has found no errors in the paper. The tight publication schedule of LNCS does not allow SPS to send reminders or search for alternative email addresses on the Internet.

Lloyd Algorithm: It is a batch centroid model. Being a batch algorithm, it applies the transformative step to all the cases at once. It is well suited for the analysis of very huge datasets, subsequently, the gradual k-means algorithm need to hoard cluster membership afterward iteration respectively, this method would save the computational expenses. Although it is slow in data convergence, it optimizes the total sum of squares and it is best for discrete data distribution.

Macqueen Algorithm: It is an iterative algorithm. It is same as Lloyd algorithm in terms of instatement, however, varies from it at the season of recalculation of centroid, i.e., done each time a case change subspace and moreover after each experience all cases. For each case straightaway, if the centroid of then subspace it beginning at now has a place with the can't avoid being the closest, no change is made. If there should be an occurrence of another centroid being the closest, the case is reassigned to the accompanying centroid and the centroid for both the old and new subspaces are recalculated as the mean of the having a place cases. This algorithm would refresh the centroid all the more frequently consequently making it more proficient. Despite the fact that this algorithm has a quick starting meeting and it improves add up to total square, however, it needs to store the two closest cluster calculations for each case, making it all the more computationally costly.

Harrington–Wong Algorithm: It uses SSE (sum of squares of errors) within the cluster to search for the partition of data space. This implies that it will dole out a case to a different subspace, regardless of the possibility that it at present has a place with the subspace of the nearest centroid, this is just done if by doing as such the aggregate inside cluster sum of square is minimized. The initialization of cluster centers is done in the same manner as that done in Lloyd and Macqueen algorithms. The cases are then allocated to the centroid closest to them and the centroids are figured as the mean of the appointed data points. Though this algorithm has fast initial convergence and it optimizes within-cluster sum of squares, but like Macqueen algorithm, it also requires to store the two nearest cluster computations for each case.

4 Methodology and Results

We have used R programming language as a platform to extract data and to do data preprocessing. On this data, we applied clustering. But before applying clustering algorithm, we have used the elbow method which gave us an appropriate number of clusters that should be used for the dataset.

Elbow Method: K-means algorithm is somewhat nave in the sense that it will generate clusters for any number of 'k,' hence we might get vague results. To tackle this problem, we need to find the right number of 'k' before applying the algorithm and hence we use the elbow method. It is a graphical method in which we run k-means clustering on the dataset for a range of values of 'k' (say from 1 to 15) and, for each value of 'k,' we calculate the SSE and then we plot the line for each value of SSE corresponding to 'k.' Our objective is to select a small value of 'k' which has a small SSE. After plotting the line, we consider the line as an arm and the place from where it is bent is the point where the elbow should be and that point gives us the number of 'k' that should be used. For example, see Fig. 7. For our data, which consisted of 270 stocks data of past 5 years having the following attributes: stock ticker, sector, opening price, closing price, adjusted close price, volume, high and low prices, we decided that we should use K-means clustering. Figure 3 shows a flowchart of K-means clustering. As discussed above in section III that judging by the dataset we can choose any of the three algorithms for k-means, so in our case, we chose Harrington–ong algorithm as it gave us the required results while computing reasonably.

This data was not ready for clustering in the means that we didn't have any solid parameters on whose basis we could have done clustering. So, we calculated the variation and %change of each stock.

$$Variation = close - open \tag{5}$$

$$\%change = ((adjclose[i] - adjclose[i - 1])/adjclose[i - 1]) * 100 \tag{6}$$

After finding out these values, we had a figure something like shown in Fig. 5. After combining the recent values of all the 270 stocks, we created a figure which is ready for clustering, see Fig. 6.

On this data, we applied the elbow method, which gave us the apt value of K in k-means clustering, see Fig. 7. In Fig. 7, we can see that the elbow bent is coming on the $k = 4$ point. Hence, the number of clusters appropriate for this data is 4. We applied k means with $k = 4$ and got the following result as shown in Figs. 8 and 9. We can see four centroids and four clusters, maximum number of stocks lying in the

Num	% change	Variation	adj close	Ticker
1	0	13.67999	43.96215	AAPL
2	4.09348534	11.910019	45.76173	AAPL
3	-1.3306534	-6.959999	45.1528	AAPL
4	0.46483053	-0.980022	45.36268	AAPL
5	-0.656899	-1.910004	45.0647	AAPL
6	-3.4959561	-13.199981	43.48925	AAPL
7	16.3285322	54.52998	50.59041	AAPL
8	-1.4469323	-12.94999	49.8584	AAPL
9	-0.912095	-4.499996	49.40365	AAPL
10	6.15231503	24.410012	52.44311	AAPL
11	-5.5783414	-15.210015	49.51766	AAPL
12	5.96546779	22.460007	52.47162	AAPL
13	12.7111039	47.079982	59.14134	AAPL
14	18.8310661	84.030014	70.27829	AAPL
15	10.5283501	51.380028	77.67743	AAPL
16	-2.5969513	-17.850021	75.66018	AAPL
17	-1.0702366	-7.16999	74.85044	AAPL

Fig. 5 Changed parameters of Apple Inc. stock

green area which is cluster number 3 from Fig. 9 consisting of 171 stocks. Red region has the second most number of stocks and it corresponds to cluster number 4 having 48 stocks. Blue and cyan regions correspond to regions 1 and 2, respectively, having 12 and 37 stocks. This is not the only way we can cluster the stocks. We can also cluster them based on the sector, so we found out the number of stocks belonging to a particular sector and we got the result which is shown in Fig. 10. From Fig. 10, we can see that the most number of stocks belong to the consumer discretionary/staples, i.e., 63 followed by financial which have 50 stocks and the least number of stocks belong to telecommunication services sector with just 2 stocks.

5 Conclusion and Future Scope

In a nutshell, we admitted about the clusters that a stock belongs to both performance wise and sector wise. From this information, applied machine learning algorithms, appropriate to cluster a stock belong to and predict the stock prices. The future work

S.No	%change	Variation	Adj Close	Stock
1	3.8186624	1.91	50.84	A
2	0.0274294	-0.120003	36.46	AA
3	12.02046	8.320007	135.35001	AAPL
4	3.8989568	2.610001	90.32	ABC
5	5.8175748	2.690003	44.2	ABT
6	6.01563	6.919998	120.72	ACN
7	4.895046	5.620002	118.93	ADBE
8	9.3808364	6.209999	81.97	ADI
9	0.360684	-0.380002	44.1	ADM
10	-1.0991197	3.849998	99.88	ADP
11	4.6717521	4.050003	85.14	ADSK
12	0	0	41.98	ADT
13	0.5508015	0.699997	52.94	AEE
14	0.3642766	0.119999	63.7	AEP
15	1.8356643	0.25	11.65	AES
16	9.172923	11.190002	129.49	AET
17	2.3433948	2.190002	71.19	AFL
18	12.810085	28.039994	246.93	AGN
19	-3.0967973	-2.48	62.27	AIG
20	3.3159977	1.32	45.17	AIV

Fig. 6 Pre-Clustering values

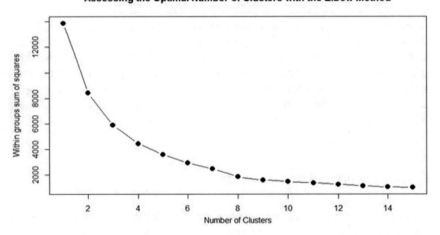

Assessing the Optimal Number of Clusters with the Elbow Method

Fig. 7 Elbow method

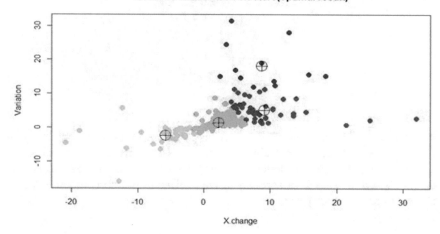

Fig. 8 K-Means clustering results

cluster	1	2	3	4
	AGN	AIG	A	AAPL
	AMGN	AKAM	AA	ACN
	AMP	AN	ABC	ADBE
	AMZN	APA	ABT	ADI
	BLK	APC	ADM	AET
	COST	ATI	ADP	ALL
	EFX	AVP	ADSK	AMD
	GOOG	BHI	ADT	AVY
	GS	BLL	AEE	BA
	HAS	CF	AEP	BAC
	ISRG	CHK	AES	BEN
	LMT	CTL	AFL	BIIB
		CTXS	AIV	BMY
		CVH	AIZ	BXP
		D	ALXN	C
		DF	AMAT	CAH
		DNB	AMT	CBG
		DNR	ANF	CL
		DVN	AON	CLF
		EOG	APD	CLX
		ESV	APH	CMA
		EW	APOL	COG
		EXPE	ARG	CSC
		FCX	AVB	CSCO
		FMC	AXP	CTSH
		FOSL	AZO	EA
		FTI	BAX	EMR
		FTR	BBBY	EXPD

Fig. 9 Cluster number of stocks

Fig. 10 Pie chart of number of stock belonging to a sector

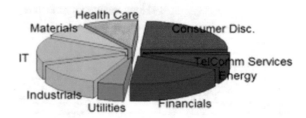

would be using the clustering results from this paper and accordingly applying algorithms of machine learning on them, like Artificial Neural Networks, Linear Regression, and Support Vector Machines. After applying these algorithms by training, the data we will find is the % of accuracy of the predicted value and propose our own method to predict stock prices.

References

1. Suresh Babu, M. et al. (2014). Clustering approach to stock market prediction. Int. J. Advanced Networking and Applications, *03*(04,NovDec), 1281–1291.
2. Martin Gavrilov, et al. (2000). Mining the stock market: Which measure is best? KDD 2000, Boston, MA USA ACM 2000 1-58113-233-6/00/08.
3. Nicolas Bertagnolli. (2015). Elbow method and finding the right number of clusters. https://www.nbertagnolli.com/jekyll/update/2015/12/10/Elbow.html.
4. Faujdar, N., & Ghrera, S. P. (2016). Performance evaluation of parallel count sort using GPU computing with CUDA. *Indian Journal of Science and Technology, 9,*. https://doi.org/10.17485/ijst/2016/v9i15/80080.
5. Neetu, F., & Satya Prakash Ghrera. (2015). Analysis and testing of sorting algorithms on a standard dataset. IEEE Fifth International Conference on Communication Systems and Network Technologies (CSNT), pp. 962–967.
6. Faujdar, N., & Ghrera, S. P. (2015). Performance evaluation of merge and quick sort using GPU computing with CUDA. *International Journal of Applied Engineering Research (IJAER), 10,* 39315–39319.
7. Neetu, F., & Satya Prakash Ghrera. (2015). A detailed experimental analysis of library sort algorithm. Annual IEEE India Conference (INDICON), pp. 1–6.
8. GopiKrishna Suvanam, & Amit Trivedi. (2013). Imbalances created because of structured products in Indian equity markets. National Stock Exchange.
9. Romero et al. (2003). Discovering prediction rules in aha! Courses: Proceedings of the International Conference on User Modelling, pp. 25–34.
10. Baldonado, M., Chang, C.-C.K., Gravano, L., & Paepcke, A. (1997). The stanford digital library metadata architecture. *International Journal on Digital Libraries, 1,* 108–121.
11. Gupta, A. (2014). nClustering-classification based prediction of stock market future prediction. *International Journal of Computer Science and Information Technologies, 5*(3), 2806–2809.
12. Gandhmal, D. P., et al. (2011). An optimized approach to analyze stock market using data mining technique. *Proceedings of International Conference on Emerging Technology Trends, 1,* 38–42.
13. Wang, J., Wu, X., & Zhang, C. (2015). Support vector machines based on K-means clustering for real-time business intelligence systems. *International Journal of Business Intelligence and Data Mining, 1,* 5464.

14. Nanda, S. R., Mahanty, B., & Tiwari, M. K. (2010). Clustering Indian stock market data for portfolio management. Expert Systems with Applications, pp. 8793–8798.
15. Kim, K. -J., & Ahn, H. (2008). A recommender system using GA K-means clustering in an online shopping market. Expert Systems with Applications, pp. 1200–1209.
16. Wu, M.-T., & Yang, Y. (2013). The research on stock price forecast model based on data mining of BP neural networks. Proceedings of the 3rd International Conference on Intelligent System Design and Engineering Applications, pp. 1526–1529.
17. Nayak, S. C. et al. (2012). Index prediction with neuro-genetic hybrid network: A comparative analysis of performance. Proceedings of International Conference on Computing, Communication and Applications, pp. 1–6.
18. Pao, Y. H. et al. (1992). Neuralnet computing and intelligent control systems. International Journal of Control, 56.
19. Boseret, B. E. et al. (1992). A training algorithm for optimal margin classifiers. Proceedings of the 5th Annual ACM Workshop on Computational Learning Theory, pp. 144–152.
20. Kim, K.-J. (2013). Financial time series forecasting using support vector machines. Neurocomputing, 55, 307–319.
21. Tan, P.-N. (2006). Introduction to Data Mining. Pearson.
22. MacQueen, J. B. (1967). Some methods for classification and analysis of multivariate observations. Proceedings of 5-th Berkeley Symposium on Mathematical Statistics and Probability, Berkeley, University of California Press, pp. 281–297.

Analysis of Lean, Green, and Resilient Practices for Indian Automotive Supply Chain Performance Using Best–Worst Method

Vernika Agarwal

1 Introduction

The Indian automotive industry is the fourth largest manufacturer of cars and the sixth largest manufacturer of commercial vehicles in 2018 [1] which is expected to reach INR 16.16–18.18 trillion (US$ 251.4–282.8 billion) by 2026 [2]. The automotive sector has been highlighted by the academic sector for this tremendous growth. Government legislation and the companies involved seek more environmental protection from manufacturers [3]. Given the growing interest in improving upon their ecological and societal image, a substantial number of automotive manufacturers are now exploring the possibility of integrating newer paradigms such as Lean, Green, and Resilient (LGR) into their traditional operations [4]. Over the last few years, much attention has been paid by the supply chain practitioners and academicians to being environmentally responsible [5]. The two major approaches in this direction are green and lean practices, which can help the industries in reducing wastes and thereby reducing the pollution levels [6]. The green aspect of the supply chain varies from managing the environmental programs externally and internally with the other channel partners [7]. The lean practice, on the other hand, can be used by the industries for reducing the waste and thereby, improving the overall efficiency [8]. Another essential paradigm in recent times is the resilient practices. This paradigm focuses on the problems responding efficiency of the supply chain (SC) in view of unexpected disturbances. Basically, the focus of the resilient paradigm is to reduce the unexpected impact of external agents in the SC, while the lean and green focus on reducing the environmental impact [9]. Together, these three paradigms can aid the manufacturing industries to overcome inefficiencies, reduce wastes, and attain

V. Agarwal (✉)
Amity International Business School, Amity University, Sector 125, Noida 201303, Uttar Pradesh, India
e-mail: vernika.agarwal@gmail.com

© The Author(s), under exclusive license to Springer Nature Singapore Pte Ltd. 2021
P. K. Kapur et al. (eds.), *Advances in Interdisciplinary Research in Engineering and Business Management*, Asset Analytics,
https://doi.org/10.1007/978-981-16-0037-1_27

349

sustainability in their operations. Since it is difficult to incorporate all the LGR practices simultaneously into the business operations, thus, the present study aids in
selecting the prominent practices that could be included in the working. The detailed
aims of the present study are as follows:

1. To identify the major LGR practices that can aid the Indian automotive supply
   chain.
2. To prioritize the LGR practices using best–worst analysis.

Discussions are held with the various academic and industry experts for the identification of the major LGR practices. BWM is used to prioritize these LGR practices,
based on discussions with various stakeholders.

The present research work is divided as follows: Sect. 2 discusses the problem
description which is followed by research methodology in Sect. 3. The illustrative
application and the result discussion are given in Sect. 4 and Sect. 5, respectively.
Finally, Sect. 6 gives the conclusion and limitations of the study.

2 Problem Description

The current study addresses the issues of inclusion of LGR practices into the supply
chain operations of Indian automotive manufacturing companies situated in the
National Capital Region (NCR), Delhi, India. The pressure from the stakeholders
demanding the inclusion of ecological and societal aspects into the operations is
making the automotive companies think about innovative strategies to sustain in
business. In this direction, the inclusion of LGR practices is a positive step that can
aid the companies to stay in the competition. However, since the inclusion of all
practices is a long-term goal, the companies need to identify the prominent LGR
practice that they can include. In the present study, companies dealing with automobile parts. The experts from these companies are consulted in structured and
unstructured interviews to understand which LGR practice is essential.

Hence, the research question addressed in the present study includes

- What are the major LGR practices that are needed for efficient working o the
  company?
- How to prioritize these LGR practices so that the objectives of the company are
  met?

3 Research Methodology

For the selection of the most prominent lean, green, and resilient (LGR) practices
from the initial list, we utilize the best–worst method (BWM). For this purpose, we
have consulted three senior DMs of the automobile part manufacturing company
situated in NCR. The DM team comprises the quality manager (with 5 years of

experience), procurement manager (with 7–10 years of experience), and the chief financial officer (with 10 years of experience). The BWM is utilized to generate weights of the LGPR using minimal comparisons, which is the highlight of this method over other MCDM techniques. It is only required that the best and the worst factors should be compared with other factors. This reduces the complexity and time of the decision-making process [10]. BWM consists of a smaller set of factors for pairwise comparisons. The comparisons matrix is made with respect to the best and the worst factors. The major advantage of using BWM is that it uses only two comparison vectors as compared to AHP (analytical hierarchy process). Thereby, reducing the complexity and time of the decision-making process.

Let the chosen LGR set be $L^C = \{L_1^C, L_2^C, ..., L_K^C\}$ practices. In this method, we evaluate the LGR practices by a set of m decision-makers (m = 1, 2, 3). Then the decision-making process of BWM comprises of the following steps [11, 12]:

Step 1. Identification of best and the worst LGR practice.

The DMs are told to identify the best and the worst LGR practices.

Step 2. Determination of the preference of best LGR practice over the decision set L^C.

The DMs are asked to give the preference of best LGR practice over the others, using a score of 1–9. The resulting vector of "Best-to-Others" is given as.

$$w = (w_{B1}, .., w_{BK}),$$

where w_{BK} gives the preference of the best LGR practice B over kth attribute and $w_{BB} = 1$.

Step 3. Determining the preference of worst LGR practice over the decision set L^C.

In this step, the preference of the others over the worst LGR attribute is determined using a score of 1–9. The resulting vector of "Worst-to-Others" is given as.

$$\overline{w} = (\overline{w}_{1W}, .., \overline{w}_{KW})^T,$$

where $\overline{w}_{kW}$ gives the preference of the kth attribute over worst economic attribute and $\overline{w}_{BB} = 1$.

Step 4. Computation of the optimal weights of LGR practices.

In this step, the optimal weighting vector $(y_1^*, ..., y_K^*)$ of the LGR practices are calculated.

The optimal weight of the kth attribute is the one that meets the following requirements: $\frac{y_B^*}{y_k^*} = w_{Bk}$ and $\frac{y_k^*}{y_W^*} = \overline{w}_{kw}$.

To satisfy the above condition, the maximum absolute difference $\left|\frac{y_B^*}{y_k^*} - w_{Bk}\right|$ and $\left|\frac{y_k^*}{y_w^*} - \overline{w}_{kw}\right|$ needs to be minimized for all LGR practices.

Thus, the optimal weighs for LGR practices can be obtained by solving the following programming problem:

$$\min_{k} \max \left\{ \left| \frac{y_B^*}{y_k^*} - w_{Bk} \right|, \left| \frac{y_k^*}{y_W^*} - \overline{w}_{kw} \right| \right\}$$

Subject to

$$\sum_{K} y_k = 1 \tag{P1}$$

$$y_k \geq 0 \, \forall k = 1, 2, ..., K;$$

Problem (P1) is equivalent to the following linear programming formulation (P2):

$$\min \, \varphi$$

Subject to

$$|y_B - w_{Bk} y_k| \leq \varphi \, \forall k = 1, 2, ..., K$$
$$|y_k - \overline{w}_{kw} y_W| \leq \varphi \, \forall k = 1, 2, ..., K$$
$$\sum_{K} y_k = 1 \tag{P2}$$
$$y_k \geq 0 \, \forall k = 1, 2, ..., K$$

The above problem (P2) is linear in nature and has a unique optimal solution. On solving problem (P2), we get the value of consistency ratio φ^* and optimal weights $(y_1^*, ..., y_K^*)$.

When the value of the consistency ratio φ^* is closer to zero value, the more consistent the comparison system provided by the DMs.

Step 5. Check the consistency of the solution.

The closer the consistency ratio is to zero value, the more consistent is the comparison system provided by the DM. We check the consistency of the solution by calculating the consistency ratio:

$$\text{Consistency Ratio} = \frac{\varphi^*}{\text{Consistency Index}}$$

Table 1 Consistency index table for BWM

v_{Bi}	1	2	3	4	5	6	7	8	9
Consistency index (max)	0.00	0.44	1.00	1.63	2.30	3.00	3.73	4.47	5.23

The value of consistency ratio closer to '0' shows more consistency, while values closer to 1 show less consistency.

4 Case Illustration

The application of the proposed methodology is divided into two steps, namely, research design and data collection. The details of each step are summarized in the below sections:

4.1 Research Design

The purpose of this step is to identify and prioritize the LGR practices for manufacturing enterprises situated in the Greater Noida region. The focus of the present analyses is to understand the impact of LGR practices in the manufacturing industries. The current study targeted experts and academicians from the Indian manufacturing sector company from NCR, Delhi, India, to identify and evaluate the prominent LGR practices. The opinion of these experts was incorporated in form of personal interview to finalize the nine (9) LGR practices as given in Table 1. Delphi methodology was used to avoid conflict of opinion (Table 2).

4.2 Application of the Best–Worst Methodology (BWM) for the Case

After the collection of data, the BWM was applied to the decision-making process. Following the steps of BWM, the experts were told to identify the best and worst LGR practices give the corresponding reference comparison values on a scale of 1–9. This data was used to generate the weights using the BWM.

The data collected from the expert panel is used to illustrate the methodologies and to develop a framework for identifying the most important LGR practices. Based on the opinion of experts the best LGR practices were found to be "LGR6: Compliance with ISO 14,001," while the worst LGR practice was "LGR 8: Business Wastage" based on the consensus of the expert panel. Following the next step, we need to give the preference rating of the best criteria over all other criteria and give the preference

Table 2 LGR practices

Notation	Criteria	Description	References
LGR1	Just-in-time (JIT) inventory	To increase efficiency and reduce waste companies use JIT to receiving goods only as they are needed in the production process	[9, 14, 15]
LGR2	Total quality management (TQM)	TQM is the process of managing the quality of the product being generated in the company. It helps to maintain the quality of the product and provide a competitive edge to them	[16]
LGR3	Strategic stock	Inventory levels that are either too short or too big are very bad for the firm. They should be set on the basis of accurate forecasting and good strategic planning	[17]
LGR4	Flexible transportation	An organization should properly utilize its transportation facility, and if it's outsourcing it, then also there should be some proper guidelines to achieve the required target	[17]
LGR5	Environmentally friendly packaging	It is the basic and most important practice which every industry can use to go green	[18]
LGR6	Compliance with ISO 14,001	By complying with ISO14001, the firm earns the respect of customers and builds a good public image which gives a major boost to their profits	[18, 19]
LGR7	Operational cost	It is the basic expenses which are incurred during the production or operation of an activity	[20, 21]
LGR8	Business wastage	It is the wastage due to over/under production	[9]
LGR9	environmental cost	The cost associated with expenses related to the ecological effect of your organization's business tasks	OWN

rating of the other criteria over the worst criteria on a scale of 1–9. These pairwise comparisons are shown in Table 3.

Weights will be calculated using P2 and coded and solved with LINGO, an optimization software. The weights are shown in Fig. 1.

Based on the value of φ, we calculate the CR. Here, we get the value of $\varphi = 0.07738$, and the value of CR is given as 0.014, which is closer to '0.' Thus, the solution is considered consistent.

Table 3 Pairwise comparison for the LGR practices	Best to others	LGR6	Others to the worst	LGR8
	LGR1	4	LGR1	6
	LGR2	3	LGR2	8
	LGR3	5	LGR3	4
	LGR4	7	LGR4	2
	LGR5	2	LGR5	7
	LGR6	1	LGR6	9
	LGR7	8	LGR7	2
	LGR8	9	LGR8	1
	LGR9	6	LGR9	3

5 Results and Discussions

The integration of LGR paradigms into the operations of the manufacturing industries can enhance their overall ecological and societal image. In this direction, the current paper asses the prominent practices that these enterprises can incorporate. With the aid of the BWM approach, these LGR practices are evaluated and found that "LGR6: Compliance with ISO 14,001" is the best in the opinion of the academicians and industry experts. In recent times, proper certifications are necessary for sustaining in the market. The legislative pressure and social awareness toward ecological concerns are putting pressure on the manufacturing sector to get necessary documentation [18]. Similarly, "Environmentally Friendly Packaging (LGR5)" is acting as the next best practice. In order to balance their business efficiencies while trying to include sustainability aspects, it is important for the organizations to realize the mitigating impacts of their SC and to redesign their operations so as to reduce the environmental impact while enhancing the social upliftment across the entire SC network [22]. The easiest step in this direction is to include green packaging and reduce the wastage due to plastic and other non-biodegradable products. On the other side, Business Wastage (LGR8) is acting as the worst practice from the expert's point of view. Business wastage is the waste due to under/overproduction, since we are focusing on JIT and TQM, and hence, this practice will be taken care of by these practices.

6 Conclusion

The concept of sustainability is increasing at a fast pace globally, specifically for automobile manufacturing companies in India. The fast varying government norms and legislations related to ecological and societal welfare are forcing the industries toward these issues. Due to the enormous development and high emissions, the automobile industry is in the spotlight. Moreover, intense competition among these sectors is another important cause for these industries to develop new ways of incorporating sustainability into their manufacturing and procurement activities. This led to LGP practices being included in their work.

A total of nine LGR practices were shortlisted for the present study based on discussions with various industry and academic experts. The present study used a recently developed methodology of BWM for prioritizing the vital LGR practices for the inclusion of sustainability based on the opinion of the expert panel.

The case of the automobile manufacturing companies situated in northern India is considered for validating the study. The best LGR practice was found to be "LGR6: Compliance with ISO 14,001"; on the other hand, the worst was "LGR8: Business Wastage" based on the opinion of the expert panel. Thus, the present study provides a framework for the analysis of LGR for the inclusion of sustainability for the automobile industry.

7   Future Scope

There are a few limitations to this study. Presently, BWM is used for the prioritization of the LGR practices, with crisp inputs; subjectivity can also be incorporated into the decision-making process by the use of Fuzzy BWM. A comparative study can also be done between various MCDM techniques like VIKOR, AHP, and ELECTRE for this study.

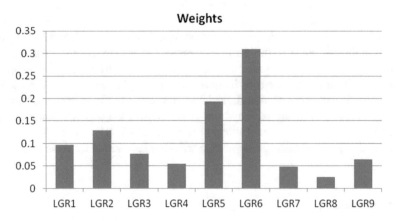

Fig. 1 Weights of LGR practices rocess

References

1. Bansal, P., Kockelman, K. M., Schievelbein, W., & Schauer-West, S. (2018). Indian vehicle ownership and travel behavior: A case study of Bengaluru, Delhi and Kolkata. *Research in Transportation Economics 71*, 2–8.
2. IBEF. (2019). Retrieved June 30, 2019, from https://www.ibef.org/industry/automobiles-presentation.
3. Gupta, V., Narayanamurthy, G., & Acharya, P. (2018). Can lean lead to green? Assessment of radial tyre manufacturing processes using system dynamics modelling. *Computers & Operations Research, 89*, 284–306.
4. Paksoy, T., Çalik, A., Kumpf, A., & Weber, G. W. (2019). A new model for lean and green closed-loop supply Chain optimization. In Lean and Green Supply Chain Management, 39–73.
5. Malik, M. M., Abdallah, S., & Hussain, M. (2016). Assessing supplier environmental performance: Applying analytical hierarchical process in the United Arab Emirates healthcare chain. *Renewable and Sustainable Energy Reviews, 55*, 1313–1321.
6. AlJaberi, O. A., Hussain, M., & Drake, P. R. (2017). A framework for measuring sustainability in healthcare systems. *International Journal of Healthcare Management, 1–10.*
7. Srivastava, S. K. (2007). Green supply-chain management: A state-of-the-art literature review. *International Journal of Management Reviews, 9*, 53–80.
8. Hussain, M., Al-Aomar, R., & Melhem, H. (2019). Assessment of lean-green practices on the sustainable performance of hotel supply chains. International Journal of Contemporary I Management.
9. Cabral, I., Grilo, A., & Cruz-Machado, V. (2012). A decision-making model for lean, agile, resilient and green supply chain management. *International Journal of Production Research, 50*, 4830–4845.
10. Rezaei, J. (2015). Best-worst multi-criteria decision-making method. *Omega, 53*, 49–57.
11. Govindan, K., Jha, P. C., Agarwal, V., & Darbari, J. D. (2019). Environmental management partner selection for reverse supply chain collaboration: A sustainable approach. *Journal of Environmental Management, 236*, 784–797.
12. Rezaei, J., Wang, J., & Tavasszy, L. (2015). Linking supplier development to supplier segmentation using Best Worst Method. *Expert Systems with Applications, 42*, 9152–9164.
13. Cabral, I., Grilo, A., & Cruz-Machado V. (2012). A decision-making model for lean, agile, resilient and green supply chain management. *International Journal of Production Research, 50*, 4830–4845.
14. Ruiz-Benitez, R., López, C., & Real, J. C. (2017). Environmental benefits of lean, green and resilient supply chain management: The case of the aerospace sector. *Journal of Cleaner Production, 167*, 850–862.
15. Ciccullo, F., Pero, M., Caridi, M., Gosling, J., & Purvis, L. (2018). Integrating the environmental and social sustainability pillars into the lean and agile supply chain management paradigms: A literature review and future research directions. *Journal of Cleaner Production, 172*, 2336–2350.
16. Hussain, M., Al-Aomar, R., & Melhem H. (2019). Assessment of lean-green practices on the sustainable performance of hotel supply chains. *International Journal of Contemporary Hospitality Management.*
17. Govindan, K., Azevedo, S. G., Carvalho, H., & Cruz-Machado, V. (2014). Impact of supply chain management practices on sustainability. *Journal of Cleaner Production, 85*, 212–225.
18. Govindan, K., Darbari, J. D., Agarwal, V., & Jha, P. C. (2017). Fuzzy multi-objective approach for optimal selection of suppliers and transportation decisions in an eco-efficient closed loop supply chain network. *Journal of Cleaner Production, 165*, 1598–1619.
19. Bai, C., & Sarkis, J. (2010). Integrating sustainability into supplier selection with grey system and rough set methodologies. *International Journal of Production Economics, 124*, 252–264.
20. Aguezzoul, A. (2014). Third-party logistics selection problem: A literature review on criteria and methods. *Omega, 49*, 69–78.

21. Prakash, C., & Barua, M. K. (2016). A combined MCDM approach for evaluation and selection of third-party reverse logistics partner for Indian electronics industry. *Sustainable Production and Consumption, 7,* 66–78.
22. Shen, C. W., & Chou, C. C. (2010). Business process re-engineering in the logistics industry: A study of implementation, success factors, and performance. *Enterprise Information Systems, 4,* 61–78.

Green Internet of Things: Next-Generation Intelligence for Sustainable Development

Mohini Jain, Gurinder Singh, and Loveleen Gaur

1 Introduction

As per United Nations Predictions, population of the world will increase to 10 billion by 2050 as compared to the existing 7.5 billion [1]. This can be regarded as a consequence of various factors like enhanced lifestyle, achievements in medical world and many others. With advancements in the field of technology brought by Internet of Things (IoT), providing anytime, anywhere access to information. On the other hand, it has affected society in negative ways like more greenhouse gas, reliance on non-renewable resources for energy generation and more raw material consumption. This has made the adoption of green technology, a very important aspect. This is how Green Internet of Things(G-IoT) comes into the picture.

As per recent trends, migration of people to cities has accelerated thus increasing urbanization. It has been anticipated that 70% of world population will move to cities by 2050 [2]. Hence, it has become important to make cities smart in order to enable economic, social and environmental wellbeing by effectively using G-IoT. To achieve the long-term objective of a cleaner and sustainable energy system, it becomes imperative to not just focus on 'from where' we generate the energy but it is equally important to allocate and handle the produced renewable energy optimally and judiciously. The fact that the energy generation and supply from renewable energy sources is not certain is the reason we must work towards its efficient utilization.

M. Jain (✉) · G. Singh · L. Gaur
Amity International Business School, Amity University, Sector 125, Noida 201303, Uttar Pradesh, India
e-mail: mohini.jain35@gmail.com

G. Singh
e-mail: gsingh@amity.edu

L. Gaur
e-mail: lgaur@amity.edu

2 Green Internet of Things (G-IoT)

IoT can be defined as a network of interconnected devices through which they can exchange data and demand. These devices include home appliances, vehicles, objects implanted with sensors, software, actuators and are characterized by connectivity with the internet. According to Gartner Research, the connected online and electronic devices will increase to 30 billion objects in 2020 from 8.4 billion in 2017. Despite numerous benefits of IoT, an increase in smart device usage in day-to-day life contributes to increased resource and energy consumption. On the other hand, it is also associated with e-waste and hazardous emissions energy. Henceforth, in order to lessen the negative influence of such waste, effective utilization of technology is necessary for human and environment. For the same reason, we are moving towards a greener future where IoT is being replaced by G-IoT, technology with green technology so that we focus more on renewable resources for energy generation.

G-IoT involves the utilization of technology without depleting the natural resources, relying on renewal resources. The idea of sustainable development leads to the evolution of G-IoT, though it is still in the early stage of maturity but if we are able to diffuse the green technology in our day-to-day life, we can move to achieve the goal of a green economy [3].

G-IoT is recognized on same communications protocols as IoT. G-IoT consists of two facets:

- Effective utilization and discarding of green devices and technologies to lessen pollution and carbon emission, thus enhancing energy efficiency.
- Designing and manufacturing green computing devices, networking designs and communications protocols with amplified power consumption and maximized bandwidth utilization.

Some of the goals of G-IoT can be summarized as effective utilization of energy, decreasing contamination and waste, utilization of non-renewable resources. In order to fulfil these goals, it is important to accomplish he following tasks [4–6]:

- Effectively use renewable sources of green power: wind, air, solar, geothermal, water, biogas sources. We can expect biodegradable batteries and wireless power.
- Reduce/save energy consumption by not using the facility when not in use (G-IoT equipment carry out sleep scheduling algorithms to switch off the appliance when not in usage).
- Use of bio-products and eco-friendly design in manufacturing G-IoT products and components and many more (Fig. 1).

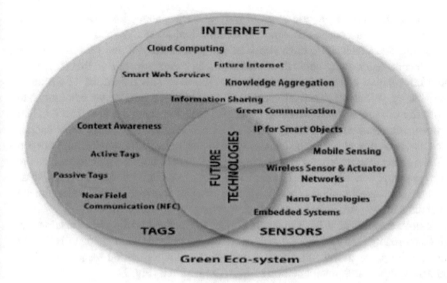

Fig. 1 G-IoT enablers [9]

3 G-IoT and Sustainable Development

The primary aim of G-IoT is to satisfy energy efficiency which is imperative for sustainable development and greener future. In order to achieve the goal of G-IoT of green design, green processing, green utilization, green communication, green disposal for sustainable development, the following things need to be considered and followed:

- Use of bio-products in manufacturing via eco-friendly design.
- Turning off G-IoT devices when not in use, and programming sleep scheduling algorithm.
- Emphasizing usage of renewable power sources such as air, geothermal, wind, solar, biogas sources.
- Executing unconventional communication methods (e.g. cognitive radio utilization);

3.1 Smart Grid

IoT acts as the first step towards self-sustainability and energy optimization at the residential level. IoT works in cooperation with smart grid technologies and helps in setting up an Energy Management System with decision-making capabilities in which all the devices such as sensors, cameras and actuators work as smart devices. In a way, IoT creates smart homes [7].

The traditional power grid allowed only one-way interaction between the grid and the consumer households which made it difficult to analyse and respond to the changing and increasing power needs of the households, whereas the new smart grid technologies along with IoT enable two-way communication so that electricity and information can be exchanged between the two. This makes the grid a more reliable and efficient source of energy. The utilities can now better communicate with the consumer to help manage and moderate the energy needs. The detailed information enables the grid operator and the consumer to manage their energy consumption in real time. The technology enables a direct communication between all the smart devices to provide information and data to both the user as well as the energy management system. The smart grid works in integration with the wind, solar and photovoltaic energy sources along with efficient consumer participation for the judicious utilization of the total energy produced.

Another aspect of IoT-based smart grid is that load scheduling for consumers is done more efficiently so that they can work on heavy loads when demand is low. In order to improve power transmission, reduce the blow of natural resources, better understanding and implementation of G-IoT is required (Fig. 2).

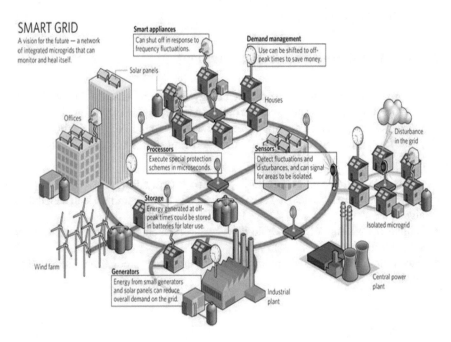

Fig.2 Smart grid

3.2   Smart City

"Technology must not only be used to upgrade and transform but also to reshape nation and its cities."[8]

IoT opens the gateway to create a smart city. To build a smart city, there has to be absolute and best utilization of all the available resources and IoT when works in integration with connected smart devices, facilitates smooth and easy functioning within the city. IoT applications improve mobility in a city and make it more citizen-friendly. Real, digital and virtual things interact with each other and make the cities smarter. This raises the standard of living for all citizens. The optimum utilization of the available electricity helps in countering the energy fluctuations as the wind and solar energy sources are variable by nature. For example, energy generation through a wind turbine will be less on a windless day and more electricity will be produced from a solar power plant on a sunny day.

IoT operates at various levels and works with embedded devices. It thereby provides valuable data about the needs of people within any community. It further presents solutions to address those needs too. IoT connectivity with Wi-Fi-enabled smart devices and sensors allow easy and real-time communication throughout the city which makes crowded cities more manageable, sustainable and affordable [9].

Proper implementation of IoT can do wonders in boosting infrastructure and thus open up a lot of opportunities for building a sustainable environment for generations to come. The number of applications of IoT in a smart city could be.

3.2.1   Standby and Specific Lighting

The street lights could act as the key to IoT connectivity because each and every street light will be used for gathering and transmitting information. A lot of electricity is wasted when street lights are left on all night-long. Standby lighting avoids unnecessary wastage of electricity as the sensors can detect the presence of individuals around and automatically switches it on/off as and when required. The specific street lighting would enable ease and safety for specially challenged people in crossing streets alone.

3.2.2   Smart Parking Spots

IoT allows smart communication between various devices. This real-time conversation tells the user if there is any free or empty parking space nearby.

A lot of fuel and time can be preserved when one does not need to waste time looking for available parking spaces across a city.

3.2.3   Easy Waste Management System

Through IoT technology, the waste collection companies would have a better under-standing of how to fill the waste containers are and they can then come and pick the garbage. This promotes a smart and efficient waste management system for all and helps making the city cleaner and more beautiful.

3.2.4   Automatic Sensors

Sensors could be established at regular intervals to provide updates and information about Air pollution, noise pollution and check river levels to avert floods. Also, IoT would manage the automatic watering system throughout the city that may as well detect leaks. There will be no requirement to regularly monitor the water or electricity meters as the consumption readings will be presented in real time through the data and information provided by IoT. This too will save resources.

3.2.5   Helps Finding Available Charging Stations

IoT connectivity all across the city could help in finding available nearby charging stations. Through easy access to information, one can easily plug their car in as soon as they need to.

There are so many other possibilities that IoT technology allows which can facil-itate smart functioning within a city. Some other benefits from IoT connectivity could be that in case of accidents, an automatic alert would go out to the concerned authorities immediately and through Remote Monitoring, a prompt update on the situation could be provided. Also, other drivers may receive cautions on GPS or through linked road signs. Traffic lights too can be adjusted to monitor the road traffic flow and minimize circulation jams. Such real-time information can lessen traffic congestion and make crowded areas more manageable. More improvements can be made in the road and public transport network through the data collected by IoT-enabled assets (Fig. 3).

4   Case Study: Irish Energy Company

The UK-based start-up—Solo Energy, launched in 2015, has deployed the IoT exper-tise to accomplish a network of energy storage units. The company was started as a facility to the renewable energy market or one can say, it started off by providing 'Energy-as-a-service'. Later, it launched a number of home energy offerings which worked on the concept of distributed energy. It includes solar and electric vehicle charging and energy storage batteries. The purpose of the company was to provide a facility for energy storage to its customers at far-reaching places. These batteries

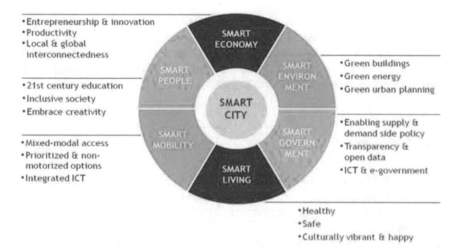

- Entrepreneurship & innovation
- Productivity
- Local & global interconnectedness

- 21st century education
- Inclusive society
- Embrace creativity

- Mixed-modal access
- Prioritized & non-motorized options
- Integrated ICT

- Green buildings
- Green energy
- Green urban planning

- Enabling supply & demand side policy
- Transparency & open data
- ICT & e-government

- Healthy
- Safe
- Culturally vibrant & happy

Fig. 3 Smart city [16]

enable the users to store the excess energy they generate and efficiently utilize that energy when needed or at times of shortage of power supply. The companies provide batteries throughout Ireland and the UK and help creating a flexible balance between energy demand and supply.

The CTO of the company believes that the storage technologies hold the potential to support the Irish Energy Management System which promotes a low-carbon economy. He further says "We want to enable deployments in remote locations". [10] By the end of the year 2018, the company plans to deploy 200 energy storage units across the entire UK and Ireland.

As of now, the customers include system operators and suppliers, as well as some renewable energy companies. The company seeks to expand its customer base and expects small business owners and domestic energy providers as its future clients.

4.1 Working

Through IoT technology, the company manages the functioning of battery at each site. The technology serves both the company as well as the customers of the storage batteries. It allows the user of these batteries to access each piece of information related to how the system is operating and how the power is being used. Data from these batteries is captured and managed through IoT technology and its sensors. Along with this, IoT reports other factors too that include weather, the time of the day and the day of the week for more detailed information. Moreover, IoT helps the company to collect all the data about storing the energy and health of its energy battery network.

The company has a Virtual Power Plant called Solo's FlexiGrid software platform that performs the task of managing all battery storage units. It is not a large industrial plant situated at one single place but serves as a connected system of batteries distributed at different locations. The platform helps in managing the battery-charging, thereby contributes to increasing the system's productivity.

4.2 Outcomes

i.   The IoT-based technology and its sensors along with the smart metering in houses could help the users of these batteries to not only generate and use their own electricity but also earn money by exporting the excess of energy to a central grid.
ii.   The energy storage batteries helped to bring a balance of management of energy supply and demand.
iii. The Virtual Power Plant assists in digitally linking various assets with each other and in turn helps in the creation of an intelligent system of renewable energy generation.

5 Conclusion

Since the traditional sources of energy generation have found to be polluting the natural environment, so these need to be replaced with newer and greener technologies that include wind and solar sources of power generation. However, these sources come up with their own battles and challenges. One of the biggest challenges is about how to effectively save this renewable energy and utilize it judiciously when the need arises. Also, it creates an obstacle on how to balance the power load on the main pillar of the electricity infrastructure, that is, the electricity grid. Henceforth, G-IoT comes as a solution. Scheming, manufacturing, usage and disposal of product and equipments in a green manner (without negatively influencing human health and environment) makes G-IoT the technology of the future. The dependence of G-IoT does not depend only on financial resources for sustainable development but also on individuals, governments and businesses taken together.

The case study of Irish company Solo Energy has been working on the mission of creating a 100 per cent renewable future. The main aim of the start-up company is to provide savings and its benefits to both the company and its clients. The storage batteries facilitate the users to store their excess of energy generated and then use that energy when the need arises. Further, in order to optimize the grid performance, Solo Energy has been working out ways on how to support the voltage on the grid at the local level. This helps in balancing and satisfying the demand on the power grid during peak times.

References

1. UNFPA (United Nations Population Fund). World Population Trends. Retrieved fromhttp://www.unfpa.org/world-population-trends
2. Jiong Jin, J. G. (2013). An Information Framework of Creating a Smart City through. *IEEE*.
3. UNEP (United Nations Environment Programme). (n.d.). Green economy initiative. Retrieved from www.unep.org/greeneconomy.
4. Leung, V. (2015). Green internet of things for smart cities. In International workshop on smart cities and urban informatics, Hong Kong. 22.
5. Pazowski, P. (2015). Green computing: Latest practices and technologies for ICT sustainability. Management, knowledge and learning. In Joint International Conference (pp. 1853–1860). Bari, Italy. 23.
6. Zhu, C., Leung, V., Shu, L., & Ngai, E. C. H. (2015). Green internet of things for smart world. *Access IEEE, 3,* 2151–2162.
7. Gapchup, A., Wani, A., Wadghule, A., & Jadhav, S. (2017). Emerging trends of green IoT for smart world. *International Journal of Innovative Research in Computer and Communication Engineering, 5*(2), 2139–2148.
8. Leung, V. (2015). Green internet of things for smart cities. International workshop on smart cities and urban informatics, Hong Kong.
9. Spring.smartcitiesconnect.org. (2019). *Smart Cities Connect - Day at Glance.* Retrieved November 19, 2018, from https://spring.smartcitiesconnect.org/program/.
10. Solo Energy. (2019). *Solo Energy.* Retrieved November 25, 2018, from https://www.solo-energy.com/.

Fabrication of Air Conditioning System Using the Engine Exhaust Gas

Rachit S. Balani, Dhirendra Patel, Subodh Barthwal, and J. Arun

1 Introduction

Cooling is the right path toward abstracting an extent of warmth from a kept or controlled space or from any substance and moving it to a spot where it is away from the object or space. From the initiation, the elucidation cooling framework gets starts by reducing the temperature of the encased space or the substance and some time later to keep it keep up at that lower temperature when veered from the circumventions of the encased space or substance. Algid is the nonappearance of warmth, thusly to lessen the temperature, one should "clear heat," instead of "checking cold." The essential focal point of stirring up a smoke upkeep refrigerant structure for movements is to make the transport cool from the space inside by using waste warmth and fumes gases from the motor. Nowadays, the cooling course of action of transports uses the "Smoke Compression Refrigerant System" (VCRS), which ingests to chop down the temperature while removing heat from inside the movement, where the cooling is required and further dispatches the shine to be somewhere else. Before long to develop the gainfulness of a movement past an all-out cutoff, the smoke pressure refrigerant structure needs to confine it as it can't use the fume gases from the motor. In a smoke pressure refrigerant framework, the structure uses control from the motor shaft as the information control needs to drive

R. S. Balani (✉) · D. Patel · S. Barthwal · J. Arun
Amity University, Noida, India
e-mail: rachitbalani9@gmail.com

D. Patel
e-mail: dpatel@gn.amity.edu

S. Barthwal
e-mail: sbathwal@gn.amity.edu

J. Arun
e-mail: jarun@gn.amity.edu

© The Author(s), under exclusive license to Springer Nature Singapore Pte Ltd. 2021
P. K. Kapur et al. (eds.), *Advances in Interdisciplinary Research
in Engineering and Business Management*, Asset Analytics,
https://doi.org/10.1007/978-981-16-0037-1_29
369

the blower of the refrigerant structure in the movement. Therefore, the motor needs to pass on additional work to run the blower of the cooling structure in this way using a supplemental extent of fuel. The loss of power of the transport for cooling gets ousted by understanding this sort of cooling framework (Figs. 1 and 2).

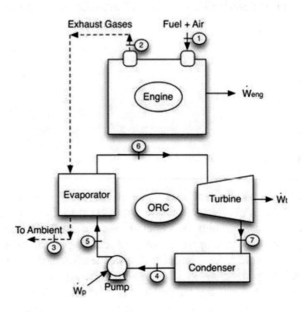

Fig. 1 A typical waste heat energy recovery system with ORC

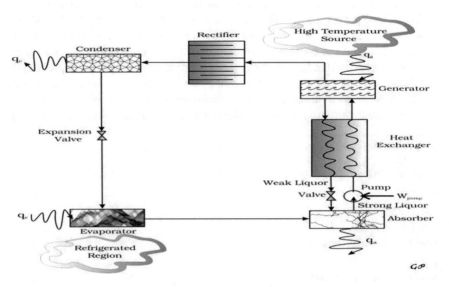

Fig. 2 Block diagram of Vapor absorption refrigeration system (VARS)

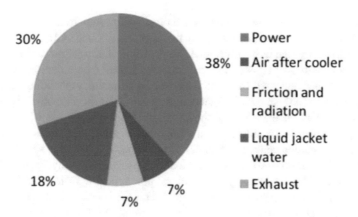

Fig. 3 Amounts of heat dissipated from engines

2 Literature Review

As many researches shown most HFC refrigerants have an appropriately very unsafe perilous environmental deviation potential. They have suggested that unending segment of the law in many community requires the utilization of refrigerants with a general temperature modification potential aftereffects of under 200–150 in all early kind developments starting in 2011 and each and each every early transport by year 2017.

In the work done by Mr. Angeloo, this engenderment contains a smoke engenderer part connected with a condenser which is itself joined, by pressure spoil valve, to an evaporator joined to a forfend which gets the fluid that is after a short time in the lower zone of the engenderer and of which fluid gets alimented to the upper zone of the engenderer, there is given in the engenderer a brilliance exchanger through which the motor cooling fluid streams. As necessities be, it utilizes the radiance of the cooling dihydrogen monoxide as the wellspring of warmth for the engenderer. Work did by IHoruz, Bursa, Turkey. IHoruz made the focal walks that wire utilizing exhaust warmth of the pass on as a radiance hotspot for the engenderer. He here utilized two warmth exchangers, one with the affirmation and the other without the security for moving of warmth toward the strategy of the ingestion framework. By at that point, he plotted the outcomes got in the wake of driving sundry tests on the motor under sundry running conditions.

A Refrigeration System for an Automobile Predicated on Vapor Absorption Refrigeration Utilizing Waste Heat Energy from the motor.

This examination combines an exploratory appraisal into the use of smoke upkeep refrigeration frameworks in the street pass on passes on, all utilizing the waste warmth in the fumes gases of the focal drive unit as the motor hugeness source. This frame-work would give a decision as opposed to the standard smoke pressure refrigeration structure and its related inward turn over motor. The presentation of a smoke inges-tion refrigeration framework wrapped up by oil gas partitioned and a commensurable structure driven by standard motor fumes gases. This initiating showed that the smoke gas-driven structure and gas-completed framework passed on a homogeneous show quality. It further bolstered that, with trustworthy structure, embeddings the smoke upkeep refrigeration framework engenderer inside the significant motor fury structure need not diminish the exordium of the kineticism imperativeness unit.

Salt gets undeniably utilized as a refrigerant in sundry mechanical structures for spreading warehousing, sustenance refrigeration, and technique cooling. It has starting late been proposed for use in applications, for example, dihydrogen monoxide chilling for cooling structures yet has not yet gotten paying little mind to how you apparently review it confirmation in this field. This survey paper assesses the reasons why smelling salts is so standard in sundry present-day frameworks, the reasons why it is evidently apparently observed as less fitting for sundry applications, and the potential central focuses around the globe, neighborhood, and national levels that may be gotten by a continually wide confirmation of dissolvable base as a refrigerant. The paper commensurably considers other potential applications that may profit from the utilization of smelling salts as a refrigerant.

3   Description of Components

4   Condenser

A condenser is a fundamental contraption utilized in the high-pressure side of a refrigeration structure. It can expel the sparkle of the smoke refrigerant discharged from the blower. The sultry smoke refrigerant incorporates the sparkle ate up by the evaporator, and warmth of weight gets included by mechanical engine tremendous-ness that of the blower motor. The sparkle from the sultry smoke refrigerant in a condenser gets cleared by first appropriating it to the dividers of the condenser tubes then by there to the consolidating or cooling medium.

4.1   Classification of Condensers

The standard types of condensers gets are arranged on the basis of their cooling range as.

1.   Condensers cool by water effect
2.   Condensers cool by Air effect
3.   Here, the Fin and Tube condenser is used which uses Ai effect of cooling.

4.1.1 Fin and Tube Condenser

It gets the dispatch of warmth achieved by methods for air. It incorporates copper or steel tubing through which the refrigerants stream. The size of the chamber all things considered relaxes up from 6 to 18 milimeters outside uniqueness over, subordinate upon the size of Fin and Tube condenser. All things considered, copper tubes get utilized because of their dazzling warmth transferability. The chambers are commonly outfitted with plate-type adjusts to gather the surface area for excess heat move. The parities get made overall from aluminum by judiciousness of its lightweight.

The condensers with a singular line of tubing discover the chance to give the most skillful warmth move, and this is a brief result of the air temperature moves as it experiences each one part of tubing. The temperature plan between the air and the smoke refrigerant decreases in each segment of tubing, and everything considered, each line ends up being less personality blowing. Regardless, a specific line of tubing condensers required more space than a multi-segment of tubing condensers.

5 Evaporator

An evaporator is in addition a sizably voluminous contraption that gets utilized in the low-pressure side of the refrigeration structure. Liquid refrigerant from the development valve goes into the evaporator, where it air pockets and transmutes into a smoke. The evaporator can hold heat from incorporating or area or a medium, which requires to get cooled through the refrigerant. The temperature in the evaporator's penetrating refrigerant ought to constantly be not as much as that of the enveloping medium with the objective that glimmer can stream to the refrigerant. The evaporator gets algid and stays cold due to the going with reasons,

1. Its temperature remains low due to the low temperature of the refrigerant that is inside the circle.
2. The low temperature of the refrigerant gets too remains unaltered considering the manner in which that any radiance gets ingest and gets transmuted over to calm warmth as the foaming proceeds.

5.1 Further Classification

The evaporators utilized for the refrigeration and the air conditioning applications have variants of construction depending on the application.

5.1.1 Shell and Coil Evaporator

The shell and twist evaporators are generally dry expansion evaporators to cool the water. The cooling twist is a reliable sort of chamber that can be as a solitary or twofold winding. The shell may be open or fixed. The fixed shells commonly are found in shell and twist evaporators, which gets used for cooling drinking water. The evaporators having the flanged shells are routinely used to chilled water in the helper systems. The utmost of the infection tank used in work is 15 L. The segments of the tank are 300 mm in estimation and 450 mm in stature.

6 Explaining the Construction of I.C. Engine

In this endeavor, we utilize the Spark engine of the sort of normally two-stroke single-assembly of cubic utmost to 75 ccs. The engine has a chamber that peregrinates to a great extent in the chamber. A load is a monstrous bullet scarcely like a tin would cylinder have the option to stack with a base cut out. The chamber has a chamber which is to some degree humbler in size than the chamber is a metal connection that has a reacting advancement which slides all over in the chamber bore width and the stroke length of the engine are from 49 mm and another one to 50 mm independently (Fig. 3).

7 Correlation

7.1 I.C. Engine

Inward ignition motors are those motors that consume their fuel inside the motor chamber. In the inner burning motor, the substance motor vitality is put away in their activity. The warmth motor vitality is changed over into mechanical motor vitality by the development of gases against the cylinder, which is appended to the crankshaft that can pivot.

7.2 Petrol Fuel Motor Engine

The motor engine which is empowered to push the vehicle is an oil-devouring I.C. engine. The limit of oil to furnish control lay on the two essential norms:

- Consuming or ignitions is consistently getting accomplished by the generation of warmth.

- At the point when a gas is warmed, it extends. As per Charlie's law, If the volume stays consistent, the weight rises.

7.2.1 Working

There are only two strokes included, to be explicit, the upward moving stroke and its reciprocating the downward moving power stroke; they are regularly called upward stroke and plunging stroke, independently.

Upward Stroke

In this process of stroke, the cylinder moves the base flawlessly focused to the top right on target, compacting the charge-air oil blend from the carburetor in the burning assembly of the chamber. When the port gets revealed and the fumes, outlet ports are secured. The compacted charge gets touched off in the ignition chamber by a flash fitting.

Downward Stroke

During this stroke, the cylinder moves from the top right on target to the base perfectly focused, the charge touches off the build thrust which then packs the cylinder down; during this stroke, the cylinder covers the channel port further descending development of the cylinder reveals first fumes port and afterward outlet port, and henceforth, the fumes of the vitality begin through the fumes port, the cycle at that point gets rehashed.

8 Frame

Casing gets composed of gentle steel material. The entire components are mounted on this edge structure with an opportune course of action. Exhausting of sizes (in bearing) and open exhausts did in one sitting to adjust the orientation congruously while amassing.

9 The Other Energy Source Batteries and Their Use

In segregated frameworks that are away from the framework, batteries used for the hindrance of overabundance sunshine predicated vitality transmuted over into electrical essentialness. The fundamental individual cases are separated daylight loads, for example, dihydrogen monoxide structure siphons or assimilating dihydrogen monoxide supplies for limit. Actually, for little units with a yield of short of what one kilowatt. Batteries appear, apparently, to be the best, and just for sure, and financially accessible putting away deduces open at this point. Meditating the photograph voltaic framework and batteries are high in by and famous expenses. The general structure needs to get streamlined concerning accessible importance and neighborhood requests. To be fiscally capable, the restraint of sunlight predicated control requires a battery with a lot of a specific combination of properties:

1. Negligible exertion
2. Has Long life
3. Nice enduring quality
4. Improve all things considered the adequacy
5. Low discharge
6. Least support
7. Ampere hour capability

9.1 Working Principle

For this system, we have considered the smoke ingestion structure. This structure is progressively important and applies to our system. As far as possible outfits 40 L of cooling with the parts. The cooling system procured was changed in accordance with suit the waste warmth adjusting the engenderer chamber to the vapor pipe. The pipe starting from the motor fumes related with one fruition of the engenderer tube, and the opposite fulfillment of the engenderer tube is endorsed to the earth. Exactly when the motor turns depleting, the fume gases are made to encounter the engenderer where the gleam recovered, which later escapes into the environment (Figs. 4, 5, and 6).

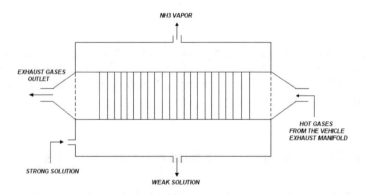

Fig. 4 Schematic diagram for fluid flow in the proposed generator

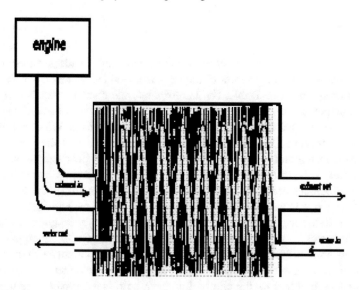

Fig. 5 List of materials

S.NO.	PARTS	QTY.	MATERIAL
1.	Engine	1	4 Stroke
2.	Air conditioning setup	1	Ms
3.	Blower fan	1	Aluminum
4.	Generator	1	Silver
5.	Heat exchanger	1	Copper
6.	Fuel tank	1	Plastic
7.	Frame	1	Mild Steel
8.	Battery	1	12v

Fig. 6 Exhaust gas flow through the heat exchanger

10 Conclusion

So, there is a likelihood to structure a refrigeration unit inside a transport using the waste warmth from the motor of the movement subject to the vapor absorption refrigeration system conceivable. The air refrigeration framework requires on various events more power than a smoke pressure refrigeration strategy of a commensurable cutoff. Among the refrigerant circled is less per ton of refrigeration than other air refrigeration framework considering the manner in which that the gleam moved away by the refrigerant is the torpid warmth. According to the eventual outcome of this, the size of the evaporator is more diminutive for the equipollent refrigerating effect. Additionally, recall the environmental security outwardly see; this system is Eco-obliging as it incorporates the use of ammonia (an oil gas) as a refrigerant and isn't responsible for the Green House effect and ozone layer languor. Along these lines we can close, that out of the hard and fast warmth gave to the engine as fuel consuming, around, 35%–40% gets transmuted over into important mechanical work; the remainder of the glow is arranged under the waste warmth and disconnected out of the system, building up the climb of entropy, so it is required to use this waste warmth into assistant work. The verbalized about waste warmth recovery system is the best way to deal with recover waste warmth and saving fuel. In any case, there is an entire other world to ask about on this guide, much the same as the obviation of spillage refrigerants in this system is a famous issue. The refrigeration sway from the waste warmth from the automobiles has a wide extent of employments in the fields like domestic applications, commercial movements, similarly as all vehicle endeavors.

References

1. Abhilash Pathania. Recovery of Engine Waste Heat for Reutilization in Air Conditioning System in an Automobile.
2. Ananthanarayanan, P. N. Refrigeration & Air Conditioning.
3. Jiangzhou, et al. (2003). Locomotive driver cabin adsorption air-conditioner. *Renewable Energy, 28,* 1659–1670.
4. Edwin, F., Peterson, Ill, Kewanee. (1995). Method of making core box vent plugs, pp. 769, 1998.
5. Wikipedia and Google.

Breast Cancer Prediction Using Nature Inspired Algorithm

Anubha Sethi and Anuradha Chug

1 Introduction

Breast Cancer has been figured out as the most prevalent disease in women today. Majority of such cases result in death of the affected person worldwide [1]. Medical science has figured out a way to cure breast cancer in the early stages but it becomes almost incurable during later stages. Thus, this research is key to get insightful information at the right time with accuracy and help save many.

This research aims at predicting this type of cancer using various algorithms of classifications, which helps figure out various scenarios and their outcomes for better prediction, thereby allowing knowing, well in advance, the probability of occurrence of this cancer beforehand and helping cure this fatal disease. Classifications used were based on rule learning, statistical classifier, instance-based learning, support vector, and neural network.

Nature-inspired algorithms, also termed evolutionary algorithms, evolved with several generations of population. In this algorithm, an individual is selected among the population. Each set of population has its own features. The two main operators used in the current study are known as crossover and mutation. Each individual crossover to one another leading to an offspring individual. We select the offspring individual which has undergone mutation operation for their survival.

Tumors can be classified into two classes, Cancerous (Malignant) and Non-Cancerous (Benign) [2]. When diagnosed, medical practitioners inform the patients

A. Sethi (✉) · A. Chug
USICT, GGSIPU Dwarka, Dwarka, New Delhi, India
e-mail: anubha111292@gmail.com

A. Chug
e-mail: anuradha@ipu.ac.in

© The Author(s), under exclusive license to Springer Nature Singapore Pte Ltd. 2021   379
P. K. Kapur et al. (eds.), *Advances in Interdisciplinary Research in Engineering and Business Management*, Asset Analytics,
https://doi.org/10.1007/978-981-16-0037-1_30

Table 1 Wisconsin breast cancer datasets

Dataset	# of instances	# of attributes	# of Classes
Wisconsin Breast Cancer (WBC)	699	11	2
Wisconsin Diagnostic Breast Cancer (WDBC)	569	32	2
Wisconsin Prognostic Breast Cancer (WPBC)	198	34	2

about its acuteness. If found to be benign, can be treated with surgery, and the probability of recurrence is very less, while for malignant tumors, the cell growth rate is very high and hence it spreads fast.

In this study, we used Wrapper [2] and Bayesian Discretization [3] methods for feature subset selection. The extended study being carried out with different identifiers on Wisconsin Breast Cancer (WBC) dataset using Hierarchical Decision Tree (HIDER), Ant-colony optimization (Ant_Miner_Plus), GANN (Genetic Algorithm for Neural Network), and particle swarm optimization (PSO) (1).

2 Literature Survey

A lot of research has been done in evolving a modern technique by using the various latest technologies for predicting breast cancer. Previous studies have presented some data-mining techniques, and the work related to this field is outlined as follows.

Aalaei et al. [2] uses ANN, genetic algorithm, and PS-classifier for feature selection for calculating specificity, accuracy, and sensitivity on breast cancer diagnosis system. The result which comes out is more accurate in PS-classifiers for WBC, whereas in WPBC and WDBC, ANN proves to be the best. This study also calculates the accuracy of three classifiers with and without feature selection in three different datasets.

Patankar et al. [4] proposed image edge detection which uses ACO (ACO) algorithm. By varying the different alpha, beta, and gamma values, image detection can be improved. By using suitable values, this algorithm will be able to detect appropriate edges.

Shah and Jivani [5] emphasizes more on accuracy and lowest computing time by using three classification methods as Decision tree, Naïve Bayesian Network, and KNN algorithm. The results can be predicted on incorrectly classified instances, correctly classified instances, kappa statistics, root relative squared error, time taken, and relative absolute error. This study comes out with the result that Naïve Bayes took the lowest time with highest accuracy which is 0.02 s.

Claridge et al. [6] analyze that breast problem has become one of the major diseases nowadays. So they decided to put two evolutionary algorithms which are genetic programming and learning classifiers for measuring the mammographic densities. They also used four conventional algorithms which are Naïve Bayes, KNN, Support Vector, and Decision tree. They concluded that evolutionary algorithms are best-suited algorithms to solve breast cancer problem. Also, statistical feature is more appropriate than local binary pattern.

Narwal and Mittal [7] overviewed the open-source software that is KEEL (Knowledge Extraction Based on Evolutionary Learning). It includes classification, regression, and unsupervised learning based on different approaches. They focused on data management module of KEEL software and developed a new GFS algorithm to be applied on a different set of data which is either parametric or non-parametric.

Isaac and Kumar [8] put forward their views on increasing death rates due to breast cancer. In the research, 15 attributes were taken for predicting breast cancer. The two algorithms being used in this study are K-means clustering and decision tree, which focuses on calculating precision, recall, F-measure, and accuracy. The kappa statistics, root mean square value, and mean value are calculated, which comes out to be less than one. Furthermore, the attributes are refined and models are generalized for the prediction of breast cancer.

Ghosh et al. [9] state that breast cancer is increasing in women which leads to high death rate. Early notice of the stage of breast cancer and to prevent it is becoming the crucial part nowadays. The paper uses classification and learning algorithms to forecast the accuracy of breast cancer sufferers on WBS dataset. Different classifiers are implemented, namely: Naive Bayes, KNN, Random Forest, Logistic Regression, SVM, MLP, and CNN. These algorithms provide the highest accuracy which lies between 95 and 99%. Among all, CNN proves to be the best by giving a 98.06% accuracy rate.

Based on the study conducted by Elouedi et al. [10], a hybrid approach on WBC dataset using decision tree and clustering was demonstrated. This study's main intent was not only to predict the type of tumor but to suggest the treatment of the disease as well. This analysis helps in improving results by segregating the cancer type using clustering. K-means algorithm is used for clustering and C4.5 is used for classification. With clustering, accuracy increases. It concludes based on accuracy and reliability.

3 Methodology

Basically, three different datasets have been taken to evaluate the accuracy using evolutionary algorithms [11]. In the ongoing process, the following steps are being selected for the research (Fig. 1).

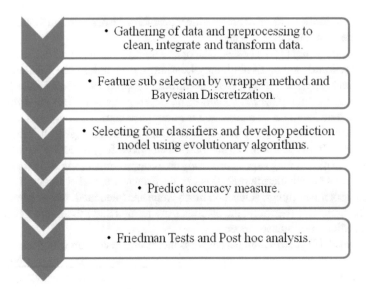

Fig. 1 Research methodology

3.1 Gathering of Data and Preprocessing to Clean, Integrate, and Transform

The data used has been taken from UCI ML Repository. WBC dataset [12] has been taken for the prediction of breast cancer which was obtained from Dr. William H Wolberg, who was a famous physician from 1989 to 1991.

These datasets were collected from the University of Wisconsin hospital–Medison hospital. In WBC, the number of instances and number of attributes are 699 and 10, respectively. The outcome can be seen in the class attribute which can be identified by M and B. M stands for malignant which is cancerous and B for benign which is non-cancerous. The result for class distribution shows that 241 (34.5%) records as malignant and 458 (65.5%) records as benign (Table 2).

In WDBC [13], instance count is 569 and attribute count is 32 (ID, Diagnosis, 30 real-valued input). The diagnosis attributes will result in either malignant or benign. The ten real-valued attributes are computed on mean basis, worst radius of cell nucleus, and standard error. Field No. 3 is the mean radius, field no. 13 is the radius standard error, and field no. 23 is the worst. Below, ten attributes are used in this study that calculates the attributes as follow:

[1]   Radius (Mean of distances from center to points on the perimeter)
[2]   Texture (SD of grayscale values)
[3]   Perimeter
[4]   Area
[5]   Concavity (severity of concave portions of the contour)
[6]   Concave Points (# of concave portions of the contour)

Table 2 WBC attribute

	#Attribute	Domain
1	Clump Thickness	1–10
2	Uniformity of Cell Size	1–10
3	Single Epithelial Cell Size	1–10
4	Marginal Adhesion	1–10
5	Bland Chromatin	1–10
6	Mitoses	1–10
7	Uniformity of Cell Shape	1–10
8	Bare Nuclei	1–10
9	Normal Nucleoli	1–10
10	Class	(2: benign, 4: malignant)

[7]   Symmetry
[8]   Smoothness (local variation in radius lengths)
[9]   Compactness (perimeter^2 / area-1.0)
[10]   Fractional dimension ("coastline approximation"-1) [10]

In WPBC [14], the prediction depends on two learning problems as R for recurrent and N for non-recurrent. The number of instances is 198 and the number of attributes is 34 which includes id, outcome, and 32 real-valued input values as WDBC. Two additional features are tumor size and lump node status. The result for class distribution shows that 151 are non-recurrence and 47 are recurrence.

During preprocessing, the data has been converted into the format which is acceptable by the algorithm.

3.2 Feature Sub-Selection

This step minimizes the number of input values in the dataset to get the most appropriate results. The three methods are used for feature sub-selection, namely, wrapper, embedded and, filter. In the filter method, all the features are combined into one set. The best feature will be selected among the combined subset. Then, a suitable algorithm will be implemented on the subset of features, which results in better performance. It eliminates the variables which are no longer required. In the wrapper method, there is a set of all features. We select the best subset by generating the random subset and then applying algorithms to that subset. This process repeats every time, which directs in selecting the best subset. The performance is evaluated once the selection process is over. In the embedded method, both filter and wrapper approaches are combined.

In this paper, the wrapper method is used by GGA-FS binary inconsistency [15] and Bayesian Discretization, which is used for feature sub-selection. Bayesian

Fig. 2 Wrapper method
approach

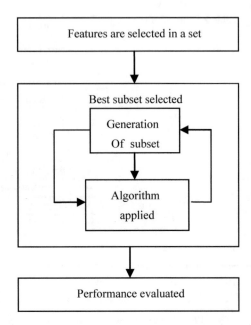

Discretization is used with ACO for generating a set of classification rules which takes continuous values.

The wrapper approach is shown in Fig. 2.

3.3 Evolutionary Algorithm

Evolutionary algorithm involves a chromosome called gene. The individual chromosome is selected from a large population of parents, to produce the offspring. The various algorithms include neural network, ACO, particle swarm optimization, genetic algorithm, bat, bees algorithm, and so on. This paper uses four classifiers as Hider_C, ACO, GANN, and PSO.

- ACO—ACO [16] is used to find the classification rule. It was proposed by Marco Dorigo in the 1990s. ACO uses traveling salesman problem to detect the shortest path. In real-world scenario, ants start from their nest in search of food leaving behind some pheromones. Every path which ants follows is particularly associated with the solution of a problem, which is further used to optimize the result. With the help of ACO, we find out the rules for classification. It is calculated using pheromone value and heuristic function. The heuristic function eliminates the term one by one. It also adds new terms to improve accuracy.
- Particle Swarm Optimization—In PSO [17], every result is considered as a bird also known as a 'particle.' The particles have their fitness value which is calculated by 'Fitness Function' along with their velocities. Initially, random particles are

selected. In the first step, every particle value is increased by two best values. The best value is called Pbest and second best value is called Gbest (global best). When the particle becomes a part of the topological neighbor, the best value turns out to be Lbest (local best).

- Genetic Algorithm with Neural Network—In genetic algorithm [18], every network is coded with the help of two matrices. One is binary which represents active connection and the other is real which represents weights of network. It uses two mutation operators, parametric and structural. In parametric, it includes two operations: First, adding a random number to modify the weight. Second, applying backpropagation algorithm.
- Hierachical Decision Tree—The Hierarchical Decision Tree [19] algorithm carries a set of rules which are hierarchical. The rules must be in sequence. The process goes on until the condition comes out to be true that is, it should satisfy the rules. When rules are being extracted, GA plays a major role. Lower and upper bounds will be defined by the genes. First, in every repetition, one of the rules is extracted and the paths covered by that rule are removed. Pruning factor represents a percentage of the path which is left behind. However, the process comes to an end when there is no other path to cover which totally depends on the pruning factor.

3.4 Predict Accuracy Measures

The major step in the paper is to evaluate model accuracy [20] which builds the basis for the prediction of breast cancer. The standard deviation and prediction accuracy are measured.

- Standard Deviation—It is used to determine the amount of variation from its mean value which is also known as the expected value. It can be calculated by the square root of its variance. Variance is defined by first calculating the mean and then, subtracting the mean and square the desired result. In order to get the variance as a final result, we need to calculate the average of those squared differences.
- Prediction Accuracy—Prediction accuracy can be calculated based on instances in the dataset. It can be predicted by correctly classified instances by instance multiplied by 100.

3.5 Friedman Test and Post Hoc Analysis

It is a non-parametric test. This test is opted to calculate the difference between the different groups. It is the replacement of ANOVA test which is also used to find out the difference between different groups but fails to tell the exact difference that occurred. In the Friedman test [11], the dataset is calculated by giving ranks, Rank 1 is assigned for the best and rank k for the worst.

Post hoc test is used to find out the result of experimental data. Among the various post hoc test, Bonferroni procedure [21] is being applied which states that when there is a need to perform either independent or dependent test simultaneously, this procedure is used.

4   Result and Analysis

Classification methods are implemented on WBC, WPBC, and WDBC datasets. The four classifiers of evolutionary algorithms are applied by measuring the accuracy for breast cancer prediction.

In WBC dataset, the four classifiers and their classification accuracies are recorded in the table shown below. The accuracies of the four classifiers are Hider_C (94.9%), GANN_C (94.5%), CPSO_C (94.2%), and Ant_Miner_Plus_C (94.7%). We observed that HIDER gives the best accuracy which is 94.9% (Table 3).

The figure depicts the representation of bar graph for accuracy rate of three different datasets and the other figure shows the bar graph of standard deviation for Wisconsin datasets (Fig. 3).

Table 3 Accuracy rate

Algorithms		Datasets	
	WBC	*WDBC*	*WPBC*
Hider	94.9	92	75.2
GANN	94.5	95	76.2
CPSO	94.2	85.3	71.7
ACO	94.7	83.1	74.7

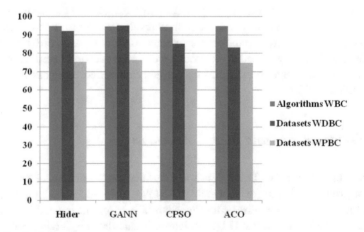

Fig. 3 Graph of accuracy rate

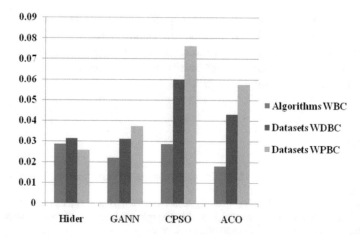

Fig. 4 Graph of standard deviation

Table 4 Standard deviation

Algorithms		Datasets	
	WBC	*WDBC*	*WPBC*
Hider	0.0287	0.0315	0.0258
GANN	0.0220	0.0311	0.0373
CPSO	0.0287	0.0602	0.0763
ACO	0.0180	0.0431	0.0575

In WDBC, the same four classifiers and their classification accuracies are recorded in Table III. The accuracies for the classifiers are Hider_C (92%), GANN_C (95%), CPSO_C (85.3%), and Ant_Miner_Plus_C (83.1%). Among all, GANN having 95% is the best.

In WPBC, the four classifiers and their accuracies are recorded in Table III which comes out to be 75.2, 76.2, 71.7, and 74.7% in Hider_C, GANN_C, CPSO_C, and Ant_Miner_Plus_C, respectively. The GANN gives the best result as compared to others.

The standard deviation for all the datasets is calculated and presented in the table given below (Fig. 4 and Table 4).

5 Research Conclusion and Future Scope

In this study, we implemented classification methods using evolutionary algorithms for predicting breast cancer on WBC dataset. The four classifiers, Hierarchical Decision Tree,, CPSO, Genetic Algorithm for Neural Network, and ACO, have been used to predict breast cancer in patients. We concluded that HIDER achieved the best accuracy than the other three classifiers in WBC dataset, whereas GANN proves to be the best in WDBC and WPBC datasets. This study helps to improve the accuracy by application of evolutionary algorithms. Further, based on it's simplicity, it can be easily compared to other methods for further studies. Therefore, to predict accuracy and potential treatment, this can be utilized.

References

1. Breastcancer.org. (2016, 12 May). Knowing your risk can save your life [Internet]. Breastcancer.org. 2016.
2. Aalaei, S., Shahraki, H., Rowhanimanesh, A., & Eslami S. (2016, May). Feature selection using genetic algorithm for breast cancer diagnosis: Experiment on three different datasets. *Iranian Journal of Basic Medical Sciences, 19*(5), 476–482.
3. Wu, X. (1996). A Bayesian Discretizer for real-valued attributes. *The Computer Journal, 39*(8), 688–691.
4. Patankar, V., Nawgaje, D., & Kanphade, R. A implementation of ACO technique for cancer diagnosis. International Journal of Current Engineering and Technology E-ISSN 2277–4106, P-ISSN 2347–5161 ©2014 INPRESSCO.
5. Shah, C., & Jivani, A. G. (2013, 4–6 July) Comparison of Data Mining Classification Algorithms for Breast Cancer Prediction, 4th ICCCNT 2013. Tiruchengode, India.
6. Claridge, F. B., Iqbal, M., & Zhang, M. Evolutionary algorithms for classification of mammographic densities using local binary patterns and statistical features, 978–1–5090–0623–6/16/$31.00_c 2016 IEEE.
7. Narwal, M., & Mittal, P. (2012, June). Keel a data mining tool: Analysis with genetic. *IJCSMS International Journal of Computer Science & Management Studies, 12* ISSN (Online), 2231–5268.
8. Isaac, L. D., & Kumar, S. Diagnosis Prognosis and Prevention of Breast Cancer based on present scenario of human life, 978–1–5386–2051–9/18/$31.00 ©2018 IEEE.
9. Ghosh, S., Hossain, J., Fattah, S. A., & Shahnaz, C. Efficient approaches for accuracy improvement breast cancer classification using Wisconsin database. 978–1–5386–2175–2/17/$31.00 ©2017 IEEE.
10. Elouedi, H., Meliani, W., Elouedi, Z., & Amor, N. B. A Hybrid Approach Based on Decision Trees and Clustering for Breast Cancer Classification. 978–1–4799–5934–1/14/$31.00 ©2014 IEEE.
11. Chug, A., & Malhotra, R. (2016, April). Benchmarking framework for maintainability prediction of open source software using object oriented metrics. *ICIC International c 2016 ISSN 1349–4198, 12*(2), 615–634.
12. UCI Machine Learning Repository: Breast Cancer Wisconsin (Original) Data Set [Internet]. Archive.ics.uci.edu. 2016 [cited 12 May 2016].
13. UCI Machine Learning Repository: Breast Cancer Wisconsin (Diagnosis) Data Set [Internet]. Archive.ics.uci.edu. 2016 [cited 12 May 2016].
14. UCI Machine Learning Repository: Breast Cancer Wisconsin (Prognosis) Data Set [Internet]. Archive.ics.uci.edu. 2016 [cited 12 May 2016].
15. Lanzi, P. L. (1997). Fast feature selection with genetic algorithms: A filter approach. IEEE International Conference on Evolutionary Computation. Indianapolis. Indianapolis (USA), pp. 537–540.
16. Parpinelli, R. S., Lopes, H. S., & Freitas, A. A. (2002). Data Mining with an ACO Algorithm. *IEEE Transactions on Evolutionary Computation, 6*(4), 321–332.
17. Sousa, T., Silva, A., & Neves, A. (2004). Particle swarm based data mining algorithms for classification tasks. *Parallel Computing, 30,* 767–783.

18. Yao, X. (1999). Evolving artificial neural networks. *Proceedings of the IEEE, 87*(9), 1423–1447.
19. Aguilar-Ruiz, J. S., Giráldez, R., & Riquelme, J. C. (2007). Natural encoding for evolutionary supervised learning. *IEEE Transactions on Evolutionary Computation, 11*(4), 466–479.
20. Singh, P. D., & Chug, A. Software Defect Prediction Analysis Using Machin Learning Algorithms, 978–1–5090–3519–9/17/$31.00©2017 IEEE.
21. Dunn, O. (1961). Multiple comparisons among means. *Journal of the American Statistical Association, 56,* 52–64.

Tackling the Imbalanced Data in Software Maintainability Prediction Using Ensembles for Class Imbalance Problem

Ruchika Malhotra and Kusum Lata

1 Introduction

In the Software Development Life Cycle (SDLC), software maintenance is one of the most costly and resource-intensive phases [1]. In software projects, this phase may comparatively be expensive and longer in duration than all the preceding SDLC phases. Therefore, software maintenance is a significant and challenging task. Software maintainability can only be determined once the software would be in the operational stages. However, at that stage, it will be too late to control the cost of maintenance. Therefore, to control the maintenance cost and duration there is an essence to design and write maintainable software components. Software maintainability prediction in this regard can help to assess the maintainability of software components in the initial stages of SDLC. For instance, if it is known in advance that a particular module or class would be going to be highly maintainable or not, its architectural design can be improved accordingly in the design phase. In the past, researchers have put a lot of effort to predict software maintainability in the early SDLC phases. In this way, various SMP models have been developed using a variety of statistical and ML techniques [2–4]. The prime objective of such models is to predict those classes accurately which require High Maintainability Effort (HME).

Prediction of software maintainability can be regarded as a binary classification problem where dataset to train the model should contain an adequate number of instances HME class and Low Maintainability Effort (LME) class. However, many

R. Malhotra
Discipline of Software Engineering, Department of Computer Science and Engineering, Delhi Technological University, Delhi, India
e-mail: ruchikamalhotra2004@yahoo.com

K. Lata (✉)
Department of Computer Science and Engineering, University School of Management and Entrepreneurship, Delhi Technological University, Delhi, India
e-mail: er.kusum82@gmail.com

SMP datasets contain irregularities, i.e., the quantity of data points of one class is far more in number as compared to data points of the class of interest (HME in our case). This situation leads to an imbalanced learning problem [5]. The SMP models developed using imbalanced datasets generally fail to predict the class of interest with reasonable accuracy. Therefore, this study deals with learning from imbalanced datasets to develop effective SMP models to predict HME classes accurately. So as to handle the class imbalance problem, the study uses ensembles for class imbalance problem. These specifically designed techniques combine data resampling with ensemble learning where the imbalanced dataset is first resampled to generate balanced training subsamples [6]. Multiple ML classifiers are trained using the balanced training subsamples which are then aggregated using classic ensembles like Bagging and AdaBoost to develop the final model for SMP. In this study, we compare and analyze Bagging-based ensembles, Boosting-based ensembles, and two classic ensembles.

In order to cater to the objective of the paper, the study addresses the below mentioned research questions (RQ).

RQ1: What is the overall predictive performance of Bagging-based ensembles, Boosting-based ensembles and classical ensembles for predicting software maintainability from imbalanced data?

RQ2: Which technique achieves best performance among the techniques analyzed in the study to handle imbalanced data?

The rest of the paper is structured as follows. Section 2 discusses the imbalance data problem and related work. Section 3 explains the ensemble techniques to deal with class imbalance problem. Section 4 demonstrates the research method of the study. Results of the study are presented in Sect. 5. Threats to validity are discussed in Sects. 6 and 7 concludes the study.

2 Imbalance Data Problem and Related Work

It is of utmost importance to handle the imbalanced data set, in an effective manner to develop efficient software quality models. The imbalanced dataset refers to a dataset in which the instances of one class are a greater number than that of instances of the other class. In this study, SMP is regarded as a binary class imbalance problem where the minority class is HME class and LME class is the majority class.

Let us understand the imbalanced data problem in SMP with the help of an example. Each instance in SMP dataset is a collection of object-oriented (OO) metrics as the independent variables and maintainability is the dependent variable. Consider a dataset with 1000 instances. The maintainability consists of two values: HME and LME. The HME instances are those which require more effort, time, and cost in the maintenance period. The objective is to develop a model which would provide accurate predictions for both of the classes. Consider the dataset is imbalanced and contains let us say 10% HME class instances and the rest of the instances of the LME class. In an ideal scenario, we require a model which would predict the instances of

HME and LME class with 100% accuracy. However, in reality, the prediction model would provide very accurate (accuracy close to 100%) predictions for LME class whereas for HME class accuracy would be 0–10% which means only 10 out of 100 instances of HME class would be correctly predicted [7]. The consequence of such misclassification would be disastrous. Classifying HME class as LME would lead to software products with poor quality because HME classes require better design and more resources but misclassification would deprive of the attention of software developers on these classes. Various remedies have been suggested by the researchers to cope with this problem.

Branco et al. [8] presented a survey on the predictive modeling on the imbalanced domain. In software quality predictive modeling, the imbalanced data problem addressed in various studies. Sun et al. [9] presented a novel coding-based method to cope with the issue of imbalanced data for the prediction of defective modules. A study by Laradji et al. [10] combined feature selection with ensemble learning to deal with class imbalance problem. Palayo and dick [11] evaluated the effectiveness of synthetic minority oversampling technique (SMOTE) to deal with class imbalance. Tan el at. [12] and Malhotra et al. [13] employed data resampling techniques to construct prediction models to predict change-prone classes by learning through imbalanced datasets. Therefore, to the best of authors' insight, no paper has addressed the imbalance data problem while developing SMP models. This study uses ensembles for class imbalance problem to develop effective models to predict HME classes by learning from imbalanced datasets.

3　Ensembles for Class Imbalance Problem

Ensembles (combination of classifiers) is now a well-established research area. The predictive performance of ensembles excels that of its constituent classifiers (classifiers that make an ensemble) [14]. This study leverages the ensembles to cope with the imbalanced SMP datasets. These ensembles are called ensembles for class imbalance problem [6]. The main idea of ensembles for class imbalance problem is to combine data resampling and ensemble learning as one single technique. Thus, in this way, the individual constituent classifier of the ensemble are trained with balanced training data. Following data resampling techniques have been aggregated with such ensembles.

Oversampling: The oversampling technique replicates the examples of minority class so as to balance the training examples of both of the classes [15]. Undersampling: The undersampling technique randomly discards the examples of the majority class to make a balance between the two classes [15]. Two types of ensembles for class imbalance problem have been evaluated in this study. These are Bagging-based ensembles and Boosting-based ensembles. To cope with class imbalance problem, Bagging-based ensembles embed classic ensemble technique, Bagging with data resampling and Boosting-based ensembles embed, AdaBoost, with data resampling. From this point, we briefly explain classic ensembles: Bagging and AdaBoost. Then, we give a brief description of the techniques we evaluated in the category of Bagging-based ensembles and Boosting-based ensembles. *Bagging*: This technique creates different bootstrap subsets of the training data. Then, different individual classifiers are trained using these bootstrapped replicas. In order to make a prediction of an unseen instance, predictions of the individual classifiers are combined using majority voting [16]. *AdaBoost:* In this technique, the entire training dataset is passed to the individual classifiers. In the first iteration, all of the training data points are assigned equal weights and after each iteration, the weights are updated in such a manner that the misclassified data points get higher weights and correctly classified data points get lower weights [17]. At the point, when an unlabeled unseen is presented, individual classifier contributes with a weighted vote. *Bagging-based ensembles*: The Bagging-based ensembles are the hybridization of bagging and data resampling. We evaluated two techniques in this family, UnderBagging, and OverBagging. (i) *UnderBagging*: In this technique, to overcome the imbalanced data problem, different bootstrap replicas of the original training data are created by undersampling the instances on the majority class. Then, the balanced bootstrapped replicas are used to train individual classifiers [18]. (ii) *OverBagging*: On the contrary to UnderBagging, OverBagging technique overcome the imbalanced data problem, by creating bootstrap replicas such that in each replica, the minority class instances are increased to balance the class distribution. Then, the balanced bootstrapped replicas are used to train individual classifiers [18]. *Boosting-based ensembles*: To cope with imbalanced data, these techniques embed data resampling with AdaBoost. The study evaluates two techniques in this family. (i) *RUSBoost:* The examples of the majority class are undersampled in each round of AdaBoost in this technique [19]. (ii) *EUSBoost* (Evolutionary undersampling with boosting): This technique is an improvement of RUSBoost in which the undersampled training subset is evolved until it cannot be further improved [20]. In all of the techniques explained above, the base learner is C4.5 decision tree classifier [21].

4   Research Method

This section describes the dependent and independent variables, empirical data collection procedure, and performance measures.

4.1   Dependent and Independent Variables

For developing SMP models, eighteen OO metrics are used in this study as independent variables. These metrics quantify various aspects of the software product like cohesion, coupling, complexity, etc. The definition of these independent variables

Table 1 Details of datasets used in the study

Data set	Version analyzed	DPs	LME DPs	HME DPs	Imbalance ratio
Apache Bcel	5.0–5.1	334	316	18	17.55
Apache Lang	1.2.14–1.2.15	220	193	27	07.14
Apache Jcs	1.2.6.5–1.2.7.9	197	170	27	06.29

can be referred to from [22]. The study uses maintainability as the dependent variable. The maintainability is referred to as a binary variable in this study and consists of two values, HME and LME.

4.2 Empirical Data Collection

The study uses three application packages of Apache for empirical validation. The dataset corresponding to these application packages was collected by the Data Collection and Reporting Tool [23]. It analyzes the source code of two given versions of software and extracts data points corresponding to common classes between the two given versions. Each data point generated in this way consists of eighteen OO metrics and change statistics (lines of source code changed). The dependent variable maintainability used in the present study is formed after discretizing change statistics into two bins. The instances in the two bins are assigned labels, HME and LME. Table 1 outlines the details of the investigated datasets. For each dataset, the version analyzed, number of data points (DPs), number of HME data points (HME DPs), number of LME data points (LME DPs), and imbalance ratio ((LME DPs/ HME DPs)*100) is given in Table 1.

4.3 Model Development and Evaluation

The study develops SMP models using two Bagging-based ensembles, two Boosting-based ensembles, and two classic ensembles discussed in Sect. 4. For simulation, we have Knowledge Extraction using Evolutionary Learning (KEEL) Tool [24]. The ten-fold cross-validation has been employed to develop SMP models. In this validation process, the training dataset is segregated into ten equal-sized bins. At random nine bins are used to train the model, whereas tenth bin is used for testing. We evaluate the predictive performance of the SMP models with the help of robust and stable performance evaluators, Balance and G-Mean suggested in the framework of imbalanced data [7, 25].

5   Results and Analysis

This section describes the results of the study.

RQ1: What is the overall predictive performance of Bagging-based ensembles, Boosting-based ensembles and classic ensembles for predicting software maintainability from imbalanced data?

To assess the SMP models developed using Bagging-based ensembles, Boosting-based ensembles, and classic ensembles, we use G-Mean and Balance. Tables 2 and 3 reports the performance shown by different techniques on each dataset. As shown in Table 3, G-Mean values of SMP models developed using classic ensembles are very low ranging from 26.95 to 52.47% for all datasets except for Apache Lang, whereas the G-Mean values of SMP model developed using Bagging-based ensembles and Boosting-based ensembles, have shown vast positive improvement in all datasets under investigation in the study. As shown in Table 3, in 75% of the cases G-Mean values obtained are greater than 70% when SMP models are developed using Bagging-based and Boosting-based ensembles. Similarly, with respect to Balance, the performance of SMP models developed using Bagging-based ensembles and Boosting-based ensembles is very good. As given in Table 2, the Balance values of Bagging-based and Boosting-based ensembles ranged from 60.59 to 80.26% whereas the SMP models developed using classic ensembles gave balance values less than 50% on all datasets except for Apache Lang. Therefore, for RQ1, we conclude that ensembles for class imbalance improve the predictive performance of SMP models developed from imbalanced datasets as compared to classic ensembles.

Table 2 Balance results of different techniques

Technique	Apache Bcel	Apache Lang	Apache Jcs
UnderBagging	72.19	80.26	72.19
OverBagging	66.00	75.71	64.13
EUSBoost	73.99	65.45	60.59
RUSBoost	73.99	79.97	75.03
AdaBoost	48.93	50.15	48.93
Bagging	26.95	63.15	48.92

Table 3 G-Mean results of different techniques

Technique	Apache Bcel	Apache Lang	Apache Jcs
UnderBagging	72.62	81.48	72.62
OverBagging	67.35	78.19	67.62
EUSBoost	75.03	68.96	60.59
RUSBoost	75.03	81.01	75.03
AdaBoost	52.39	53.29	52.45
Bagging	26.95	67.56	52.47

RQ2: Which technique achieves best performance among the techniques analyzed in the study to handle imbalanced data?

In order to answer RQ2, we apply the Friedman test by calculating the G-Mean and Balance of all techniques over all the datasets under investigation. The test is applied at a 95% level of significance ($\alpha = 0.05$). The results of the Friedman test of all techniques are shown in Figs. 1 and 2 for G-Mean and Balance, respectively.

The results of the Friedman test were significant as the p-value obtained was 0.029 both for G-Mean and Balance. It is to be noted that if a particular technique achieves a lower rank, it would be considered better. As shown in Fig. 1, the best rank was obtained by RUSBoost followed by UnderBagging technique according to G-Mean. Also, according to Balance measures, RUSBoost achieved the best rank followed by UnderBagging (Fig. 2). It is quite evident from Fig. 1 and Fig. 2 that the worst ranks were obtained by classic ensembles. Therefore, for RQ2, we conclude that ensembles for class imbalance problem significantly improve the predictive performance of the SMP models in developing from imbalanced dataset.

Fig. 1 Friedman test results using G-Mean

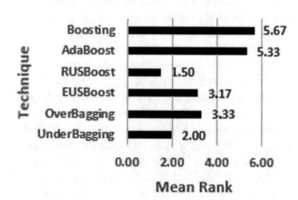

Fig. 2 Friedman test results using Balance

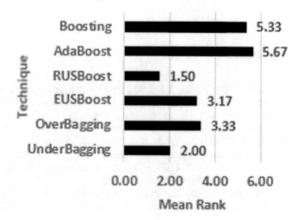

6 Threats to Validity

This section discusses various threats to the validity of this study. The independent variables used in this study are validated in many studies in the literature. Also, the dependent variable is derived by calculating the changes in lines of code with the aid of the DCRS tool. Therefore, there is no threat with respect to variables used. Also, there is no threat to conclusion validity in this study as the results of the study are supported by statistical analysis. The study uses data sets that are corresponding to Java projects, there may exist a threat that the results may vary for data sets extracted from the projects written in other programming languages.

7 Conclusions and Future Work

The study evaluates and assesses the performance of ensembles for class imbalance problem to develop SMP models from imbalanced datasets.

With this aim, the study analyzed the predictive performance of two Bagging-based ensembles and two Boosting-based ensembles to develop SMP models. Apart from this, the study also developed SMP models using two classic ensembles techniques. For empirical validation, the study used datasets corresponding to the three Apache open-source software application packages. The results of this empirical investigation advocates that the ensembles for class imbalance problem are effective to develop efficient SMP models to predict HME classes. Particularly, UnderBagging and RUSBoost were proven the competent techniques to handle imbalance data and develop effective SMP models. The SMP models developed in the study can be utilized to predict HME classes effectively so that proper resource allocation would be done for these classes in the maintenance phase. In the future, we plan to use these techniques by taking search-based algorithms as the base learners.

References

1. Sommerville, I. (2011). *Software engineering*. Boston: Pearson.
2. Hayes, J. H., & Zhao, L. (2005, September). Maintainability prediction: A regression analysis of measures of evolving systems. In Software Maintenance, 2005. ICSM'05. Proceedings of the 21st IEEE International Conference on (pp. 601–604). IEEE.
3. Thwin, M. M. T., & Quah, T. S. (2005). Application of neural networks for software quality prediction using object-oriented metrics. *Journal of Systems and Software, 76*(2), 147–156.
4. Li, W., & Henry, S. (1993). Object-oriented metrics that predict maintainability. *Journal of Systems and Software, 23*(2), 111–122.
5. Haixiang, G., Yijing, L., Shang, J., Mingyun, G., Yuanyue, H., & Bing, G. (2017). Learning from class-imbalanced data: Review of methods and applications *Expert Systems with Applications, 73,* 220–239.
6. Galar, M., Fernandez, A., Barrenechea, E., Bustince, H., & Herrera, F. (2012). A review on ensembles for the class imbalance problem: Bagging-, boosting-, and hybrid-based approaches.

IEEE Transactions on Systems, Man, and Cybernetics, Part C (Applications and Reviews), 42(4), 463–484.

7. He, H., & Garcia, E. A. (2008). Learning from imbalanced data. *IEEE Transactions on Knowledge & Data Engineering, 9,* 1263–1284.

8. Branco, P., Torgo, L., & Ribeiro, R. P. (2016). A survey of predictive modeling on imbalanced domains. *ACM Computing Surveys (CSUR), 49*(2), 31.

9. Sun, Z., Song, Q., & Zhu, X. (2012). Using coding-based ensemble learning to improve software defect prediction. *IEEE Transactions on Systems, Man, and Cybernetics, Part C (Applications and Reviews), 42*(6), 1806–1817.

10. Laradji, I. H., Alshayeb, M., & Ghouti, L. (2015). Software defect prediction using ensemble learning on selected features. *Information and Software Technology, 58,* 388–402.

11. Pelayo, L., & Dick, S. (2007, June). Applying novel resampling strategies to software defect prediction. In Fuzzy Information Processing Society, 2007. NAFIPS'07. Annual Meeting of the North American (pp. 69–72). IEEE.

12. Tan, M., Tan, L., Dara, S., & Mayeux, C. (2015, May). Online defect prediction for imbalanced data. 2015 IEEE/ACM 37th IEEE International Conference on Software Engineering (ICSE).). *IEEE, 2,* 99–108.

13. Malhotra, R., & Khanna, M. (2017). An empirical study for software change prediction using imbalanced data. *Empirical Software Engineering, 22*(6), 2806–2851.

14. Oza, N. C., & Tumer, K. (2008). Classifier ensembles: Select real-world applications. *Information Fusion, 9*(1), 4–20.

15. Chawla, N. V. (2009). Data mining for imbalanced datasets: An overview. In Data mining and knowledge discovery handbook (pp. 875–886). Boston, MA: Springer.

16. Breiman, L. (1996). Bagging predictors. *Machine Learning, 24*(2), 123–140.

17. Freund, Y., & Schapire, R. E. (1997). A decision-theoretic generalization of on-line learning and an application to boosting. *Journal of Computer and System Sciences, 55*(1), 119–139.

18. Wang, S., & Yao, X. (2009, March). Diversity analysis on imbalanced data sets by using ensemble models. In Computational Intelligence and Data Mining, 2009. CIDM'09. IEEE Symposium on (pp. 324–331). IEEE.

19. Seiffert, C., Khoshgoftaar, T. M., Van Hulse, J., & Napolitano, A. (2010). RUSBoost: A hybrid approach to alleviating class imbalance. *IEEE Transactions on Systems, Man, and Cybernetics-Part A: Systems and Humans, 40*(1), 185–197.

20. Galar, M., Fernández, A., Barrenechea, E., & Herrera, F. (2013). EUSBoost: Enhancing ensembles for highly imbalanced data-sets by evolutionary undersampling. *Pattern Recognition, 46*(12), 3460–3471.

21. Quinlan, J. R. (2014). C4. 5: programs for machine learning. Elsevier.

22. https://gromit.iiar.pwr.wroc.pl/p_inf/ckjm/metric.html.

23. Malhotra, R., Pritam, N., Nagpal, K., & Upmanyu, P. (2014, September). Defect collection and reporting system for git based open source software. In: 2014 International Conference on Data Mining and Intelligent Computing (ICDMIC). IEEE. pp. 1–7.

24. https://www.keel.es.

25. Menzies, T., Greenwald, J., & Frank, A. (2007). Data mining static code attributes to learn defect predictors. *IEEE Transactions on Software Engineering, 1,* 2–13.

26. Zimmerman, D. W., & Zumbo, B. D. (1993). Relative power of the Wilcoxon test, the Friedman test, and repeated-measures ANOVA on ranks. *The Journal of Experimental Education, 62*(1), 75–86.

Petri Net Modeling of Clinical Diagnosis Path in Tuberculosis

Gajendra Pratap Singh, Madhuri Jha, and Mamtesh Singh

1 Introduction

Tuberculosis (TB) is one of the highly contagious diseases which are mainly caused by the *Mycobacterium tuberculosis* bacteria (*Mtb*) [9]. The infection is mainly initiated by the inhalation of *Mtb* present in air droplets, which may be discharged by the person who is infected by the active TB bacteria. By the WHO report one can observe that the maximum case of mortality due to TB is from People Living with HIV (PLHIV) cases. The immune system weakens in the case of HIV so the bacteria of TB, which may be present in the latent form inside the body, can act actively which leads to intense infection [6]. Mortality in the case of Multi-Drug Resistant TB (MDR-TB) is highly in consideration which can have several reasons behind, possibly the late diagnosis or no diagnosis is one of the factors. Several diagnosis centers are being organized by the government to detect this disease in early stage but the diagnosis of TB is still a matter of discussion. Due to all problems still facing in India, the researches on TB prevention is always being motivated. In [34], authors designed a mathematical model on the reviewed major contribution in TB transmission. TB in children is also coming into focus from the last 6–7 years, this may be because of lack of nutrients or living with an infected person. The authors discuss the diagnosis and treatment of TB in children in [4]. In [32, 33], the WHO report

G. P. Singh (✉) · M. Jha
School of Computational and Integrative Sciences, Jawaharlal Nehru University, New Delhi 110067, India
e-mail: gajendra@mail.jnu.ac.in

M. Jha
e-mail: jhamadhuri81@gmail.com

M. Singh
Department of Zoology, Gargi College, University of Delhi, Delhi, India
e-mail: s.mamtesh@yahoo.com

for diagnosis pathways is discussed by which we motivated to model the whole process with the help of a graphical approach dealing with at most the important aspects behind this. Here, a clinical diagnostic path covering all the possible steps speculative TB patients will go through to have a suitable treatment is discussed with modeling by a Discrete Event System (DES), Petri net (PN). PN gives a descriptive analysis of the diagnosis system to get an optimal diagnostic algorithm.

Figure 1 gives the prior idea of the diagnosis process for Tuberculosis. The person showing TB symptoms must go through the prescribed diagnostic process. For children, it's recommended for Interferon-γ Release Assay (IGRA) blood test, and accordingly, the treatment is suggested either Fixed-Dose Combination (FDC) or Self-Administrated Therapy (SAT). In adults, the treatment for PLHIV and NON-PLHIV patients is different. If a PLHIV patients CD4 count is lower than the average, then its recommended to go through Antiretroviral Therapy (ART), and if CD4 count is more than average, then they should get TB treatment which is INH Preventive Therapy (IPT), where Isoniazid (INH) is the key drug in the treatment of TB. With NON-PLHIV patients, they should undergo the same diagnostic process called TB Skin Test (TST) and Chest X-ray. These two tests are recommended to all general patients. After these tests, four cases arise: (i) + ve TST with + ve chest X-ray, (ii) –ve TST with + ve X-ray, (iii) + ve TST with –ve X-ray, and (iv) –ve TST with –ve X-ray.

In the first two cases, the patients are supposed to have active TB bacteria, so they will be recommended to have a Sputum test and Drug-Susceptible Test (DST) to confirm the active bacteria and to identify the drugs which are resistant or susceptible. If Drug-Susceptible (DS), then recommended to take First-Line Drug (FLD) treatment, while in another case, if Drug-Resistant (DR), then patients are recommended Second-Line Drug (SLD). Again in case (iii), it is supposed to have Extra Pulmonary Tuberculosis (EPTB) in the patients and recommended Anti-tuberculosis Treatment

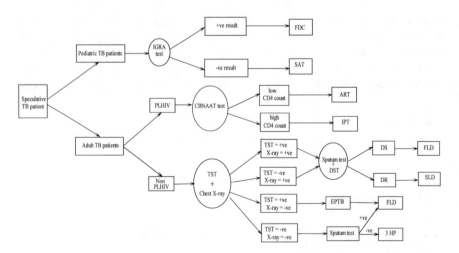

Fig. 1 Diagnosis Process and treatment recommended for Tuberculosis

with the use of Adjunct Corticosteroids (ATT with AC). The last case (iv) infers the confusion of inactive TB bacteria or not, so the patient, in this case, is recommended to go through the sputum test to confirm the disease. If in case it detects + ve, then patients are recommended to undergo ATT with AC; otherwise, it is considered that the immune system is very weak and, because of that, TB symptoms were detected in those patients. Thus, they are recommended for the treatment of Latent TB Infection (LTBI) which is 3HP, the treatment which is the fixed combination of Isoniazid (INH) and Rifapentine (RPT).

Modeling this diagnosis process is important, as it will help understanding the complexity of different treatments in TB patients [31].

2 Related Work

Pathology and epidemiology of TB is a vast area of research in which several researchers have worked already from the early eighteenth century. Whether it is about the diagnosis or the treatment, several mathematical models have also been generated to have a clear concept of increment or decrement in the rate of an affected person from TB. In [24], authors have shown the time delay from latent infection to active infection by making a model using some epidemiologic data. Several diagnostic tools have been invented and proven beneficial in the early diagnosis of TB. A summarized report is presented on the diagnosis process in [19] which proposed an organized algorithm for the rapid investigation of TB in European countries.

3 Methodology

PN is a special type of directed bipartite graph in which two types of nodes are present; one is Place (P) and another is Transition (T). Here, the arcs are also characterized by incoming flow (I^-) and outgoing flow (I^+), where I^- represents the arc from place to transition and I^+ represents the arc from transition to a place. These are directed and can have some weight associated with them depending upon the flow. Also, places can be assigned with some number from whole numbers which is called tokens on that place, and this will decide the initial marking (μ_0) of the net. Formally, PN is defined as a 5-tuple set (P, T, I^-, I^+, μ_0). With the initial marking μ_0, it is also called a Marked PN. Any transition fires at the given marking if and only if the outgoing arcs from all the places are less than and equal to the present marking of that place. After firing, the tokens are moved from the input places and deposited on the output places accordingly.

A transition t $\in$ T fires at the marking μ iff, for all the places, t satisfies the following condition:

$$I^-(p, t) \leq \mu(p)$$

Table 1 Description and firing ability of some transitions taken in the Model

Transition	Description	Transition	Description
	The normal transition follows the usual enabling condition		AND-join, enable only when all the incoming places have tokens
	XOR-split follows usual enabling condition but after firing transfer token to only one outgoing place		XOR-join, enable if any of the incoming places have tokens

After firing at μ, the new marking μ' is given by

$$\mu'(p) = \mu^0(p) - I^-(p, t) + I^+(p, t)$$

It is shown as $\mu \xrightarrow{t} \mu'$. Thus, we can say μ' is reachable from μ through t [16, 18, 21, 23].

PN is being widely used to model the biological system to explain the structural properties which can help to observe all the possible reachable states coming in the system [15, 26–28]. The reachability graph is one of the major tools in the PN which helps to analyze any network at every stage. PNs can also behave as a recommender system so to deal with some complex networks [11, 30].

In this paper, we have used some different types of transitions, as dealing with the biological process certain times the resulting events can be more than one and are independent of each other. Sometimes more than one independent event can result in one event by combining the simultaneous result. Therefore, we are using some special structure of transition instead of using a normal transition. Table 1 shows different transitions that have been used in this paper.

The graphical approach to model any network provided by the PN is not the only way to do all the analysis of that network, but it has several other tools to describe the network properties. Some of the properties are defined below which is used further for the analysis of the diagnostic path.

3.1 Some Analysis properties (power) of Petri Net

Reachability Graph: The *Reachability Graph* of a PN is denoted by $R(PN, \mu)$ where all the markings μ are taken as the vertices and the transition responsible for the corresponding marking is taken as edges. It helps in drawing the results of the reachability of one state to another state. One state is reachable from another state if there is a directed connected path found between them.

Invariant Property: *Invariant analysis* plays an important role when PN modeling deals with any metabolic network or genetic network. There are two kinds of invariant one is *p-invariants* and another is *t-invariants*. The former is the set of places where the sum of the tokens remains constant throughout the process. In a biological network, it helps to find the conservation relation. The t-invariants are the sequence of transition which can produce the initial marking after a certain number of intermediate steps. In a biological network, it helps to conclude in the cyclic behavior of the process.

Liveness and Boundedness: *Liveness* of any PN model assures that the process can eventually occur. A PN is life in a marking μ if there exists a marking sequence from any other marking μ^1. *Boundedness* is the property of a PN which can assure the bounded value of tokens in any place. One can assure that the accumulation of tokens is not taking place which is beneficial in almost every metabolic pathway.

State Machine: Any system that stores the information or resources at any time that can be used to make changes further.

Marked Graph: A PN is called a marked graph where each place edge has exactly one incoming and outgoing arc.

Here, as we are validating from the WoPeD software for PN, we will discuss some more important properties of the net related to the system modeled.

Net Density (ND): It is the ratio of the number of arcs in the net to the maximum number of arcs possible for the same number of nodes. Any net is considered well structured if it comes between $(0 - 0.1)$ and if it exceeds the value 0.5 then the net is not well structured.

Cyclicity (Cyc): It is the ratio of the number of nodes in the cycle to the total number of nodes. Any model with high cyclicity is more likely toward error probability.

These powers of the PN can help to describe the modeled system in a better way. An advanced PN analysis of techniques provides systematic and qualitative results of the modeled system. PN model analysis is getting attention because of its simple graphical approach and certain properties. Some high-level PNs like timed PNs, stochastic PNs, fuzzy PNs, and colored PNs are also finding focus because of their adaptability with any real-time value systems [3, 10].

The PN shown in the paper is created by WoPeD software [7] and the basic structural properties also have been performed with the same software. The behavioral properties of the net shown in the paper are obtained with the help of PIPE software [2].

4 Petri net Modeling and Analysis of Diagnosis Process in TB

While modeling any biological network with PN the first thing is to decide the places and transition to being assigned. In this paper, we have modeled the clinical diagnostic path for TB and recommended treatment accordingly.

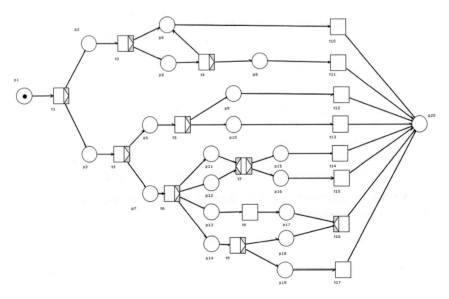

Fig. 2 Petri net model for the Diagnostic Process

For the diagnosis process explained in Fig. 1, modeling is shown in Fig. 2, where places are taken as different stages in the diagnosis process and the treatment recommended while transitions are taken as the prescribed clinical test like sputum test, chest X-ray, etc., required in diagnosis at different stages. The tokens on the place show the active form of that place or happening of that event. This diagnosis process gives us a clear idea about the variety of TB disease is present in the population and how the treatment differs in each case. Table 2 shows the description of places and transitions taken in the model of Fig. 2.

In the model, the person coming with the symptoms of TB for diagnosis is taken as a token in the first place. For the generosity of the model, we have taken one token on p_1 and zero on all the other places.

$$i.e., \quad \mu^0(p_1) = 1 \ and \ \mu^0(p_i) = 0, \ \forall \, p_i \in P$$

Next, t_1 is enabled and as it is an XOR-split transition so token will be transferred to any one of the next places either in pediatric patients (p_2) or adult patients (p_3).

$$\mu'(p_2) = \mu^0(p_2) - I^-(p_2, t_1) + I^+(p_2, t_1) = 0 - 0 + 1 = 1$$

$$Also, \mu'(p_3) = \mu^0(p_3) - I^-(p_3, t_1) + I^+(p_3, t_1) = 0 - 0 + 1 = 1$$

For each section the next stage diagnostic process is different. If t_2 will fire, it shifts the token to either p_4 or p_5. If the token is on p_5, then the children are supposed not to affect by active TB bacteria, but in this case, a higher test is recommended which is the sputum test (t_4). If the child gets—ve result in this test too then they are treating with SAT (t_{11}). This is required to increase immunity among kids having

Table 2 Description of places and transitions taken in the model

Places	Description	Transition	Description
p_1	Speculative TB patients	t_1	Clinical separation based on age
p_2	Pediatric TB patients	t_2	IGRAs test
p_3	Adult TB patients	t_3	Separation based on the first information
p_4	Children with −ve IGRAs (LTBI)	t_4	Sputum test
p_5	Children with + ve IGRAs or + ve TB test	t_5	CB-NAAT test
p_6	PLHIV TB patients	t_6	TST + chest X-ray
p_7	Non-PLHIV TB patients	t_7	Sputum + DST
p_8	-ve sputum test in children	t_8	PCR (Polymerase Chain Reaction)
p_9	Low CD4 count	t_9	Sputum test
p_{10}	High CD4 count	t_{10}	FDC
p_{11}	Patients with + ve TST and + ve chest X-ray	t_{11}	SAT
p_{12}	Patients with -ve TST and + ve chest X	t_{12}	ART
p_{13}	Patients with + ve TST and -ve chest X	t_{13}	IPT
p_{14}	Patients with -ve TST and -ve chest X	t_{14}	FLD
p_{15}	Patients with DST + ve	t_{15}	SLD
p_{16}	Patients with DST -ve	t_{16}	ATT with AC
p_{17}	EPTB patients	t_{17}	CDC recommended regimens, 3HP
p_{18}	+ ve Sputum test		
p_{19}	LTBI patients		
p_{20}	End of the treatment		

TB symptoms. While, If the token flows to p_4, then the children diagnosed with active TB bacteria and FDC (t_{10}) is recommended to the patients. The fixed-dose combination (FDC) is the improved grouping of drugs recommended in first-line treatment. Once t_{10} or t_{11} fires, then the process leads to the end of treatment (p_{19}). Likewise in adult patients, they are separated according to other clinical challenges at the initial stage. HIV is the most common in the patients having TB symptoms So, we have considered this case only. PLHIV case (p_6) or NON-PLHIV case (p_7). The diagnostic and treatment process are different in both cases. The PLHIV patients are recommended to diagnose by the CB-NAAT test (t_5) where CD4 count is measured. Accordingly token shifts to either p_9 or p_{10}. Then transition t_{12} and t_{13} can fire to lead the process to the end of the treatment. Similarly, for NON-PLHIV (p_7) patients, the

TST and X-ray (t_6) will fire to shift the token to any one place p_{11}, p_{12}, p_{13}, p_{14} which enable t_7, t_8, t_9, accordingly. This firing of t_7 leads to either DS (p_{15}) or DR (p_{16}) which then enable t_{14} or t_{15}, at last, it will fire to attain the end of treatment place. If t_8 fires then token shifts to p_{17} (EPTB) which enables t_{16} to bring the process to end. In the last, if t_9 will fire, token shift to either p_{18} or p_{19} to make t_{16} and t_{17} enabled again, leading to the end of the process. The model behaves as per the requirement of the diagnostic path shown in Fig. 1. Thus, it can help to analyze the structural and behavioral properties of the process.

5   Results

In the model, we have covered almost all the different cases and one can validate this by using real data values.

By observing the structural analysis of the net, we get the results shown in Fig. 3.

In Fig. 3, we can observe that the net is bounded which infer that the sample is to be diagnosed is moving toward the right path and there is no accumulation of tokens at any place other than the last place that is the end of treatment which is the desired result.

Also, we observed the Ratio Metrics for the net in Fig. 2. The properties like Net Density (ND), Coefficient of nets connectivity (CNC), average number of arc per node (DeN), average number of arcs per place (DeP), average number of arcs per transitions (DeT), degree of sequentiality (Seq), degree of net routing (RR), and cyclicity (Cyc) are shown in Fig. 4. These metrics are explained to check the error probability of the net. From the analyzed data, one can infer that the net proposed in Fig. 2 is executing with less error probability.

Fig. 3 Structural analysis of the net shown in Fig. 2

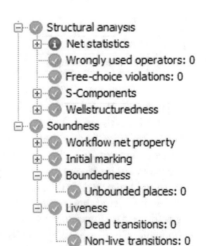

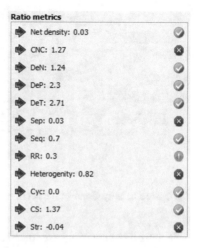

Fig. 4 Ratio metrics of the net in Fig. 2

While validation through real data, one can assign weights on the arc according to the data range and can find the shortest path as well as the longest path of the treatment. Without weight on the arcs, the longest path which is also called the diameter of the net is shown in Fig. 5. We can observe that the longest path contains the node where the skin test comes −ve and chest X-ray test comes + ve, and patients are drug-resistant and recommended to SLD. From the data of WHO, we can observe that this case is most contributing toward MDR-TB or XDR-TB as it takes maximum time to diagnose the actual situation of the disease.

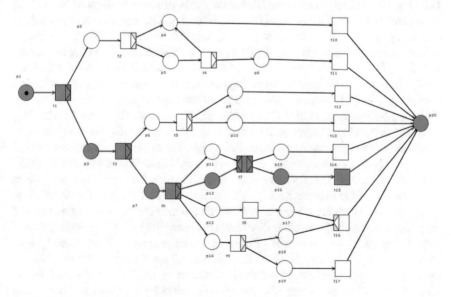

Fig. 5 The Longest path from Diagnosis to Treatment (Diameter)

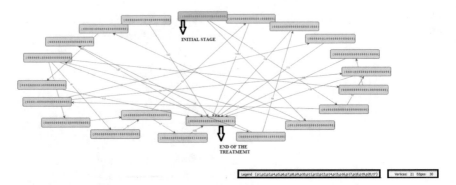

Fig. 6 Reachability Graph of the net in Fig. 2

When all the markings are traced as a directed graph taking responsible transition as arcs we get the reachability graph of the net in Fig. 2 is shown in Fig. 6. This a directed graph of 21 vertices and 30 edges. The nodes are the marking obtained by the respective firing of all transitions. This graph is a clear picture of all the reachable states in the diagnostic process in TB also shows the relation between them about the number of steps to reach on End of treatment from Initial Marking.

6 Discussion and Future Work

TB is one of the highly contagious diseases which is still a big threat worldwide. Lots of researches are still going on in the field of curing TB. Yes, the mortality rate is decreasing but still, about one-third population is affected by this disease. The major ratio in mortality due to TB is the PLHIV case which is generally coming in the case of MDR-TB or XDR-TB case. Several types of research have been done and are being motivated in this field to either diagnose it early or can get proper medication accordingly. In [25], authors presented a summarized report on the work of the European Network for Global Cooperation in the field of AIDS and TB (EUCO-Net). The two main problems in the way to stop TB is the late diagnosis and the drug resistance of the first-line and second-line drugs. There are several reasons behind the late diagnosis of this disease. In this paper, the clinical diagnosis path is being modeled with a DES, PNs for better execution of the path. It is very important to start the treatment of the patients having symptoms of TB according to the actual stage of the disease. There are several tests discovered which can help in the treatment of the patients accordingly and can slow down the mortality rate. PN modeling allows considering each and every stage involving in the process as it has such type of modeling tool. Earlier some authors have shown the application of PN modeling in diseases like cancer and cardiovascular dysfunction [1, 14, 22, 29]. PN modeling is very much efficient in dealing with biological systems by analyzing and simulating their structural behavior by explaining certain advanced properties like a siphon, trap, invariant properties, and many more [5, 12]. Certain high-level PN is also defined to explain some complex biological networks like Colored PN, Time PN, Fuzzy PN, and many more [8, 13, 17, 20]. In this paper, also some important aspects have been focused on regarding the diagnosis of TB. The longest path is traced which is possibly one of the reasons behind the multi-drug resistance in TB. Structural properties have been analyzed which shows the net is well structured, bounded, and live. Metric analysis has been shown which proves the error probability of the net is very low and can be used to draw beneficiary conclusions from the path. The proposed net can be used to validate the real data as it gives a systematic and optimal diagnostic algorithm in TB. In the future, one can think of applying Time PN by bounding the places with a certain time limit so as to estimate the actual duration of the process from diagnosis to treatment.

Acknowledgements This work is funded by the Department of Science and Technology (DST)-Science and Engineering Research Board (SERB) project id (File No: ECR/2017/003480/PMS).

References

1. Behinaein, B., Rudie, K., & Sangrar, W. (2018). Petri net siphon analysis and graph-theoretic measures for identifying combination therapies in cancer. *IEEE/ACM Transactions on Computational Biology and Bioinformatics (TCBB), 15*(1), 231–243.
2. Bonet, P., Lladó, C. M., Puijaner, R., & Knottenbelt, W. J. (2007). PIPE v2. 5: A Petri net tool for performance modeling. In *Proc. 23rd Latin American Conference on Informatics (CLEI)*.
3. Brauer, W., & Reisig, W. (2009). Carl Adam Petri and "Petri nets." *Fundamental Concepts in Computer Science, 3*(5), 129–139.
4. Britton, P., Perez-Velez, C. M., & Marais, B. J. (2013). Diagnosis, treatment and prevention of tuberculosis in children. *New South Wales Public Health Bulletin, 24*(1), 15–21.
5. Cherdal, S., & Mouline, S. (2018). Modeling and simulation of biochemical processes using petri nets. *Processes, 6*(8), 97.
6. Ford, N., Matteelli, A., et al. (2016). TB as a cause of hospitalization and in-hospital mortality among people living with HIV worldwide: A systematic review and meta-analysis. *Journal of the International AIDS Society, 19*(1), 20714.
7. Freytag, T. (2016). WoPeD—Workflow Petri net designer. *University of Cooperative Education, 279–282.*
8. Gilbert, D., Heiner, M., Ghanbar, L., & Chodak, J. (2019). Spatial quorum sensing modelling using coloured hybrid Petri nets and simulative model checking. *BMC Bioinformatics, 20*(4), 173.
9. Goldman, R. C., Kevin, V. P., & Barbara, E. L. (2007). The evolution of extensively drug-resistant tuberculosis (XDR-TB): History, status, and issues for global control. *Infectious Disorders-Drug Targets (Formerly Current Drug Targets-Infectious Disorders), 7*(2), 73–91.
10. Gupta, S., Kumawat, S., & Singh, G. P. (2019). Fuzzy petri net representation of fuzzy production propositions of a rule-based system. In *International Conference on Advances in Computing and Data Sciences*, pp. 197–210.
11. Gupta, S., Singh, G. P., & Kumawat, S. (2019). Petri net recommender system to model metabolic pathway of polyhydroxyalkanoates. *International Journal of Knowledge and Systems Science (IJKSS), 10*(2), 42–59.
12. Herajy, M., Liu, F., & Heiner, M. (2018). Efficient modelling of yeast cell cycles based on multisite phosphorylation using coloured hybrid Petri nets with marking-dependent arc weights. *Nonlinear Analysis: Hybrid Systems, 27,* 191–212.

13. Herajy, M., Liu, F., Rohr, C., & Heiner, M. (2018). Coloured hybrid petri nets: An adaptable modelling approach for multi-scale biological networks. *Computational Biology and Chemistry, 76,* 87–100.

14. Jung, J., Kwon, M., Bae, S., Yim, S., & Lee, D. (2018). Petri net-based prediction of therapeutic targets that recover abnormally phosphorylated proteins in muscle atrophy. *BMC Systems Biology, 12*(1), 26.

15. Kansal, S., Acharya, M., & Singh, G. P. (2012). *Boolean Petri nets.* IntechOpen: Petri Nets-Manufacturing and Computer Science.

16. Kansal, S., Singh, G. P., & Acharya, M. (2010). On Petri nets generating all the binary n-vectors. *Scientiae Mathematicae Japonicae, 71*(2), 209–216.

17. Liu, F., Heiner, M., & Gilbert, D. (2018). Fuzzy Petri nets for modelling of uncertain biological systems. *Briefings in bioinformatics.*

18. Murata, T. (1989). Petri nets: Properties, analysis, and applications. *Proceedings of the IEEE, 77*(4), 541–580.

19. Migliori, G. B., Giovanni, S., Senia, R. -K., Rosella, C., D'Ambrosio, L., Abubakar, I., Bothamley, G. et al. (2018). ERS/ECDC statement: European union standards for tuberculosis care, 2017 update. *European Respiratory Journal, 51*(5), 1702678.

20. Olszak, J., Radom, M., & Formanowicz, P. (2018) Some aspects of modeling and analysis of complex biological systems using time Petri nets. *Bulletin of the Polish Academy of Sciences. Technical Sciences, 66*(1).

21. Peterson, J. L. (1977). Petri nets. *ACM Computing Surveys (CSUR), 9*(3), 223–252.

22. Rovetto, C., Cano, E., Ojo, K., Tuñon, M., & Montes, H. (2018). Coloured petri net model for remote monitoring of cardiovascular dysfunction. *In Memorias de Congresos UTP*, pp. 405–411.

23. Russo, G., Pennisi, M., Boscarino, R., & Pappalardo, F. (2018). Continuous Petri Nets and microRNA analysis in melanoma. *IEEE/ACM Transactions on Computational Biology and Bioinformatics (TCBB), 15*(5), 1492–1499.

24. Salpeter, E. E., & Salpeter, S. R. (1998). Mathematical model for the epidemiology of tuberculosis, with estimates of the reproductive number and infection-delay function. *American Journal of Epidemiology, 147*(4), 398–406.

25. Martina, S., Giehl, C., Mcnerney, R., Kampmann, B., Walzl, G., Cuchí, P., Wingfield, C. et al. (2010). Challenges and perspectives for improved management of HIV/Mycobacterium tuberculosis co-infection. *European Respiratory Journal, 36*(6), 1242–1247.

26. Singh, G. P., Kansal, S., & Acharya, M. (2013). Construction of a crisp Boolean Petri net from a 1-safe Petri net. *International Journal of Computer Applications, 73*(17).

27. Singh, G. P., Kansal, S., & Acharya, M., Embedding an Arbitrary 1-safe Petri Net into a Boolean Petri Net. *International Journal of Computer Applications, 70*(6).

28. Singh, G. P., & Kansal, S. (2016). Basic Results on Crisp Boolean Petri Nets. *Modern Mathematical Methods and High-Performance Computing in Science and Technology*, pp. 83–88.

29. Singh, G. P., & Gupta, A. (2019). A Petri Net Analysis to Study the Effects of Diabetes on Cardiovascular Diseases. IEEE Xplore. ISBN: 978–93–80544–36–6. (accepted).

30. Singh, G. P., & Singh, S. K. (2019). Petri net recommender system for generating of perfect binary tree. *International Journal of Knowledge and Systems Science (IJKSS), 10*(2), 1–12.

31. Storla, D. G., Yimer, S., & Bjune, G. A. (2008). A systematic review of delay in the diagnosis and treatment of tuberculosis. *BMC Public Health, 8*(1), 15.

32. World Health Organization. (2018). *Global tuberculosis report 2018.* World Health Organization.

33. World Health Organization. (2009). Pathways to better diagnostics for tuberculosis: A blueprint for the development of TB diagnostics by the new diagnostics working group of the Stop TB Partnership.

34. Zwerling, A., Shrestha, S., & Dowdy, D. W. (2015) Mathematical modeling and tuberculosis: advances in diagnostics and novel therapies. *Advances in medicine.*

Selecting Optimal Areas of Feedwater Exchangers in Steam Turbines for Cost Saving Using Dynamic Programming

Saeib A. Alhadi Faroun, Asaad A. M. Al-Salih, Vikas Rastogi, and Rajesh Kumar

1 Introduction

1.1 Towards Cost-effective Design in Feedwater Exchangers

In steam turbine-based power plants, and in order to achieve 'optimal' designs of feedwater exchangers, a variety of specific problems may arise and thus require to be handled. The motive behind solving those problems is through minimizing the power production cost besides for efficiency maximizing. In other words, the need for finding such optimal solutions for any of these problems comes notably from either of the two main reasons; solution design with minimal permissible fabrication cost, or achieving maximal probable dependability, or both [1–9]. In due course, multi-process industries have reportedly utilized the technique of 'dynamic programming,' in diverse processes, aiming to achieve optimization. Owing to reasons clarified in Sect. 1.3, the optimization policy in this work has adopted the top-down design approach in the relevant system modeling. This has involved the selection of optimal areas of feedwater exchangers in a four bleeding points single steam turbine-based power plant besides its cost saving.

S. A. A. Faroun
Senior Expert, Ministry of Electricity, Government of Iraq, Baghdad, Iraq
e-mail: saafa436@yahoo.com

A. A. M. Al-Salih (✉)
Department of Electrical Engineering, University of Baghdad, Baghdad, Iraq
e-mail: alsalih1996@yahoo.co.in

V. Rastogi · R. Kumar
Department of Mechanical Engineering, Delhi Technological University, Delhi, India
e-mail: Rastogivikas@gmail.com

R. Kumar
e-mail: dr.rajeshmits@gmail.com

© The Author(s), under exclusive license to Springer Nature Singapore Pte Ltd. 2021
P. K. Kapur et al. (eds.), *Advances in Interdisciplinary Research in Engineering and Business Management*, Asset Analytics,
https://doi.org/10.1007/978-981-16-0037-1_33

This paper is keen to highlight the main related aspects of applying the technique of dynamic programming in the pertinent design process.

1.2 Some of the Other Works in the Field

Within the confines of designing feedwater exchangers in steam turbines, with an aim for cost saving, there has been a wide spectrum of related works in the literature. Among these is the work presented by Chan and Kantamaneni (2015) which is about efficient energy-saving achievement with the confines of industry besides the available relevant mechanisms. Hereby, comprehensive bottom-up modeling initiatives concerning the evaluation and estimation of the trends regarding the consumption of energy were in fact conducted including eight different sectors of industry. Spanning the range up to 2050, this comprehended the issue of the potentials toward the saving of energy [10]. Alobaid et al. (2016) have delineated a study about their progress in the dynamic simulation of power plants with thermal basis. This study comprehended the relevant necessary mathematical models, the diverse essential constituents to describe the basic processes, in addition to the automation, mechanization, and all electrical organizations of power plants with thermal basis. This was associated with realistic models of available paradigms [11]. In a power plant that is a supercritical unit, Wołowicz et al. (2012) exhibited the notion of utilizing the exhausting hot gases coming out of the turbine in such a way to preliminarily heat up the plant feedwater. The turbine was hereby combined with the system, such as to bypass a heat exchanger with regenerative action of heat transfer due to the hot gases coming under high pressure [12]. Cziesla et al. (2009) published their work regarding some highly developed power plants on steam basis 800 + MW, with their upcoming combined cycle system alternatives (abbreviated as CCS). This work has actually delineated the construction of some sophisticated USC (ultra-supercritical) 800 Mega Watt power plant (steam basis) depending on the indication plant SSP5/6000 (that is a 1 × 800 ultra-supercritical) [13]. Towler and Sinnott (2012) published a book about their design in chemical engineering involving the fundamentals, practicalities, and economical affairs of the whole design of the plant besides the pertinent processes. It has dealt with the design of chemical procedures and apparatus preceded by the application of chemical engineering fundamentals [14]. In the context of liberalized markets of power, Rong and Lahdelma (2013) have presented their power systems based on the best possible procedures of joining power and heat (generally referred to as *Combined Heat and Power*, which is denoted as CHP). On such CHP systems, they discussed the effects of the markets of the liberalized power. They further presented a distinctive unit assurance form for such system with diverse kinds of constituents [15].

1.3   Method, Approach, and Achievement

Recently, among the fast advancement in machinery and data processing equipment, the notion of how to attain 'optimization' has received considerable attention. Based on that, and as mentioned in the literature, 'optimization theory' and optimization methods that deal with selecting the best possible alternatives for a given object function, start to form the best convenient resort [1–9]. In this work, 'dynamic programming' has been selected then utilized as an adequate 'technique' for optimization due to its uniqueness in solving diverse problems faced in the context of power plants.

Two different approaches for the "optimization design modeling" are available; the Top-Down Design (TDD) and the Bottom-Up Design (BUD) approaches. Due to its step-wise refinement property, the TDD approach for modeling assessment has been chosen, rather asserted, then adopted by the optimization policy in this work. It thereby states that the key feature in dynamic programming is first to set the power plant 'overall' state, then to set the intermediate states [2]. Hence, and as a practical paradigm, a pertinent problem related to a thermal power plant, operating on organic fuel, with single steam turbine chamber having four bleeding points, as illustrated in Fig. 1, has been considered and articulated. Different areas of feedwater exchangers (within the bleeding points enclosures) have whereby been analyzed with an aim to attain maximum saving (in \$/sec). To optimize a specific solution in such area distribution of the enclosures, this has practically involved distributing $1000m^2$ total area of heat exchange in this steam turbine power plant. The said area has therefore been empirically distributed as several different area combinations;

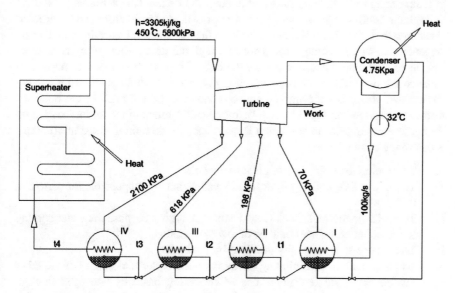

Fig. 1 Selection of optimal areas (I, II, II, IV) of feedwater heaters

each of which is allocated to the four subsequent feedwater heaters, in order to form four concurrent stages of heat exchange. Heat saving at all the respective area combinations of feedwater heaters has whereby been calculated for each of these combinations, yielding to calculate the cost of steam used at each combination. The combination with maximum saving has thus been indicated showing the minimum cost of steam used.

The achievement in this work is represented in attaining the 'maximum' saving of 1.186 Cent/sec through distributing the 1000m^2 total area 'A' of the four consecutive feedwater heaters as in the partial area combinations of A I $= 200$m^2, A II $= 200$m^2, A III $= 300$m^2, and A IV $= 300$m^2, respectively, where A I represents the lowest bleeding pressure chamber (right side of turbine in Fig. 1), and A IV represents the highest bleeding pressure chamber (left of Fig. 1). Therefore in order to 'minimize' the cost of steam used, the conclusion here is to allocate the total area 'A' of the four consecutive feedwater heaters as 0.2A, 0.2A, 0.3A, and 0.3A, respectively.

2   Background and Basic Concepts

2.1   Feedwater Exchangers in Steam Turbine

As demonstrated in Fig. 1, the main two units in any thermal power plant are the steam boiler (with superheater) and the turbine. The provided water to the boiler is named as feedwater. Feedwater is hereby heated in advance up to the saturation level of temperature to be thereby prepared to enter the boiler. This action is to be done through the feedwater pumps under the high-pressure heater in order to be forwarded ahead into the boiler. In high-pressure heaters, the feedwater is basically heated by the "regenerative water heating," that is done through the steam gotten from the turbine. This technique would cause to reduce the loss of heat in the condenser besides to enhance the overall steam turbine efficiency among the whole plant. After leaving the turbine, steam is concentrated and condensed to be fed back to the boiler as feedwater as shown in Fig. 1. Being "economically" crucial in warming/heating up the cycling water, prior to supplying it again back to the boiling container (boiler), is due to the causes below:

i.   To develop the efficiency of plant as a whole.
ii.   To remove CO_2 and O_2 (dissolved in water) that are responsible for corrosion in the boiler.
iii. To avoid the thermal shock (i.e., the thermal stress) because the water coming to the boiler container could be of low temperature.
iv.   To enhance steam quantity provided by the boiler.
v.   To participate in precipitating (away from the boiler tank) some of the other diverse tiny particles (impurities of corrosion) basically conveyed through steam to condensate in the condenser and the boiler [3].

In the recent 'big' power stations/plants, turbine chambers have inevitable steam hemorrhage that can be used in feedwater warming up. The feedwater is thus warmed up/heated, set in pressure, more heated as well to keep its pressure and temperature above the boiler water temperature prior to the superheating process. This will increase the power production efficiency of the power plant as a whole.

2.2   The Concept of Dynamic Programming and the Top-Down Design Policy

The concept of *Dynamic Programming* (also denoted as *dynamic optimization*) is a technique that follows the optimization policy of TDD assessment modeling mentioned in Section I.3. The TDD is used to resolve the single multifarious or multiside problem through analyzing such problem into an assortment of uncomplicated smaller problems/sub-problems. Then to resolve all of these smaller problems each one alone, with a goal to sort out the smaller elucidations under some final stage of integration.

Instead of computing the same smaller problem, once it takes place again, it is merely possible to call on the formerly calculated solution. Hence, this will save the precious calculation period of time with a simple cost of storage. It is moreover feasible to make an index for the solutions of every smaller problem. In order to expedite the calling on process, this index could be generally constructed depending on the pertinent parameter values [15].

The pertinent formulation of dynamic programming in this paper envisages distributing the total area of heat exchange in the feedwater side in any power plant (with a specified bleeding points steam turbine) into a certain number of parts (with different combinations) to form concurrent chambers of heat exchange. Each chamber is thus coupled to an individual steam bleeding point of the turbine. Therefore, the pertinent data (of each of the said different combinations) involving the respective physical parameters such as the area besides the specific local temperature and pressure values at the relevant positions have all to be applied in the related equations of "Dynamic Programming" available in the literature, hence to estimate the individual and total "Saving" (in cent/sec) yielding to reduce the overall cost of heating of the steam used in the plant. This will eventually lead to maximize plant efficiency and minimize cost.

2.3   Optimal Areas Selection and Cost Saving Using Dynamic Programming

In any steam turbine-based power plant, it is deemed to be essential, rather crucial, for the plant design engineers to allot heaters thermal regions in such a way to keep the "running cost" of the plant *minimal* under *maximal* savings. This can be achieved by utilizing some technique that is known as Dynamic Programming. It is actually a technique that is appropriate to be applied and used in a broad range of decision problems with multi-stage structure. It is basically built on an optimization basis with a nonlinear nature [4]. Mechanical parts are usually designed by mechanical technologists and engineers to accomplish either the utmost possible element lifetime or the lowest manufacturing price or could be both [5]. From the design point of view, any problem generally comprehends numerous factors in design, some of which are exceedingly quite responsive to design manners. It is thus important to take into account that dynamic programming comprehends the inherent development of all probable encountered cases in all possible paths. That is because the design should inevitably depend on the basis of seeking and attaining optimality. For this reason, dynamic programming often achieves the global system optimum. Thus, changing the issue of optimal control into an issue of optimizing parameters is hereby a quite relevant issue and needs comprehensively and deeply to be considered [6].

2.4   System Mathematical Model

As is typical of dynamic programming, temperatures can be computed by the use of Eq. 1.

$$t_0 = t_i + (t_c - t_i)(1 - e^{-UA/wcp}),\tag{1}$$

where t_o = outlet feed water temp. from heater °C, t_i = inlet feed water temp. to heater °C, t_c = bleeding steam temp. to heater °C, U = coefficient of thermal conductivity Kw/m²/k, W = flow rate of feed water Kg/s, C_p = heat transfer coefficient Kg/k. Heat can be calculated from Eq. 2.

$$Q = m_w C_{pw}(T_0 - T_i)\,10^{-6}\,[GW].\tag{2}$$

where Q = heat transfer to water, m_w = mass flow kg/s, C_{pw} = constant of heat transfer kg/k. The "saving" hereby represents the price of *heat saved* at the boiler *lesser extraction cost* of the used steam.

Likewise, W_{fw} denotes the cost of heat at boiler (60cent/Gj) and W_{ex} denotes extraction steam cost (cent/Gj), *Saving* (*S*) is therefore defined here as in Eq. 3.

$$S = (W_{fw} - W_{ex})\,Q.\tag{3}$$

2.5   *Number of Transfer Units NTU's*

Considering Fig. 2, the effectiveness (ε) of the heat exchanger thermal expression in the counterflow direction might be articulated as a functional relation of two dimensionless sets; UA/W_{min} besides W_{min}/W_{max} [2], whereby U = coefficient of thermal conductivity $K_W/m^2/$ Ko, A denotes various areas of feedwater heat exchangers in m^2, W_{min} denotes the rate of heat transfer (K_W/Ko) at minimum temperature, while W_{max} denotes the rate of heat transfer at maximum temperature. The possibility to articulate this effectiveness as a functional relation of UA/W_{min} and W_{min}/W_{max} yields to graphic arrangements of the effectiveness of heat exchangers of various configurations. The group UA/W_{min} is called the number of transfer units (NTUs). Typical graphical presentations for the basic configuration of heat exchangers are shown in Fig. 2.

3   Methodology

The methodology of this work is based on the concept of optimization. This general concept is appropriately viable to be applied in both continuously operational functions besides staged progressive processes. With a rather restricted resolution, the term.

'Optimization' denotes the attempting action to attain the finest possible resulting outcome within specifically set conditions. In the managerial/technical sides of any engineering system, within the confines of processes of planning, designing, or even constructing then maintaining stages, all the pertinent engineers are required to consider quite diverse relevant judgments [7]. The final target of all of those judgments is to either make the best possible limit of the wanted advantage or diminish the endeavors needed for the achievement. Optimization is generally considered of

Fig. 2 Typical graphical presentations for the basic configuration of heat exchangers [2]

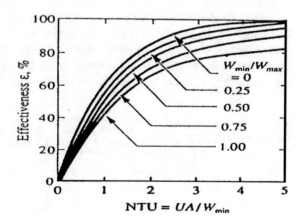

many types based on the difficulty and intricacy of the faced geometrical complexity. For optimal control achievement besides parameters optimization, it is convenient to resort to provide some all-purpose problems as well as some specialized problems. To seek optimization, the respective physical system can be modeled under the best possible elucidation problem. When this action comprehends merely single-target functional process, then the issue of developing/solving that optimization is denoted as of a single target/objective type. Whereas, in the case of an optimization problem that has more than single-target functional process, the issue of obtaining another optimal resolution is denoted as of multi-target/objective type [8]. The technique of dynamic programming albeit faces some foremost weakness referred to as the nuisance of dimensionality that might require some suitable cure under the concept of *dimensionality reduction*. This in fact is quite applicable and convenient in solving broad scope of multifaceted and compound problems in diverse fields of judgment reaching [9].

3.1 Engineering Perspective

In diverse states of affairs concerning engineering sides, the dynamic programming problem is generally represented in the stages breakdown into the relevant distinct stages. This involves calculating the possible distinctions aiming to attain a fairly accurate outcome. Through applying dynamic programming, one can notice that Table 1 is merely a necessary routine, yet Table 2 shows that the application of dynamic programming is effectual and crucial. Resolving the problem under treatment, that seems in constrains, among applying dynamic programming is clearly evident through optimizing the process of heating up the feedwater. Actually, it is a widespread application in the big vital power plants to preheat up the main boiling container (the superheater) by the turbine bleeding steam extracted to boost the heating temperature of the feedwater. This consequently yields to upgrade the effectiveness of the overall power of steam circulation in the plant. In order to perform extraction on the turbine bleeding steam (at diverse values of pressure), several heaters need to be utilized in most of the available power stations. These feedwater heaters improve the efficiency of the whole cycle as indicated through the calculation of each specific case [2]. Accordingly, at the turbine duct, for each joule of the outcome work, it has been deemed to provide the superheater boiler a steam power

Table 1 Characteristics of the conclusive model	Extraction & heater number	Saturation temperature (C°)	Worth of extraction cent/GJ
	1	90	23
	2	120	28
	3	160	38
	4	215	47

Table 2 Extraction steam data for stage I

Total area (m²)	Area for stage I (m²)	T 1 °C	Q [GW]	Saving cent / second
0	0	32.00	0	0
100	100	60.27	0.0118448	0.0043826
200	200	74.76	0.0179164	0.006629
300	300	82.19	0.0210287	0.0077806
400	400	86.00	0.0226241	0.0083709
500	500	87.95	0.0234419	0.0086735
600	600	88.95	0.0238611	0.0088286
700	700	89.46	0.024076	0.0089081
800	800	89.72	0.0241862	0.0089489
900	900	89.86	0.0242426	0.0089698
1000	1000	89.93	0.0242716	0.0089805

circulation heat of about 3j. Hereby, the 2j disparity represents the loss quantity discarded in the vicinity of the condenser. It is however hopeless to think to utilize a part of the discarded heat in the vicinity of the condenser area to heat up the water in the boiler container. The reason behind that is due to the fact that heating up the boiler container through the steam exhausted out of the enclosure of the turbine won't be possible due to simple physical reasons. As hereby we won't gain any difference in temperature between the boiler enclosure and the exhausted steam out of the enclosure of the turbine in order to create the necessary force to transfer that heat. The exhausted steam in the vicinity of the condenser area is actually having a lesser temperature than that bleeding from the turbine (extraction steam); the last may therefore be indirectly utilized in assisting for heating up the boiler container. As a matter of fact, 1 kg of the steam bleeding from the turbine (extraction steam) after departing the container of the boiler will accomplish in the turbine chamber a work. It shall thereby utilize the rest of the energy left onto the soaked liquid at the condenser aiming in heating up the feedwater. Effectively, the provided heat (in overall) to the 1-kg steam within the container of the boiler is thus fully transferred/changed to work. It is further found out that among heating up the feedwater via the bleeding steam (extracted) from the turbine chamber; the overall efficiency of the steam circulation is clearly increased in comparison with discarding the possibility of two joules of heating up the boiler for each turbine work of 1j. It is also postulated that the steam with higher pressure level is forceful and more useful than steam with lower pressure level since the turbine bleeding steam that is extracted at high pressure should therefore be more powerful to convey supplementary work in the duct of the turbine. Hereby, it might be helpful to start with a *virtual* practical exemplary depicted case in order to realize the best possible way to use dynamic programming for the purpose of applying "optimization" to resolve this type of issue. Once we know the pertinent characteristic properties of some heat exchanger, i.e., as its boiler temperature, condenser fluid temperature, inlet temperature, are all identified, then it would be

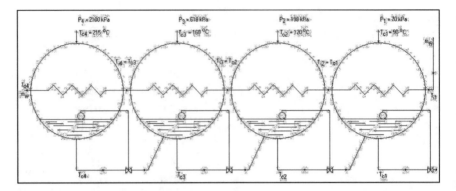

Fig. 3 Feedwater heat exchangers

possible to consider Eq. 1 on the acting invariant fluid (with no phase change) in order to estimate its outlet temperature.

Considering Fig. 1, a proposed economical analysis suggested an overall 1000 m^2 of heat exchanging locality to be utilized in the four feedwater heat exchangers (heaters). Choosing increments of 100 m^2, the assigned 1000 m^2 is envisaged for many different possible versions of distribution among the said four heaters. For all of the four heaters, the coefficient of the overall heat conveying (i.e., heat transfer) is 2800w/m^2.k. The value of steam extracted (bleeding from the turbine) can be estimated through some relevant computations in thermodynamics. This is simply displayed in Table 1. As per each Giga Joule, the boiler heat cost is 60 cents. As illustrated in Fig. 3 [2], and in order to find out the distribution of the pertinent area in optimal manner through the use of dynamic programming, 100 kg/s should be the flow rate for feedwater heat exchangers.

3.2 Analytical Treatment

Prior to starting with some required resolution, a proper advice in this regard is perhaps to speculating then anticipating the optimum resolution temperament. Considering Figs.1 and 3, that illustrate the concept of selecting optimal area stages of feedwater heat exchangers, one can notice that the four stages from the right (the lowest pressure and lowest temperature side) are A I, A II, A III, and A IV, respectively. As will be seen through inspecting the following five tables (Tables 1, 2, 3, 4 and 5), the number of examined alternatives for areas distribution among the four stages (which have been investigated to attain better cost savings) is 787 alternatives. Through inspecting these tables, one can observe that this number represents 11 areas from the first stage, 66 areas from the second stage, 245 areas from the third stage, and 465 areas from the fourth stage. For each arrangement of areas, the overall economical save for the entire number of disseminated areas, besides the

Table 3 Stages I and II

Total area [m²]	A II [m²]	A I [m²]	T2[c°]	Saving II [Cent/s]	Total saving [Cent/s]
0	0	0	32	0	0
100	0	100	62.27	0	0.0043826
100	100	0	74.89	0.0057508	0.0057508
200	0	200	74.76	0	0.0066291
200	100	100	89.38	0.0039034	0.008286
200	200	0	96.88	0.0086987	0.0086987
1000	0	1000	89.93	0	0.0089805
1000	100	900	104.55	0.0019698	0.0109395
1000	200	800	112.04	0.0029928	0.0119417
1000	300	700	115.89	0.0035432	0.0124513
1000	400	600	117.86	0.003876	0.0127046
1000	500	500	118.87	0.0041455	0.012819
1000	600	400	119.38	0.0044766	0.0128475*
1000	700	300	119.65	0.0050227	0.0128033
1000	800	200	119.78	0.0060369	0.0126659
1000	900	100	119.85	0.0079891	0.0123717
1000	1000	0	119.89	0.0117843	0.0117843

Table 4 Stages I to III

Total Area [m²]	A III [m²]	A II [m²]	A I [m²]	T3 [c°]	Saving III [Cent/s]	Saving II [Cent/s]	Saving I [Cent/s]	Total saving [Cent/s]
0	0	0	0	32	0	0	0	0
100	0	100	0	74.89	0	0.0057508	0	0.0057508
200	0	200	0	96.88	0	0.0086987	0	0.0086987
200	200	0	0	126.037	0.0086987	0	0	0.0086987
1000	0	600	400	119.38	0	0.0044766	0.0083709	0.0128475
1000	100	400	500	138.36	0.0018966	0.0040009	0.0086735	0.014571
1000	200	500	300	149.14	0.0028093	0.0048904	0.0077806	0.0154803
1000	300	300	400	154	0.0035559	0.0039452	0.0083709	0.015872
1000	400	400	200	157.02	0.0037007	0.005647	0.0066291	0.0159767*
1000	500	200	300	158.23	0.0044401	0.0037377	0.0077806	0.0159585
1000	600	300	100	159.13	0.0043485	0.00693	0.0043826	0.015661
1000	700	200	100	159.48	0.0050862	0.0059043	0.0043826	0.0153731
1000	800	0	200	159.59	0.00782	0	0.0066291	0.014449

Table 5 Stages I to IV

A total (m²)	A IV (m²)	A III (m²)	A II (m²)	A I (m²)	T4 C°	Saving IV c/s	Saving III c/s	Saving II c/s	Saving I c/s	Total saving c/s	Max
1000	200	0	100	700	185.92	0.0044436	0	0.0019958	0.0089081	0.0153475	
1000	200	0	600	200	189.82	0.0038479	0	0.0059557	0.0066291	0.0164327	
1000	200	0	200	600	187.89	0.0041426	0	0.0030695	0.0088286	0.0160407	
1000	200	0	300	500	188.9	0.0039883	0	0.0037188	0.0086735	0.0163806	
1000	200	0	500	300	189.69	0.0038687	0	0.0048904	0.0077806	0.0165397	
1000	200	0	400	400	189.42	0.0039092	0	0.0042445	0.0083709	0.0165247	
1000	300	700	0	0	207.43	0.0026484	0.0116893	0	0	0.0143378	
1000	300	600	100	0	207.38	0.0026651	0.007703	0.0057508	0	0.016119	
1000	300	600	0	100	207.35	0.0026776	0.0090264	0	0.0043826	0.0160866	
1000	300	500	200	0	207.29	0.0026976	0.0056127	0.0086987	0	0.0170091	
1000	300	500	0	200	207.19	0.0027345	0.0075793	0	0.0066291	0.0169429	
1000	300	500	100	100	207.26	0.0027101	0.0062792	0.0039034	0.0043826	0.0172753	
1000	300	400	0	300	206.87	0.0028456	0.0066775	0	0.0077806	0.0173037	
1000	300	400	300	0	207.11	0.0027611	0.0044498	0.0102098	0	0.0174207	
1000	300	400	100	200	207	0.002798	0.0054227	0.0029565	0.0066291	0.0178062	
1000	300	400	200	100	207.07	0.0027736	0.0047795	0.0059043	0.0043826	0.01784	
1000	300	300	400	0	206.76	0.0028848	0.0036752	0.0109844	0	0.0175444	
1000	300	300	0	400	206.25	0.0030621	0.0059029	0	0.0083709	0.017336	

(continued)

Table 5 (continued)

A total (m²)	A IV (m²)	A III (m²)	A II (m²)	A I (m²)	T4 C°	Saving IV c/s	Saving III c/s	Saving II c/s	Saving I c/s	Total saving c/s	Max
1000	300	300	300	100	206.72	0.0028973	0.0038323	0.00693	0.0043826	0.0180422	
1000	300	300	100	300	206.51	0.0029693	0.0047366	0.002471	0.0077806	0.0179576	
1000	300	300	200	200	206.65	0.0029217	0.0041387	0.0044719	0.0066291	0.0181615	***
1000	300	200	500	0	206.07	0.0031263	0.00293	0.0113815	0	0.0174378	

*** "Maximum Saving" alternative among several other different combinations of area distribution

individual cost savings in every area, has been calculated. It is desirable to use the lowest possible steam cost which might propose a bigger region in stage I. However, every further piece of area in the said stage would be lesser effectual compared with the preceding piece of area. The reason behind that is owing to the less disparity in temperature between the feedwater and the steam. Thus, there must be a compromise between trying to use the low-cost steam and maintaining a high-temperature difference. Because the entire area has hereby been already identified, this issue appears a bit constrained. None the less, through using as a 'state variable' the entire dedicated area, this problem would be changed into a proper unconstrained type.

Actually, it might hereby be advantageous to embark on treating this problem from its front (right) side concerning the flow at the feedwater (i.e., stage A I side), because the inlet temperature of the feedwater is known (32 °C).

4 Results

There are five tables for justifying this method. Table 1, in dynamic problem solution, is merely a routine, yet as mentioned earlier, "dynamic programming" becomes effective starting from the second table upward. Results of the final solution steps are evidently exhibited in Table 5, where the solution consolidates. Table 2 demonstrates the temperature of the outlet starting at Stage I for diverse included stage areas (A). Table 3 uses the state variables as the total regions (areas) dedicated in the initial 2 stages (A I and A II). So for instance, if 1000 m² is offered for the initial two stages, the allotment in optimal will be allocating 400m² for the first stage besides 600 m² for the second, yielding the (US $) economical save (saving) of 0.0128475 Cent per second.

5 Conclusion and Future Scope

Dynamic programming works efficiently in the distribution of areas on turbine feedwater heat exchangers for getting better cost saving and efficiency. And once there is a necessity to optimize a feed-forward a systematic structure formed through a sequence of procedures involving thermomechanical constituents in which the outcome state becomes the income to the next, the need may arise for dynamic programming to be applied, carried out, and experienced. The crucial issues usually appear are in specifying the status (state) variables besides managing to arrange the pertinent tables. As well as that, it is further important, rather essential to calculate the flow rate and heat transfer to feedwater heat exchangers (in GW) for all stages with given time t_i. For each area (A), the saving in each stage has been calculated in addition to the total saving for all distributed areas. A point of significance that deserves hereby to be highlighted is that when comparing the work presented in this paper with some other similar work, one can notice the improvement. Table 5 hereby shows that the full area of 1000m^2 committed in the presented approach of "Dynamic Programming" signifies an optimal allocation of regions of 200,200,300,300 that has achieved an overall saving in US $ of 1.816 Cent per second. Whereas another work [2] within some different application of optimization has signified an optimal allocation of regions (for the same given area) was 100,300,300,300 achieving an overall saving in US $ of 1.804 Cent per second. This means that the comparable improvement achieved by the work presented in this paper is 0.012 Cent per second. The future scope may involve to study how to reduce the losses in the turbine aiming for an upgrade in the overall efficiency in the plant power generation.

References

1. Chong, E. K. P., & Zak, S. T. H. (2004). An introduction to optimization. 2nd edn. New York.
2. Stoecker, W. F. (1976). *Design of thermal systems* (3rd ed.). New York: Mc Graw-Hill.
3. Nag, P. K. (2001). Power plant engineering. 2nd Ed. New Delhi.
4. Uijlings, J. R. R., Van de Sande, K. E. A., Gevers, T., & Smeulders, A. W. M. (2012). Selective search for object recognition.
5. Dandy, G. C., & Warner, R. F. (1989). *Planning and design of engineering systems* (1st ed.). London: Unwin Hyman Ltd.
6. Kalyanomy, D. (1995). Optimization for engineering design algorithms & examples. New Delhi.
7. Hull, D. G. (2003). Optimal control theory of applications. New York.
8. Rao, S. S. (1984). Optimization theory and application. 2nd Ed. New Delhi.
9. Kalyanmoy, D. (2002). *Multi objective optimization using evolutionary algorithms.* Kanpur: John Wiley & Sons ltd.
10. Rao, S. S. (1995). Engineering optimization theory and practice. 3rd ed. New Delhi.
11. Chan, Y., & Kantamaneni, R. (2015). Study on energy efficiency and energy saving potential in industry and on possible policy mechanisms. Contract No. ENER/C3/2012439/S12. 666002. A report submitted by ICF Consulting Limited, ENER/C3/2012–439/S12. 666002, 1 Dec.
12. Alobaid, F., Mertens, N., Starkloff, R., Lanz, T., Heinze, C., Epple, B. (2016). Progress in energy and combustion science. Technische Universitat Darmstadt, Institute for Energy Systems and Technology, Otto-Berndt-Straße2, 64287 Darmstadt, Germany, 59, 79–162.
13. Wołowicz, M., Milewski, J., Badyda., K. (2012). Feedwater repowering of 800 MW supercritical steam power plant. *Journal of Power Technologies, 92*(2), 127–134.
14. Cziesla, F., Kremer, H., Much, U., Riemschneider, J. -E., & Quinkertz, R. (2009). Advanced 800+ MW steam power plants and future CCS options. Siemens AG, Energy Sector, Copyright © Siemens AG 2009. COAL-GEN Europe—Katowice, Poland. September 1–4.
15. Towler, G., & Sinnott, R. K. (2012). Chemical engineering design: principles, practice and economics of plant and process design. 2nd Ed, Butterworth-Heinemann, p. 1320.
16. Rong, L. (2013). Optimal operation of combined heat and power based power systems in liberalized power markets. ©Encyclopedia of Life Support Systems (EOLSS).

Conservation of Forests Using Satellite Imaging

Ahmed Majid Bahri and Asaad A. M. Al-Salih

1 Introduction

1.1 The Blessed Trees in the Quest Toward Green World

Natural forests and those cultured through human plantation are obviously composed of trees. As postulated, trees during the metabolism processes of photosynthesis naturally absorb agents of carbon dioxide (CO_2) and provide Oxygen (O_2). With a rather restricted resolution, Oxygen is vital, in fact inevitable, for creatures' respiration (including humans) in order to survive. Trees thus naturally help in the removal of carbon dioxide and other harmful gases from the air and in refreshing ambience with oxygen instead. Trees further provide food, medicines, and livelihood to hundreds of millions of people. They also provide firewood, lumber, and other products like furniture. They moreover ensure habitat for animals, birds, insects, and mammals. In addition, they prevent the land from flooding, conserve soil and water, regulate climate, and preserve natural beauty. Thus it is required to conserve and protect trees forests despite the worldwide environmental changes.

Unfortunately, day by day, "trees" are being cut for many reasons such as the need of biofuel, furnishing, paper manufacturing, urbanization, industries, and so on. With this clearing of trees, the ecological balance in the environment involving different plants and creatures' species besides humans is getting disturbed. In addition, fertile

A. M. Bahri
Section of Electronic Computer, Ministry of Health and Environment, Baghdad, Iraq
e-mail: bahry3@yahoo.com

A. A. M. Al-Salih (✉)
Department of Computer Science, Jamia Millia Islamia, New Delhi, India
e-mail: alsalih1996@yahoo.co.in

Department of Electrical Engineering, University of Baghdad, Baghdad, Iraq

© The Author(s), under exclusive license to Springer Nature Singapore Pte Ltd. 2021   429
P. K. Kapur et al. (eds.), *Advances in Interdisciplinary Research
in Engineering and Business Management*, Asset Analytics,
https://doi.org/10.1007/978-981-16-0037-1_34

soil, without trees' roots, will get loose and be washed away by rainwater or sand storms that promote desertification. This effect in turn makes changes in irrigated crops causing many species to be near to extinction and different types of diseases to be observed. The natural reaction is making the entire world take notice and pay attention toward this severely serious issue; the trees cutting and deforestation.

1.2 Greenhouse Effect

As motivated earlier, trees during the process of photosynthesis are naturally absorbing agents of carbon dioxide CO_2 and refreshing ambience through emitting Oxygen O_2. Nonetheless, emission of CO_2 is being done on large scale due to industries and vehicles consuming non-eco-friendly types of fuel like petrol, gasoline, kerosene, diesel, etc. This causes severe environmental pollution besides enhances and promotes Earth's surface temperature. This further leads to an increase in global warming causing the phenomenon called the "greenhouse effect" that boosts the harmful phenomenon of desertification. Hence, with deforestation, the level of absorption of CO_2 by trees is reduced making climatic and environmental changes in a negative direction.

1.3 World's International Initiatives

Under such circumstances of environmental changes, and since the seventies of the past twentieth century, all countries of the world have started to look into the problem of deforestation and the problem of the greenhouse effect with utmost priority. Ever since. Green Pease and Environment Supporters organizations have been established then embarked their activities in different parts of the world under the slogan of "SAVE OUR PLANET SAVE OUR SOULS". As such, a trend is being nowadays lead by the United Nations organization aiming to promote conservation of forests, diminish deforestation, and defeat desertification.

This paper is keen to show that it is quite viable to use "satellite imaging" technologies for "remote sensing" applications through utilizing the free "Google Earth" in order to obtain a "strong tool" for the *conservation of forests* in this world; this gives a low-cost application with an effective unique outcome.

2 Related Works

Under the worldwide environmental changes, a broad spectrum of diverse works has been achieved in the field. Hereby, Almeer [1] has presented his work about plants-vegetation removal/deforestation via deserts images taken from Google Earth (free) through utilizing some strong Back Propagation Neural Networks (BPNN) approach in Hue, Saturation, and Value (HSV) coloring space. He set up a scheme that has considered some neural networks to effectively identify plants-vegetations in addition to differentiating their regions out of barren regions, metropolitan, and land-transportation regions in their neighborhood. He afterward exploited some perceptron (multi-layer), some sort of learning algorithm (supervised), in order for the scheme to be trained within the confines of the functional bond joining plants-vegetation and barren region areas/deserts that has been initiated merely on color basis. Al-Wassai and Kalyankar [2] presented their work regarding the major limitations faced with satellite images in remote sensing. They clarified that the available technologies cannot be considered of appreciable hi-tech level because the spectral, temporal, and spatial resolutions are still not combined altogether in an optimal level by a single sensor. They concluded that the mutual swapping in position (trade-off) in resolution (spatial and spectral) would stay for a while due to the lack of software tools. They reached that in order to snatch the information that is quite valuable among image data, aiming to better utilize the available remote sensing apparatus in an optimal level, some novel advancements in data gathering techniques are inevitably required. Rai [3] published his work about the application of remote sensing in land mapping, forestry, and ecology. He has adopted some information system with a geographical perspective built on a remote sensing basis. The work provides an overview of a case study that is based on the model system. In some Indian zones of the Himalaya mountains, this work is applicable for practice in the North-East Indian regions. Saito and Sakaguchi [4] exhibited their work about resolutions regarding satellite imagery in forestry with an aim to monitor ecosystems and forests. The monitoring is based on the analysis of satellite imagery for which time-series data exists from the past is an effective way of quantitatively evaluating these conservation activities. Devabalan [5] demonstrated his satellite remotely sensed work focussed on an image processing computing atmosphere that is grid-based. In accordance with Grid conventionality with diversified inhomogeneous sources for computing, a Grid atmosphere was contracted for the required processing of the said satellite images. Duijvenbode et al. [6] showed their efforts regarding forest stratification in Fiji using very high-resolution satellite imagery. It is a project work aiming toward climate protection through forest conservation in pacific island countries. Miettinen et al. [7] have revealed their work concerning the remote sensing of woods depletion in forestry at the south-east side of Asia. It tried to achieve provincial sight outlook among satellite data of around 5–30 m. It aims for a better greenish global ecology and forests' conservation. Khare and Ghosh [8] have demonstrated some satellite remote sensing technologies dedicated to monitor and conserve biodiversity type of life on earth. This has comprehended the classification of the most significant and highly developed sensor devices regarding the essentially required parameters pertinently considered in biodiversity. Donaldson and Storeygard [9] presented their economical perspective work about some satellite application on pertinent areal data. It seeks to give a flavor of how remotely sensed data have been used by economists so far and what might be done in the future. Dixit et al. [10] have delineated their work focussed on some categorization plan applied on satellite areal images which is based on characteristic features related to texture. The referred plan involved segregating the work into three significant stages. The first is pre-processing of the areal image, the second is extracting the relevant image features, and the third is features categorization. It was based on using some artificial neural network (ANN) with support vector machine (SVM) basis. From a different perspective, Adjognon1 et al. [11] have presented their remarkable work about satellite-based tree cover mapping for forest conservation in the drylands of Sub Saharan Africa (SSA). Its application field was in Burkina Faso gazetted forests. They presented a tree cover assessment under a resolution of high level concerning 12 forests over there. They have utilized a supervised categorization algorithm that is based on Random Forest areas. It was actually based on satellite imagery of Sentinel-2 detected remotely in the interval starting from March and ending at the end of April 2016. Sample points with ground truth have represented the utilized methodology (under manual labeling) through utilizing images with a resolution of 10 m. They displayed images with a combination of three bands; green, red, and near infrared (NIR). These images were taken from the satellite Sentinel-2 capable to provide multispectral data. This ensures to assess the sensed tree cover with an accuracy level of about 80% in average. Last but not the least, Goldblatt et al. [12] have introduced their review about the general current perspectives regarding geospatial data for research within an economic development basis. They scanned the whole issue of deforestation/desertification by highlighting five different case studies. Focusing on the developing countries, they have illustrated the recent modern versions of High Tech equipment besides the latest resources of the geospatial big data. Generally, they have participated in an appreciable attempt in the field with a goal to devise some innovative thoughts applicable in communal activities.

3   Basic Concepts

3.1   Satellite Imaging

Emerging satellite imaging technologies cover various disciplines such as biotechnology, transportation, robotics, information technology, material science, applied mechanics, and so on. Though this wide spectrum of technologies represents new and significant developments within a field, converging technologies which previously represent distinct fields are nowadays moving toward stronger interdisciplinary fields with similar goals. Generally, the expression satellite imagery denotes photos/images of planets (including our dear Earth) taken through different satellites equipped on-board with specialized imaging equipments operating under worldwide states' governments and business organizations. There are also elevation maps, usually made by radar images. Many of those organizations are selling images to different governments and business houses like Apple and Google. Among these, imaging satellites are as follows: Image Sat International, Spot Image, Black Bridge,

Meteosat, Geo Eye, ASTER, Hyperion, Digital Globe (Quickbird), IKONOS, World-view, SPOT-5, Orb View-3, etc. [4, 5]. Table 1 exhibits some data sources derived from different satellites and sensors [9]. Satellite images can thereby be used for remote sensing in different fields of application involving: surveying, agriculture, oceanography, forestry, desertification, fishing, warfare, intelligence, etc. Satellite image analysis is thus being conducted by diverse remote sensing applications. The crucial part of satellite imagery concerning forestation/deforestation is the issue of segmentation that is generally based on *Edge Detection* concept clarified in Sect. 4.1.

3.2 Remote Sensing

The expression *Remote Sensing* denotes all specialized proce-dures/methods/techniques for data acquirement and information analysis by means of electromagnetic technologies [3]. Such procedures wrap up the entire span of the electromagnetic range/spectrum starting at low-frequency radio waves, passing by high-frequency radio waves, ultra high frequency (UHF)/super high frequency (SHF) radio waves, microwaves (millimetric waves), sub-millimetric waves, far infrared waves, near infrared waves, visible light waves, ultraviolet waves, X-ray waves, and terminating with gamma-ray waves within the said spectrum. Remote sensing, as a science, denotes to acquire, process, analyze, interpret, and utilize all types of information/data gotten through resources situated above-ground (aerial) besides in space such as balloons, satellites, and airplanes [6]. When the talk is about remote sensing that is space borne, sensors are hereby commonly installed on-board in rocket ships rotating in orbits around the earth. In this concern, several satellites for remote sensing are usually available for launching into particular orbits that are of sun-synchronous basis or geostationary basis. Actually, in geostationary basis type, satellite looks immobile compared to the surface of earth [7]. For instance, the trajectories of meteorological satellites' orbits allow a satellite to scan for all time the same regions of scope on earth. In simple words, geostationary satellites are following the movement of earth around itself within the same earth rotational velocity around its axis. Whereas sun-synchronous satellites look as if they are "stationary" with the sun in relevance to earth movement around itself. In fact, the trajectory of the sun-synchronous satellite is close to the earth polar orbit where its altitude being one that the satellite shall for all time bypass above a position at certain latitude at the very limited time [7]. For example, Landsat, SPOT, IRS, … and so on. Thus, satellites for observing earth generally track orbits of the sun synchronous type. In accordance with the energy source, the *active* and *passive* systems are the two main essential sorts of systems available for remote sensing.

The active satellites system for remote sensing (for instance RADAR-SAT and ERS-2) is generally self-equipped with its individual source for radiating electro-magnetic waves. The radiated beam of electromagnetic waves is orientated toward the surface side of earth terrains while the reflected part of energy of the said elec-tromagnetic waves shall be directed back from the earth surface and immediately

Table 1 Data sources [9] derived from different satellites and sensors

Source	Economics applications	Highest resolution	Pricing	Availability by year	Examples	For more information
Landsat	Urban land cover, beaches, forest cover, mineral deposits	30 m	Free	1972-(8 satellites)	Foster and Rosenzweig (2003), Faber and Gaubert (2015)	https://landsat.usgs.gov/
MODIS	Airborne pollution, fish abundance	250 m	Free	1999- (Terra); 2002-(Aqua)	Foster et al. (2009), Burgess et al. (2012)	https://modis.gsfc.nasa.gov/
Night lights (DMSP-OLS, VIIRS)	Income electricity use	~1 km	Free digital annual (DMSP-OLS) and monthly (VIIRS) composites	Digital archive 1992–2013+(VIIRS 2012-; film archive 1972–1991)	Henderson et al. (2012), Chen and Nordhaus (2011)	https://ngdc.noaa.gov/eog/
SRTM	Elevation, terrain roughness	30 m	Free	2000 (static)	Costinot et al. (2016) via Global agro-ecological zones	https://lta.cr.usgs.gov/SRTM

recorded aboard [6]. With a rather restricted resolution, this energy encompasses a broad assortment of wavelengths which in overall composes the electromagnetic spectrum. Usually, micrometer (μm) hereby represents the unit of wavelength; 1 μm is equal to 1×10^{-6} m. And as preceded earlier in this section, the distinct separated sets of uninterrupted wavelengths, based on the corresponding frequency bands, are named as "wavebands". These wavebands were agreed to be called (from higher to lower frequency bands) as: the gamma-ray band, the X-ray bands, the ultraviolet band, the visible light band, the infrared band, the microwave band, and the radio waves band: SHF, UHF, HF (Short wave, Medium wave, and Long wave). Thereby, when considering the electromagnetic spectrum bands, it is obviously postulated that the waveband of the visible light (that spans merely 0.4$-$0.7 μm and detected/sensed through the mere human eyes) spans solely a too short division of that spectrum. In the design of some remote sensing appliance, it is common to consider the number of included wavebands (1 or even more) that the appliance is going to function on. This point should be clearly articulated in the design specifications of the pertinent perspective [8]. That is since the mentioned electromagnetic wavebands are going to penetrate and navigate through the diverse particles$-$materials composing the atmospheric envelope around the Earth.

From another corner of focus, when considering a passive satellites system (e.g., SPOT-3 HRV, and Landsat TM), one may see that it utilizes the sun rays as its source for electromagnetic waves. When sun rays hit Earth terrains, an interaction with the Earth face through ray's reflection will take place. Sensors that on-board the satellite remote sensing system will detect and quantize the reflected energy as measured quantities.

3.3 Google Earth

The application *Google Earth* can be deemed as a well-liked appliance on the Internet at no cost. It presents a documentary collection of aerial images and satellite photos that integratedly covers the surface of the whole Earth for efficient monitoring. Prior to the development of satellite imaging with appreciable declaration/resolution, it was quite difficult, somehow restricted, and very costly to get metadata. Nonetheless, satellite pictures are at the present time at no cost besides provided on the Internet in some locations. So that is why Google Earth has been resorted by this work as an active source for research [1].

As some images are in limited access for the public, besides the current unavailability of high speed in downloading, both facts form the main obstacles to download high-resolution satellite images. In spite of all that, the availability of "fast" internet has facilitated the act of image downloading and has made it quite quicker for users in general cases. Figure 1a$-$c hereby illustrates some satellite Google Earth images.

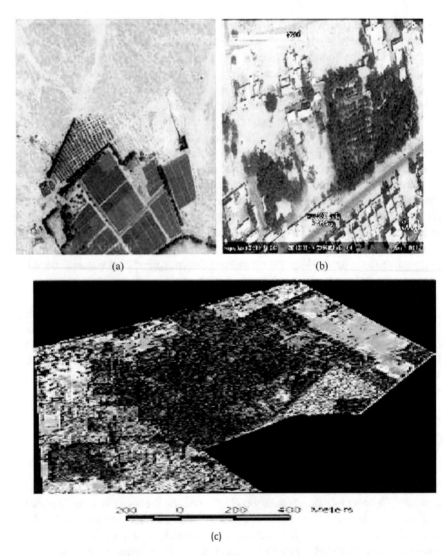

Fig. 1 Satellite *Google Earth* Images for different aerial pictures showing the landscape and topographic terrains involving scattered and manmade plantation: **a** and **b** panchromatic pictures [1], **c** multispectral picture [3]

3.4 The REDD+ Forest Conservation Program

Under the U.N. umbrella, several forest conservation activities in the developing economies have participated to combat climate change. In this context in 2005, the Reducing Emissions from Deforestation and Forest Degradation in Developing Countries Plus (REDD+) program has embarked [4]. It is a program that seeks to

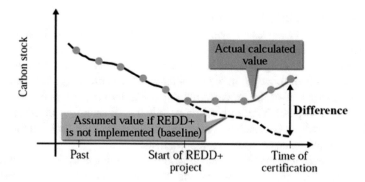

Fig. 2 Increasing carbon stock. By implementing REDD+, carbon stock increases over the baseline predicted value. Credits are issued once the difference is calculated and certified. After the REDD+ project starts, increases in carbon stock due to the implementation of forest conservation are evaluated on a regular basis. As with the baseline formulation procedure, the total amount of biomass in the project region is calculated using a combination of satellite imagery and field surveys, and the change in carbon stock is derived [4]

reduce the harmful emissions caused by depletion of forests, desertification, and deforestation in third world countries. It further seeks to manage forests in a sustainable manner. In other words, it aims in conserving forest carbon stock, i.e., carbon in the Earth's ecosystem involving biomass in soil, trees, forests, woods, and vegetation cover. Thus the REDD+ program is keen to sustain, maintain, and improve the global forests' carbon stock using some convenient certain methodologies. It starts to incorporate some pertinent concept whereby increases in carbon stock stemming from forest conservation are rewarded with credits. The REDD+ program concept is clarified in Fig. 2.

4 Research Methodology

4.1 Edge Detection

Recalling Sect. 3.1, the term *edge detection* denotes to detect or sense an "edge" separating two surfaces differing in brightness values. It actually involves an assortment of analytical/mathematical techniques targeting to specify positions in a digital image with discontinuities or whose brightness alters sharply [2–4]. These positions are normally arranged into non-straight line segments group named as edges. In this concern, *step detection* in general denotes the issue of discovering discontinuities in the 1D signal, whereas *change detection* refers to identifying discontinuities in the 1D signal over time. In the fields of computer vision, image processing, machine vision, and mostly in the sides of features detection and extraction, the "basic" action mostly considered hereby is edge detection. The simplest form for detecting edges is in general through using *Central Differences Gradient approach* clarified hereunder.

Considering a digital image with a pixel "L" represented in the Cartesian coordinates as L(x, y), a creeping filtering mask can be applied on that pixel in order to estimate the gradient (change) in pixel intensity through creeping, horizontally and vertically, among half of the previous pixel, passing by the pixel itself, and halting at the half of next pixel. The implication here is that at each step of discretization, the estimated gradient is considered to exclusively set an EDGE pixel at the steps of "changing" in the value of the estimated gradient.

In order to compute the *first-order* gradients for an image or for its "smoothed" copy, several acting operators may thus be applicable. Hereby, the motivated *Sobel* operator with central differences basis, as clarified in Fig. 3, represents one of the straightforward techniques in this regard. Further, the literature includes some other edge detection techniques that are highly sophisticated being operating through *sub-pixel* precision besides adopting *second-order directional derivative* approach for the gradient *zero*-crossing detection. This yields an expedient routine action for the detection of edges. Edge detection in colored pictures can be easily delineated in the literature.

$$L_x(x,y) = -\frac{1}{2}L(x-1,y) + 0 \cdot L(x,y) + \frac{1}{2} \cdot L(x+1,y)$$

$$L_y(x,y) = -\frac{1}{2}L(x,y-1) + 0 \cdot L(x,y) + \frac{1}{2} \cdot L(x,y+1),$$

corresponding to the application of the following filter masks to the image data:

$$L_x = [+1/2 \quad 0 \quad -1/2]L \quad \text{and} \quad L_y = \begin{bmatrix} +1/2 \\ 0 \\ -1/2 \end{bmatrix} L.$$

The well-known and earlier Sobel operator is based on the following filters:

$$L_x = \begin{bmatrix} +1 & 0 & -1 \\ +2 & 0 & -2 \\ +1 & 0 & -1 \end{bmatrix} L \quad \text{and} \quad L_y = \begin{bmatrix} +1 & +2 & +1 \\ 0 & 0 & 0 \\ -1 & -2 & -1 \end{bmatrix} L.$$

Given such estimates of first-order image derivatives, the gradient magnitude is

$$|\nabla L| = \sqrt{L_x^2 + L_y^2}$$

while the gradient orientation can be estimated as

$$\theta = \text{atan2}(L_y, L_x).$$

Fig. 3 Central differences masking approach to compute first-order gradients for edge detection in images [2]

4.2   The Proposed Model

The introduction and literature review discussed above have inspired to propose a unique model in Satellite imaging for the conservation of forest and protecting trees from being cut. There are satellites like IKONOS or SPOT-5 which send images to earth after a relatively short interval like five minutes. These images can be processed to locate the position of trees within this reasonably short period of time. Figure 4 illustrates the data flow diagram proposed in this forests' conservation model. The pertinent hierarchical details can thereby be summarized as in the following stepwise algorithm.

(1)   Images from the satellite can be collected.
(2)   Study area should be marked and "gridded" to get the edges of trees.
(3)   Edges of trees should be marked.
(4)   Location coordinates of trees, i.e., altitude, longitude, and latitude values of each tree to be stored in the database against the image.
(5)   Images received should be put to the plotted grid and checked for the empty space for all the locations previously marked.
(6)   If the location is empty, an output message needs to be sent to the concerned authority with location and map to reach them using Google map service.

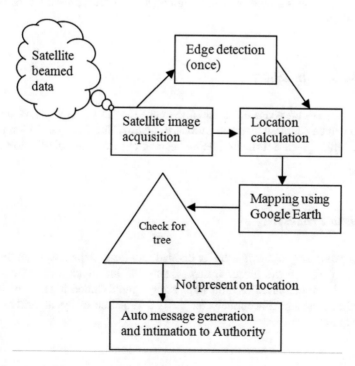

Fig. 4 The proposed forests' conservation model

In this way if a tree is cut illegally, it could be traced out and culprits shall be caught. The spatial resolution of image data may be measured by utilizing satellite data taken with high level (5 m) to medium level (30 m) of resolution [1]. Thus, in order for the remote sensing satellites to reliably have a rich land cover mapping of Earth terrains information, the pertinent image data should enjoy spatial resolutions of around 0.6–1 m [4]. In general, satellites systems get multispectral data with a lesser resolution degree of about 2.8–4.0 m. Within the electromagnetic spectrum, they further get panchromatic data in a sole distinct band among 450–900 nm, commonly demonstrated as monochromatic (gray-scale) with a higher resolution degree of about 0.6–1.0 m. Generally, the multispectral detectors/sensors, located on these satellites systems, gather data in single spectral bands. We can use a green band with 520–600 nm spectrum, which provides important spectral color information about the target [9].

5 Model Realization Remarks

Due to some relevant technical limitations, the realization and validation of the "forests' conservation model" proposed in this work will be introduced in some next work. Pertinent model realization remarks are hereby presented for possible consideration.

5.1 Data Acquisition

The data acquisition phase needs satellite image data in the green band spectrum. The data can be acquired by a geostationary satellite. The data acquisition phase is purely satellite given information being beamed out from the satellite to the earth station.

5.2 Data Processing

The data processing phase is actually divided into three steps, i.e., edge detection, checking of tree on the location, and passing the information to the concerned authority. A sample-based classification using a multi-dimensional technique can be tested and evaluated accurately for the determination of the algorithm for the proposed model [5].

5.3 Trees Mapping

Once the processed data are available, they can be plotted on the map. All the trees' maps are available online to be used by the Geographic Information System (GIS) professional with the application on Google Earth. Thus with these maps, new evidence of disturbances can be distinguished [3].

5.4 Governmental Role of Parallel Processing

It might not be possible to process such a large amount of data at one central location. Hence the data can be distributed locally by the respective state government to divide the locations and act accordingly. This can be one of the steps in working for the United Nations Framework Convention on Climate Change (UNFCCC) or Reduction of Emissions from Deforestation and Forest Degradation (REDD+). For countries with very little forests land, it becomes obviously necessary to protect the forest.

6 Conclusion and Future Scope

The greenhouse effect causes global warming besides increasing environmental pollution due to the harmful carbon dioxide (CO_2) emission. It is therefore important to control such an effect that is worldwide deemed to be crucial in maintaining life on earth. U.N. and world countries are now seeking the formulation of substantial strategy outlines to set the sustainable management of these forests. Satellite imaging can hereby be a very efficient means for this purpose; it can practice a crucial part in the conservation of forests, maintaining the ecosystem, and diminishing desertification. Satellite remote sensing provides forest conservation strategies by using almost freely available resources. High-resolution data giving minute details over a very small area with the date to date updates are nowadays available, making it easy to process. These resources help the researcher to assist in mapping and conserving the forest. The proposed model is using satellite imagery for the conservation of forest and protecting trees from being cut. A more efficient edge detection approach is required for the tree classification under the high-resolution satellite imagery. Forest conservation is the inevitable need of life on the Earth planet in today's changing world.

References

1. Almeer, M. H. (2012). Vegetation extraction from free Google Earth images of deserts using a robust BPNN approach in HSV space. *International Journal of Advanced Research in Computer and Communication Engineering, 1*(3), 134–140.
2. Al-Wassai, F. A., & Kalyankar, N. V. (2013). Major limitations of satellite images. *Journal of Global Research in Computer Science, 4*(5), 51–59.
3. Rai, P. K. (2013). Forest and land use mapping using remote sensing and geographical information system: A case study on model system. *International Academy of Ecology and Environmental Sciences (IAEES) - Environmental Sceptics and Critics, 2*(3), 97–107.
4. Saito, K., & Sakaguchi, H. (2013). Satellite imagery solutions for monitoring of forest and ecosystems. *Hitachi Review, 62*(3), 204–208.
5. Devabalan, P. (2014). Satellite image processing on a grid based computing environment. *International Journal of Computer Science and Mobile Computing, 3*(3), 1039–1044.
6. Duijvenbode, J., Reiche, J., & Forstreuter, W. (2014). Forest stratification in Fiji using very high resolution satellite imagery. In *Project work for 'Climate Protection through Forest Conservation in Pacific Islands Countries'*, Aug 2014 (pp. 11–33).
7. Miettinen, J., Stibig, H.-J., & Achard, F. (2014). Remote sensing of forest degradation in Southeast Asia—Aiming for a regional view through 5–30 m satellite data. *Global Ecology and Conservation, 2,* 24–36.
8. Khare, S., & Ghosh, S. K. (2016). Satellite remote sensing technologies for biodiversity monitoring and its conservation. *Cloud Publications International Journal of Advanced Earth Science and Engineering, 5*(1), 375–389.
9. Donaldson, D., & Storeygard, A. (2016). The view from above: Applications of satellite data in economics. *Journal of Economic Perspectives, 30*(4), 171–198.
10. Dixit, A., Hedge, N., & Reddy, B. E. (2017). Texture feature based satellite image classification scheme using SVM. *International Journal of Applied Engineering Research, 12*(13), 3996–4003.
11. Adjognon1, G. S., Rivera-Ballesteros, A., & Soest, D. V. (2019). Satellite-based tree cover mapping for forest conservation in the drylands of Sub Saharan Africa (SSA): Application to Burkina Faso gazetted forests. *Development Engineering, 4,* 100039, Elsevier, Journal homepage: www.elsevier.com/locate/deveng. https://doi.org/10.1016/j.deveng.2018.100039.
12. Goldblatt, R., Jones, M., & Bottoms, B. (2019). Geospatial data for research on economic development. *Development Engineering, 4,* 100041, Elsevier, Journal homepage: www.elsevier.com/locate/deveng. https://doi.org/10.1016/j.deveng.2019.100041.

Exploration of Impact of Lean Six Sigma and TRIZ in Banking Sector

Nishant K. Tripathi, Ishit Sheth, and R. P. Mishra

1 Introduction and Literature Survey

Every organization irrespective of their business field needs a source of income to stay in the business. Everyone associated with the business has their own needs which need to be satisfied by the business in a very effective manner to be profitable and keep on doing what they are doing. Over the past centuries, companies have developed many different techniques to keep themselves attracting customers to retain their business. These studies started after the development of quality control charts by Walter Shewart in the late 1920s. In the late 1990s, Motorola developed a technique called six sigma which has several statistical tools already developed by other practitioners previously. This methodology is proven to be a structured and disciplined process to deliver the consistent and precise outputs on a continuous basis.

In the 1950s, Toyota Motors developed a system now widely known as Lean Manufacturing Systems or the Lean Management. This system was constructed on the basis of removing those processes which do not add value to the system. This system focuses on work standardization and value stream mapping. In the past years, these principles were considered as totally unrelated and entirely different methodology compared to six sigma. In recent years, these two methodologies which were considered completely different have been integrated into one methodology widely

N. K. Tripathi · I. Sheth · R. P. Mishra (✉)
Mechanical Engineering, Birla Institute of Technology and Science (BITS), Pilani, Rajasthan, India
e-mail: rpm@pilani.bits-pilani.ac.in

N. K. Tripathi
e-mail: nkt.bcpl@gmail.com

I. Sheth
e-mail: sheth.ishit@gmail.com

© The Author(s), under exclusive license to Springer Nature Singapore Pte Ltd. 2021 443
P. K. Kapur et al. (eds.), *Advances in Interdisciplinary Research*
in Engineering and Business Management, Asset Analytics,
https://doi.org/10.1007/978-981-16-0037-1_35

known as Lean Six Sigma as any process improvement project demands the use of both methods as an integrated approach.

The Lean Six Sigma approach uses a five-step method to solve any problem or bring improvement into the process; widely known as DMAIC model which stands for Define phase, Measure phase, Analyze phase, Improvement Phase and Control Phase. An important difference between other quality improvement programs and LSS is that in each phase of LSS, there are several tools and techniques required to be implemented, and top management is required to allow necessary time and resources in each phase to endeavor continuous improvement. There are so many studies conducted across the globe in the field of business management on the effect of implementation of this systematic approach. Kumar et al. [1] presented a case of an Indian SME, according to their findings applying the systematic framework of LSS increased overall equipment effectiveness, and reduction in defect rate was also observed. They reported annual savings of $46,500 by implementing the LSS framework.

In a retail bank, Vijaya [2] showed that using the LSS framework one can reduce the rate of rejection of applications received for opening a new account by a great extent along with boosting employee morale. In their study of a particular retail bank headquartered in India having three sub branches, rejection of applications submitted was reduced to 3.4% from the previous 10% and reported saving of 1.6 million INR. This study showed how much impact the use of LSS framework can bring in the financing industry.

Vashishth et al. [3] presented a detailed review of the implementation of LSS methodology in the field of finance. They conducted a systematic review of 30 articles published in the finance field in between the years 2002 and 2015. In their study, they found that usage of LSS in the finance sector gave major 12 benefits out of which 9 were operational benefits which included reduction in operational costs, reduction in cycle time and improved customer satisfaction among others. Further, they showed that among many reasons for limited application of LSS framework in this sector, one of the major reason is the lack of awareness of potential benefits that can be achieved in financial sector among organizations and these limited resources do not provide guidelines for implementation of advanced tools in this sector which limits the usage of LSS framework.

One major finding from the literature survey was that in the DMAIC model, there are plenty of tools and techniques available for each of the phases except Improvement phase as in every study improvement phase is either carried out using Design of Experiments (DOE) or using brainstorming activities. This approach does not provide a systematic methodology to implement in similar studies. To address this gap in the studies, a method developed by Altshuller [4] known as TRIZ, which stands for Theoria Resheneyva Isobretatelskehuh Zadach in Russian and Theory of inventive problem-solving in English language, can be implemented in the Improvement phase of DMAIC model. TRIZ is a systematic approach that allows to solve the problem using a predefined structured method. In this method, users have to identify their problem from already defined 39 features; according to their philosophy whenever one tries to improve one problem it contradicts some other feature in the organization which can be resolved using 40 inventive principles provided in the TRIZ framework. This method is based on the belief that every problem has a solution available for a similar kind of problem solved by someone else, somewhere else in the world. A lot of research has proven the integration of TRIZ tools with six sigma methodology significantly increases the potential of Six Sigma for existing systems as well as in the creation of new products/services and processes that allows for the company to stay competitive [5]. Chao-Ton Su [6] provided a framework for applying TRIZ to improve the service quality level in an e-commerce setup. According to their study, 29 significant determinants of quality have been found which are closely related to customer satisfaction. Wang et al. [7] applied LSS and TRIZ methodology to reduce cycle time in the operation of a bank and reported a reduction of almost 33% in cycle time leading to annual savings of $828,000.

2 Research Use Case

We have taken the practical use case of a public sector undertaking bank of India working in a distant village where majority of its customers are illiterate or have very little knowledge of using sophisticated systems of a bank. As this bank has its presence in metro cities as well as remotest areas of India, it has a very huge customer base. Opening an account in a bank is the first interaction of a customer with bank; so that time taken in opening a new account is a major source of customer's impression about bank. Additionally in the recent past, Indian government mandated to have a bank account in order to obtain various financial aids like subsidies and scheme benefits so the requests to open a new savings account increased in the public sector undertaking banks. Besides opening an account, customers do visit regularly for depositing money, withdrawing money, passbook printing, loan enquiry, knowing about government offers, etc.

According to Voice of Customers (VOC), one of the primary issue they faced is the time taken in the processing of a new savings account and as many had very less knowledge of the documents required for opening a new account and procedure to be followed for the same, they found the process to be very complicated and difficult which needs to be made more customer-friendly, and time required in processing should be reduced. To address this problem, LSS approach has been applied with the integration of TRIZ methodology as it is a framework that provides stepwise approach needed to implement and it has shown positive results in previous research studies.

2.1 Define Phase: A Three-Step Process

1. **Project Identification**: Interacting with employees, shadowing the customers and the observational studies suggested that the cycle time of the savings

accounts opening process is higher than the observed time in other bench-marks. Additionally, the Process needs to be simplified for both employees and customers to follow easily. Hence the project was selected as "Reducing cycle time and simplifying the procedure of account opening process in a public sector undertaking bank". In this study, the variable critical to quality is taken as the cycle time of savings account opening.

2. **Determine the vital few factors**: A flow chart is prepared mapping the high-level activities involved in the project. As in the below chart, the sections marked in red are identified as the main bottlenecks of the project.

3. **Determine defects and interpret the successful indicators**: According to benchmarks, there is almost a difference of 40% between required and actual cycle time. As shown in the above flowchart, major indicators in cycle time were document verification and waiting time in the queue to deposit money in the account (Fig. 1).

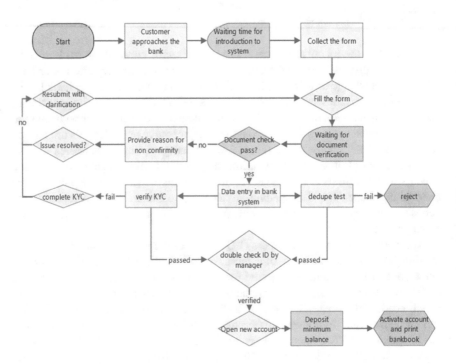

Fig. 1 Process map

2.2   Measure Phase: A Two-Step Process

Here the effort is put down to verify and establish the performance standard of the process so as to obtain a baseline for future improvements. The two steps involved are as follows:

1. **Data collection plan**: In order to understand the process capability before improvement, a data collection plan is deployed to gather sample data from actual time collection. Sample data is collected through observation and shadowing techniques. Random samples of people from different demographics and at different points of time of day were chosen to consider all possible variabilities in the process. A sample size of five people was considered at a specific time period and in total, sample data for 50 people were collected. The details can be referred from Table 1.
2. **Measuring as-is process capability**: Using the sample data, we plot the Individual control chart, Moving range chart, Normal probability plot, Capability plot and get the following results. The control chart of 50 taken samples shows the mean waiting time of 32.158 min and process capability as −0.50. For a process to be in control and capable of meeting the target effectively, the process capability index is required to be 1.33 as per standard 3σ. While conducting the capability analysis, the upper specification limit is considered to be 30 min as discussed with the bank manager and found from various benchmarks available. So our process is not capable enough to meet the target (Fig. 2).

2.3   Analyze Phase: A Three-Step Process

In this phase, the performance objective is expounded and the key sources of variations are pointed out.

1. **Root cause Identification**: The cause and effect matrix has been utilized to identify the root causes of the problem of variation and waste between target time and actual time (Fig. 3).
2. **Failure Mode and Effects Analysis (FMEA)**: There are factors that can cause function failure in the key process which can be identified in advance using FMEA and a Risk Priority Number (RPN) can be allocated. The associated factors with a high RPN will be selected and corrective actions will be recommended. Failure Mode and Effects Analysis (FMEA) is an organized way to deal with finding potential failure modes that may exist inside the structure of the procedure. Failure modes are the manners by which a procedure can come up short. Effects are the manners in which these disappointments can prompt waste or harm customers. Failure Mode and Effects Analysis is intended to recognize, organize and limit these failure modes.
    The following procedure is adopted for FMEA analysis:

Table 1 Measured sample time data involved in various activities of the process

Sr. no.	Waiting time for introduction to system	Fill the form	Document check passed	Data entry in bank system	Double ID check by manager	Verify KYC	Dedupe test	Deposit Minimum balance	Activate account	Total time
1	10.8	6.4	2.4	1.7	1.3	2.1	1.5	4.6	1.1	31.9
2	10.9	7	2.1	2.7	0.9	1.7	2.9	4.3	0.7	33.2
3	10.5	6.4	1.7	1.4	1.1	1.3	1.4	4	1.2	29
4	9.1	6.2	1.6	2.8	0.8	0.8	3.2	5.5	1.7	31.7
5	10.9	7.8	2.4	2.1	0.9	1.5	2	5.5	1.7	34.8
6	9.3	6.9	1.4	3	1	0.8	1.6	4.9	0.9	29.8
7	10.5	8.4	2.2	1.9	1	1.9	1.5	4.2	1.4	33
8	10.2	7.3	3	1.8	1.3	0.8	1.9	4.6	0.6	31.5
9	9.2	6.4	2.8	1.8	0.9	0.8	2.7	5.4	0.9	30.9
10	10	6.3	1.9	2	1	2	2.9	5.4	1.9	33.4
11	10.6	6.1	1.9	2.2	1.4	0.7	2.7	5.9	1.8	33.3
12	9.4	8.2	1.4	1.9	1	1.6	1.2	5.1	1.1	30.9
13	10.3	6.5	1.6	1.9	1.1	1.5	2.1	4.2	0.7	29.9
14	9.3	7.2	2.3	1.4	1	1.1	2	5.6	1.5	31.4
15	8.6	6.3	2.7	1.8	1.4	0.9	3.1	4.4	1.2	30.4
16	10.4	7.4	2.2	1.5	1.5	1	2.9	5.3	0.7	32.9
17	10.3	7.9	2.4	1.4	1.1	1.5	1.9	4.6	0.9	32
18	10.9	6.4	2.5	2.5	0.8	0.9	1.6	5.7	0.6	31.9
19	9	6	2.8	1.5	1	1.6	2.4	5.9	1.5	31.7

(continued)

Table 1 (continued)

Sr. no.	Waiting time for introduction to system	Fill the form	Document check passed	Data entry in bank system	Double ID check by manager	Verify KYC	Dedupe test	Deposit Minimum balance	Activate account	Total time
20	9.3	6.5	2.5	2.3	1.5	1.5	2.5	5.6	1	32.7
21	9.5	8.1	3	2	0.9	2	1.6	5.6	1	33.7
22	10	6.7	2.3	3	1.3	1.5	1.4	4.7	1.5	32.4
23	8.9	8.5	2.9	2.1	1.4	0.5	1.6	4.9	1.3	32.1
24	9.5	7.8	1.4	2	1.4	1	1.9	5.9	0.5	31.4
25	9	7.1	1.8	2.6	1.5	0.8	2.9	5.1	1.8	32.6
26	9.8	6.2	2.9	2.7	1.2	0.7	2	4.7	1.4	31.6
27	9.5	6.5	1.6	2.2	1.1	0.5	2.4	5.4	1.9	31.3
28	10.5	6.3	1.9	1.8	1.4	2.1	2.4	4.5	1.4	32.3
29	9.7	6	1.9	2.6	1.4	0.8	1.9	4.1	0.9	29.3
30	10.8	7.3	1.3	2.2	1	1.3	3.1	4.2	1.7	32.9
31	9.5	7.7	3	2.9	1.5	2	2.7	5	1.5	35.8
32	9.7	7	2.2	2.7	1.4	0.8	1.7	4.5	1.2	32.2
33	11	7.9	1.4	3.2	0.8	1.9	2	4.8	1.4	34.4
34	8.9	6	2.7	1.9	0.9	2	2.9	5.1	1	31.4
35	9.7	7.9	2.3	2.3	1.2	1.3	2.1	4.7	0.7	32.2
36	10.4	7.8	2.3	2.6	1	1.1	1.5	4.4	1.5	32.6
37	10.4	7.7	1.7	1.7	1.3	1.5	2.4	4.9	0.5	32.1
38	9.5	7.2	2	3.2	1.3	1.5	2	5.3	1.3	33.3

(continued)

Table 1 (continued)

Sr. no.	Waiting time for introduction to system	Fill the form	Document check passed	Data entry in bank system	Double ID check by manager	Verify KYC	Dedupe test	Deposit Minimum balance	Activate account	Total time
39	10.5	7.2	1.6	1.7	1.3	1.9	3.1	5.3	0.7	33.3
40	9.4	6.7	2.2	1.8	1.3	2.2	1.8	4.9	0.6	30.9
41	8.7	7.2	2.8	3.2	1.3	1.6	2.6	4	1	32.4
42	9.3	8	1.5	1.6	0.9	1.5	2.7	5.1	1.6	32.2
43	8.7	6.3	2.8	3	1.1	2	1.2	5.6	1.1	31.8
44	9.5	6.8	1.8	2.4	1.3	2	2.7	4.1	0.8	31.5
45	9.1	7.6	2.7	2.2	1.3	0.8	3	5.3	0.9	32.9
46	10	6.4	1.7	1.7	1.3	1.4	2.3	5.6	0.7	31.1
47	10.S	6.1	1.4	2.2	1.3	0.6	3.1	5.4	0.8	31.7
48	10.3	7.6	2.6	3	1.2	0.9	3	5.8	1.2	35.6
49	9.5	6.8	2.9	1.9	1.4	2	2.8	5.6	1.4	34.4
50	9.3	7.6	2	1.8	1.4	0.6	1.6	5.2	0.7	30.2
Average	9.824	7.034	2.168	2.196	1.182	1.356	2.248	5.008	1.142	32.158

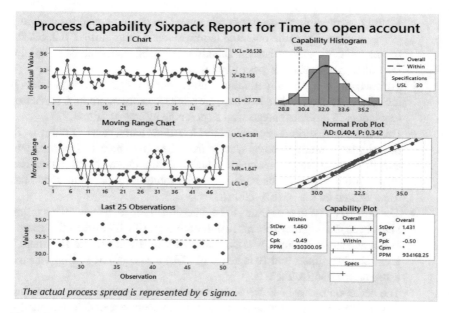

Fig. 2 Process capability six pack

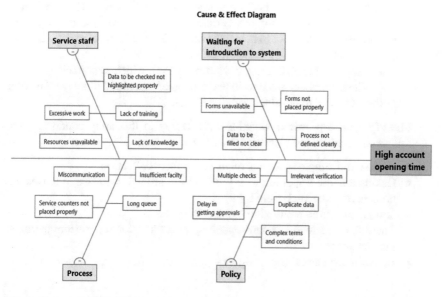

Fig. 3 Cause and effect diagram

- Identify all possible modes/types of failure that could happen for each and every function. These are potential failure modes. Going further, if necessity occurs, go back and rewrite the function with a bit more details to be sure the failure modes show a loss of that function.
- For each and every identified failure mode, identify all the consequences on the system, process, related system, processes, customer, product, service or regulations. These are categorized as potential effects of failure.
- Determine how serious each effect is. Determine using severity rating or S. Severity is normally rated on a scale from 1 to 10, where 1 is unimportant and 10 is disastrous. In a scenario where a failure mode has more than one effect with different S, the highest severity rating for that failure mode has to be written on the FMEA Table 2.
- The occurrence rating or O has to be determined for each cause. This rating helps in estimating the probability of failure occurring because of that reason during the entire lifetime. We rate Occurrence on a scale of 1 to 10, where 1 is extremely unlikely and 10 is inevitable. List the occurrence rating for each cause on the FMEA Table 2.
- The detection rating or D has to be determined for each control parameter. This rating helps in determining how well the controls can detect either the cause or its failure mode after they have happened but before the customer is affected. Normally Detection is rated on a scale from 1 to 10, where 1 means the control is absolutely certain to detect the problem and 10 means the control is certain not to detect the problem (or no control exists). List the detection rating for each cause on the FMEA Table 2.
- Calculation of the Risk Priority Number(RPN), which is defined as $S \times O \times D$. These numbers guide us in the ranking of potential failures in the order of their severity to be addressed first to last.

3. **Identify vital few initial variables**: According to the study conducted, the following variables are identified which need major rectification.

   - A clear and simple system is required for form collection.
   - Details that are important to fill in the form are not clear which makes the process more time-consuming.
   - Too much data to be filled in by an employee.
   - There is no defined standard operating procedure for supervisors to check the documents.
   - No defined standard operating procedure for staff.

2.4 Improve Phase: A Three-Step Process

Here, the process performance has been improved using TRIZ methodology.

1. **Development of solutions**: The analysis phase highlighted the significant causes based upon which the improvement activities using TRIZ methodology are

Table 2 FMEA analysis

Specific functions of the process	Probable failure modes	Probable effects of failures	Severity	Probable causes of failure	Occurrence	Detection	Risk priority no (RPN)
1. Getting the application form	Unnecessary movement of man and material	Unnecessary Increase of waiting time	6	Forms not kept in correct place	7	10	420
2. Completing the application form	Incomplete application form	Stress, confusion, seek help or incorrect form completion	7	Same data to be filled multiple times. Complicated form design	7	10	490
3. Verification and acceptance	Application form incomplete/ incorrect data filled	Customer revisit and re-filling of the application form	7	Data to be checked not highlighted properly	4	10	280

Table 3 The contradiction matrix with suggested inventive principles

			Worsening feature				
			Weight of the stationary object	Stress	Stability of object	Reliability	Adaptability
			2	11	13	27	35
Improvving features	3	Speed	–	6, 18, 38	1, 28, 33	11, 27, 35	10, 15, 26
	22	Loss of energy	6, 9, 18, 19	–	14, 26, 39	10, 11, 35	–
	25	Loss of time	10, 5, 20, 26	4, 36, 37	3, 5, 22, 35	4, 10, 30	28, 35
	33	Ease of operation	1, 6, 13, 25	2, 12, 32	1, 3, 32, 35	8, 17, 27, 40	1, 15, 16, 34
	36	Device complexity	2, 26, 35	1, 19, 35	2, 17, 22, 19	1, 13, 35	15, 28, 29, 37

suggested. TRIZ analyzes problems through the unique perspective of contradiction.TRIZ utilizes 40 inventive principles based upon 39 contradiction parameters [4, 8, 9]. In order to gain a competitive advantage in the operation of its business, the improvement of customer satisfaction becomes the primary goal of the company. Different issues incorporate additional expenses because of holding up time of procedures and working mistakes. 40 inventive principles are applied to resolve the above contradictions (Table 3).

2. **Implementation of improvement plan**:
   The observation data, identified issues and suggested improvements were shared with the bank authorities and a request was made for the implementation of the suggested improvement plan as per innovative principles. This can be referred from Table 4.

3. **Redesigning the process map and identification of the new process capability**:

- After the proposed improvement plan has been implemented and progressed for approximately 3 months, again sample data is taken from the revised process map for recalculating the process capability.
- After implementing improvement action from Table 4, a significant improvement in the waiting time of opening an account and the operational cost was observed, and internal failure costs are also predicted to reduce.
- The process capability after the improvement has been calculated and can be seen in Fig. 4. The results of the analysis show that the average waiting time was reduced dramatically from 32.16 min to 22.346 min for each operation. Furthermore, the process capability of Ppk and Cpk is enhanced from 0.66 and 0.65 to 2.56 and 2.41, respectively (Table 5).

Table 4 The proposed action plan combined with the inventive principles

Identified Issues in the chosen process	Suggested improvement actions
Most of the time customers have to wait for collecting the required application form	Implement visual management tools using 5S (Sort, Set In order, Shine, Standardize and Sustain). This will lead to proper categorization of forms and ease of accessibility by customers
Form data/information not easily understood by customers and hence they look for help	Redesign the application form by reducing redundant data and provide visual cues for filling in the required information (Inventive principle 35)
Sometimes staff has lack of knowledge required to resolve queries involved in the process	Conduct regular knowledge training sessions at particular intervals of time. Prepare standard operating procedures to avoid confusions
The key staff member involved in the process has a low operating speed	Consider operating speed as one of the key metrics for performance appraisals. Assign time-bound goals and reward on achieving those milestones (inventive principle 28, 33)
Lot of cognitive load on staff while checking and verifying the application form	Streamline the information to be checked by staff by highlighting and sequencing those required sections (inventive principle 32)

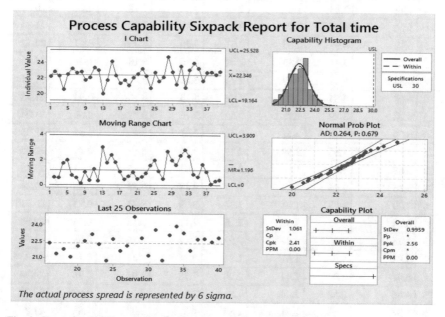

Fig. 4 Six pack capability after the improvement plan is implemented

Table 5 Measured sample time data involved in various activities of the process after solution implementation

Sr no	Waiting time for introduction to system	Fill the form	Document check passed	Data entry in bank system	Verify all details	Deposit minimum balance	Activate account	Total time
1	6.19	3.69	2.74	2.25	3.04	3.48	0.81	22.2
2	5.49	4.17	2.28	1.95	4.14	3.77	0.98	22.78
3	6.4	3.68	2.66	1.96	3	3.61	0.91	22.22
4	4.5	4.38	1.64	1.54	4.38	3.45	0.61	20.5
5	6.43	3.98	1.89	1.89	3.48	3.67	1.12	22.46
6	6.3	4.71	2.7	1.66	3.11	3.79	0.94	23.21
7	6.24	5.32	1.72	1.61	3.04	3.57	1.17	22.67
8	5.65	4.97	1.96	2.48	3.37	3.37	0.95	22.75
9	5.69	4.37	1.93	1.68	3.96	3.33	0.77	21.73
10	5.24	4.17	1.54	2.32	4.16	3.57	1.06	22.06
11	5.68	3.85	2.73	2.07	4	3.98	1.01	23.32
12	6.52	4.32	2.15	2.83	2.84	3.47	0.87	23
13	4.78	3.85	1.6	2.16	3.5	3.18	0.94	20.01
14	5.26	3.66	2.54	2.08	3.43	3.97	0.8	21.74
15	6.12	4.95	2.22	2.51	4.29	3.36	0.67	24.12
16	4.93	4.21	2.23	2.62	4.16	3.18	1.01	22.34
17	4.58	5.43	1.79	2.24	3.36	3.31	0.6	21.31
18	6.08	4.55	2.07	1.88	3.13	3.1	0.92	21.73
19	5.28	3.79	1.78	2.52	3.76	3.22	0.69	21.04
20	6.09	3.76	2.15	2.19	3.81	3.4	0.65	22.05
21	6.11	3.58	2.59	2.73	3.15	3.58	0.79	22.53
22	5.68	5.26	1.66	2.63	3.02	3.85	1.05	23.15
23	5.15	3.94	2.64	2.97	3.18	3.53	0.83	22.24
24	4.99	4.49	1.91	1.98	3.39	3.22	0.73	20.71
25	6.14	3.71	2.52	2.26	4.11	3.18	0.76	22.68
26	4.75	4.65	1.84	2.52	3.42	3.75	0.68	21.61
27	5.84	5.02	1.61	1.8	3.74	3.37	0.7	22.08
28	6.98	3.8	2.44	2.98	3.73	3.95	0.86	24.74
29	6.66	3.52	2.66	1.83	3.4	3.74	0.97	22.78
30	5.36	3.88	2.01	1.88	4.3	3.09	0.68	21.2
31	5.38	5.21	1.75	2.97	3.95	3.13	1.15	23.54
32	4.75	4.06	2.45	1.73	3.19	3.58	1.01	20.77
33	5.01	5.5	1.56	2.84	3.43	3.84	0.88	23.06

(continued)

Table 5 (continued)

Sr no	Waiting time for introduction to system	Fill the form	Document check passed	Data entry in bank system	Verify all details	Deposit minimum balance	Activate account	Total time
34	6	4.91	2.46	2.37	4.12	3.33	0.69	23.88
35	6.64	4.36	2.01	2.23	3.53	3.53	0.96	23.26
36	5.91	3.67	2.62	1.78	3.07	3.9	0.69	21.64
37	6.08	3.97	1.92	2.19	3.72	3.71	1.1	22.69
38	6.32	3.7	1.82	2.82	3.45	3.93	0.71	22.75
39	6.76	3.54	1.83	2	4.3	3.31	0.71	22.45
40	6.32	4.57	1.99	1.87	3.27	3.69	1.13	22.84
	5.757	4.27875	2.11525	2.2205	3.58575	3.52475	0.864	22.346

2.5 Control Phase

The sample data after implementation of the action plan was used to draw the control chart, shown in Fig. 5. Taking the aid of these control charts, a resilient control plan of risk management to prevent system failure is proposed. The possibility points and influence points, ranking from 1 to 9 points individually are used to determine the final risk score, which is formed by possibility of multiplying influence.

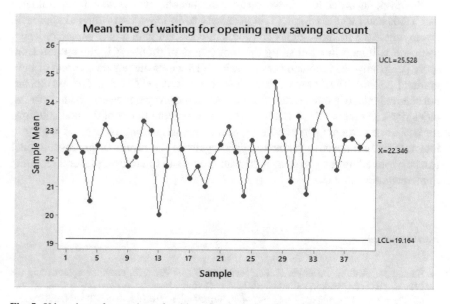

Fig. 5 X bar chart of mean time of waiting after the improvement plan

3   Conclusion

The presented research work has tried to demonstrate the impact of usage of Lean Six Sigma with TRIZ methodology in the service sector. Taking the banking sector as an example, the impact of these methods in increasing the effectiveness of various operations was analyzed through sample data. Real-time data for account opening operation was collected and after the analysis, it was observed that the process mean waiting time of savings account opening in Indian public sector undertaking banks can be reduced to 22.35 min from 32.16 min by implementing the above-mentioned improvement plan. Process capability can also be improved dramatically using this approach.

Six Sigma plays an important role in effective management strategy programs for quality management and has always delivered the best-in-class customer quality experience. But if we talk about innovative problem-solving or innovation regarding new products and services, it fails sometimes. While the original TRIZ was mostly used in analysis and delivery of innovative problem-solving for manufacturing processes. Nowadays, it could be said that businesses are converging on TRIZ for innovation; the way they have converged on Lean and Six Sigma for operational excellence. As previous research has proven the integration of TRIZ tools with lean six sigma methodology, this study has also showed how significantly this integration increases the potential of Lean Six Sigma for existing systems as well as in the creation of new products/services and processes that allows for the company to stay competitive.

However, in order to achieve measurable benefit by adapting methodologies like TRIZ and Lean Six Sigma it is necessary to follow a user-centered approach. Deploying these methodologies and tools in a system with a user-centered design approach will help in eliminating the root causes of problems in the system from every perspective be it manpower or machine. However the practical aspect of integrating LSS with TRIZ or any other innovative tool like UX Design process etc. has not been examined in other industrial sectors like the process industry, manufacturing industries and others. There is a lot of scope of arriving to out of the box solutions for problems pertaining in service as well as manufacturing industry if we try to integrate these innovative methodologies together. The results of proposed model can be explored more on various industrial grounds in future using several real time implementation studies.

References

1. Kumar, M., Antony, J., Singh, R. K., Tiwari, M. K., & Perry, D. (2006). Implementing the Lean Sigma framework in an Indian SME: A case study. *Production Planning & Control: The Management of Operations, 17*(4), 407–423. https://doi.org/10.1080/09537280500483350.
2. Vijaya, S. M. (2016). Rejects reduction in a retail bank using Lean Six Sigma. *Production Planning & Control*. https://doi.org/10.1080/09537287.2016.1187312.

3. Vashishth, A., Chakraborty, A., & Antony, J. (2017). Lean Six Sigma in financial services industry: A systematic review and agenda for future research. *Total Quality Management & Business Excellence, 30*(3–4), 447–465. https://doi.org/10.1080/14783363.2017.1308820.
4. Altshuller, G. (2000). *The innovation algorithm: TRIZ, systematic innovation and technical creativity.* Worcester, MA: Technical Innovation Center Inc.
5. Averboukh, E. (2005). Six sigma trends: Six sigma leadership and innovation. www.isixsigma.com.
6. Chao-Ton, Su., Lin, C.-S., & Chiang, T.-L. (2008). Systematic improvement in service quality through TRIZ methodology: An exploratory study. *Total Quality Management & Business Excellence, 19*(3), 223–243. https://doi.org/10.1080/14783360701600662.
7. Wang, F.-K., & Chen, K.-S. (2010). Applying lean Six Sigma and TRIZ methodology in banking services. *Total Quality Management, 21*(3), 301–315. https://doi.org/10.1080/147833609035 53248.
8. Domb, E., & Tate, K. (1997). 40 inventive principles with examples. *TRIZ Journal.* www.triz-journal.com
9. Silverstein, D., DeCarlo, N., & Slocum, M., (2008). *INsourcing innovation - How to achieve competitive excellence using TRIZ.* New York, USA: Auerbach Publications, Taylor & Francis Group.

Information and Communication Technologies for Sustainable Supply-Chain—A Smart Manufacturing (SM) Perspective

U. C. Jha and P. Siano

1 Introduction

The sustainability is being incorporated in the corporations' strategy, not only by its importance for the future of the society, but as a competitive priority [28]. The adoption of information and communications technologies for sustainability objectives has been largely investigated by the scholar literature in recent years and its impact in the sustainability programs have been quite positive in many scenarios [19, 25, 30].

The possibility to use new ICT technologies for manufacturing companies to achieve competitive advantages has generated a lot of research interest. Technologies like cloud computing, internet of things, machine learning, and new sensors are subjects of governmental and private programs to improve the regional industrial parks competitiveness. Some of these initiatives are Industrie 4.0, in Germany, Industrial Internet of Things and Smart Manufacturing, in United States, and Factory of the Future in European Union [6, 22, 40].

A more connected supply-chain, with available information about materials, products and energy enabling more refined decisions about productions and logistics is the common scenario among these initiatives. Improvements in the sustainability of the operations are expected by the program developers and researchers [6, 43].

Buchholz [6], in the Factories of the Future roadmap, presents opportunities to impact the economic, social, and environmental sustainability. Stock and Seliger [43] propose a totally connected industrial system for the Industrie 4.0, with a better energy management scheduling renewable energy sources efficiently. Kang et al. [22] define

U. C. Jha (✉)
Lovely Professional University (LPU), Punjab, India
e-mail: udai.22511@lpu.co.in

P. Siano
University of Salerno, Salerno, Italy
e-mail: psiano40@gmail.com

© The Author(s), under exclusive license to Springer Nature Singapore Pte Ltd. 2021   461
P. K. Kapur et al. (eds.), *Advances in Interdisciplinary Research
in Engineering and Business Management*, Asset Analytics,
https://doi.org/10.1007/978-981-16-0037-1_36

Smart Manufacturing as a "future growth engine" that can improve manufacturing considering societies, human, and environment.

Information Systems and Supply Chain researchers have studied for more than one decade different models and applications of ICT for sustainable objectives. Information systems for reverse logistics [25], packaging reuse [30], and green product development [10] are examples of subjects already explored.

Since the Smart Manufacturing research is still elaborating its operational model, and how the new proposed technologies can impact in more sustainable supply chain is not clear yet, a systematic literature review is required to evaluate how the ICT can be used to improve the sustainability and how the new technologies can change it.

This article aims to answer the following questions:

- What are the ICT applications for the sustainable supply chains already studied in the scientific literature?
- How these proposals can be incorporated or modified by the Smart Manufacturing technologies?

This document is organized by the sections: smart manufacturing, whit a bibliographic research about the smart manufacturing programs; research design, with the methodology followed; systematic review, with all steps of the review; discussion, integrating the review data with Smart Manufacturing; and conclusion.

2 Smart Manufacturing

The extensive offshoring of the industrial function of western companies during the 1990s resulted in twenty years of decline of manufacturing in Europe and USA [35]. Due to the important role of this sector for job creation, drive research and innovation, and economical competitiveness, several public and private initiatives was created to recover the western regional industrial parks [6, 35]. Such programs have in common the extensive research in the applications of modern ICT to create a more connected and sustainable supply chain [6, 18, 21].

The most influential programs are the Industrie 4.0, from Germany, the Smart Manufacturing and Industrial Internet of Things, both from USA, and the Factory of the Future, from the European Union, all of them with collaboration agreements [7, 47]. Other important regional programs are Nouvelle France Indutrielle, from France, and the east initiatives made in China 2025, and Manufacturing Innovation 3.0, from Korea [22, 47].

Despite some regional particularities, all of them have as the central proposal a more integrated industrial system, with three integration axes: horizontal integration, vertical integration, and life-cycle integration. The horizontal integration in the inter-company network that coordinates the entire supply-chain, including materials and energy suppliers, factories, and logistics. The vertical integration is in the factories, between the machines systems, also called Operation Technologies (OT), and business or IT systems. The life-cycle integration is the information flowing since the product engineering to this disposal, including its manufacturing and post-sales services [1, 21, 38].

This integration enables the digitization of the entire manufacturing system, working as a coordinated but decentralized linked physical and virtual elements. This cross-linked intelligence is called Cyber-Physical System (CPS) [43]. The main technologies present in the Smart Manufacturing are [7, 12, 21, 22]:

- Internet of Things: Internet of Things is the technology that enables a network of physical objects and its integration with software. It is one of the core technologies of Smart Manufacturing, responsible for communication with sensors and controllers.
- Big Data: Big Data is the data set with wide range, complex structure, and size, that cannot be processed by the traditional methods. Due to the large amount of data generated by the sensors and systems integrated in the Smart Manufacturing, the infrastructure has to be designed to deal with Big Data.
- Cloud Computing and Cloud Manufacturing: Cloud computing is a structure to outsource data processing and IT infrastructure. Cloud Manufacturing is the applications of this technology to communicate with distributed sensors and allocate manufacturing resources on demand.
- Additive Manufacturing: Additive Manufacturing is the technology that converts 3D models into physical objects. It is basically implemented in the 3D printers, that can be part of a highly flexible manufacturing plant, or a distributed manufacturing system.

Sensors Technology: The Sensors are the most important hardware technology in Smart Manufacturing, responsible for the real-time data collection and control. To make the Smart Manufacturing viable, sensor has to be cost effective, resilient and offer open communication standards.

- Machine Learning and Data Mining: Machine Learning and Data Mining are technologies that can extract knowledge from the data and apply intelligence for optimized decisions. They are intelligent elements of Smart Manufacturing that can use all the data captured by the system to generate computational models and coordinate its elements.

Davis et al. [12] suggest five operational categories impacted by the Smart Manufacturing technologies: smart machine line operations, in-production high-fidelity modeling, dynamic decisions, enterprise and supply chain decisions, design, planning and model development Table 1.

A system that can capture real-time data about production, logistics and products, and coordinate its resources, can improve the sustainability of the supply chain. An intelligent allocation of resources, like products, materials, energy, and water, can cause a good impact in the environmental management [43].

Buchholz [6], for the Factory of the Future, suggests opportunities for environmental, economic, and social sustainability. Best performance across the supply chain, reconfigurable factories capable of small-scale production, high-performance

Table 1 Operational categories of smart manufacturing opportunities [12]

Smart machine line operation	In-production high-fidelity modeling	Dynamic decisions	Enterprise and supply chain decisions	Design, planning and model development
Integrated process machine and product management	Enhanced management complex behaviors	Performance management global integrated decisions	Smart grid interoperability	Design models in production
Benchmarking machine-product interactions	Rapid qualification components products materials	Untapped enterprise degrees of freedom in efficiency, performance, and time	In situ measurement and integrated value chains	Product/material in production quality
Machine-power management	Integrated computational material engineering	Enterprise analytics and business operational tradeoff decisions	Tracking, traceability, and genealogy	New product, material technology insertion
Adaptable machine configurations		Configurable data and analyses for rapid analytics and model development	External partner integration and interoperability	

production and resource efficiency in manufacturing are opportunities for the economical sustainability. For the social, he lists increasing the human achievements, creating safe and attractive workplaces, and create sustainable care and responsibility for employees and citizens in global supply chains. For environmental, reducing the consumption of energy, increasing the usage of renewable energy, reducing the consumption of water and other process resources, near-to-zero emissions in manufacturing processes, optimizing the exploitation of materials.

3 Research Design

The literature review is an important tool to map an intellectual territory and, oriented by the research question, develop the existing body of knowledge [49]. According to Tranfield et al. [49], the adoption of a systematic review helps to counteracting bias, enhancing the legitimacy and authority of the resultant evidence.

According to Seuring and Müller [41], the literature review model has four steps:

- Material collection: Definition of the unit of analysis such as single paper, and the limits for the material to be collected.

- Descriptive analysis: Summarization, usually quantitative, of the formal aspects of the material. For example, the number of publications per year.
- Category selection: Selection of structural dimensions to be applied to the selected material. This step can be reviewed after the material evaluation, adding new categories proposed by the literature.
- Material evaluation: Analysis of the collected material using the categories selected.

The result of the literature review is going to answer the first research question. The second research question is going to be dealt by the discussion of the results considering the Smart Manufacturing concepts, presented in Sect. 2 (Fig. 1).

4 Systematic Review

The Web of Science, the Thompson Reuters scientific publications repository, was used to collect the material considered by this research. Web of Science and Scopus, from Reed Elsevier, are considered the most comprehensive scientific knowledge bases, and the best options to bibliographic and bibliometric studies [4]. According to Archambault et al. [4] both bases are highly correlated, so only one of them was selected for this study. It was only considered peer-review papers published in English.

The search terms were selected based on Fahimnia et al. [14] review about green supply chain, with an additional expression about information systems. The query was composed for three groups of keywords. The first "Information Technology(ies)" and "Information System(s)", about ICT, the second "Manufacturing" and "Supply Chain", and the third "Green", "Ecological", "Environmental" and "Sustainable", and "Environmental" and "Sustainability". The logical expression used in search, using "*" as wildcard character, was:

("Information Technolog*" OR "Information Syste*") AND (Ma nufacturing OR "Supply Chain") AND [Green OR Ecological OR (Environmental AND Sustainable) OR (Envi ronmental AND Sustainability)]

The search, applied to the title, abstract and keywords fields, and restricted by peer-reviewed papers and English language, returned 93 results. All of them had the abstracts read to eliminate the documents that didn't have the ICT applications for sustainable supply chain as an important subject. Only 37 of them was kept in the final collection.

4.1 Descriptive Analysis

The 37 articles collected is distributed between 2000 and 2016, with a great concentration in recent years. Around 60% of them was published after 2013.

The distribution of the papers in journals is very disperse, with only five journals with more than one publication. They are Journal of Cleaner Production and International Journal for Production Economics, both with 3 papers, MIS Quarterly, Supply Chain Management and Expert Systems with Applications, with two. Journals from different areas illustrates the multidisciplinary nature of the subject.

We have a diversity of research methodology adopted for this material. They are theoretical and conceptual (11), case study (9), survey (9), modeling and simulation (3), design science (3), and literature review (2). Besides the existence of new researches dedicated to theory development and case studies, that characterizes a body of knowledge still in development phase, the number of surveys shows that we already have some mature sub areas. The multidisciplinary nature of this research can be perceived too by the presence of modeling and simulation works, very common in operations management field, and design science, from the software development field.

We have two literature reviews, one about cleaner supply chain management practices [44] and another studying the corporate sustainability development in China. Both works have the technology as one of the resulting categories of the review.

4.2   Category Analysis

As initial category analysis was selected the three Smart Manufacturing integration axes, vertical, horizontal and life-cycle. The vertical integration category grouped all researches that proposes the integration with machines inside de manufacturing plants. The horizontal integration is the set of technologies that communicate between different companies of the supply chain. The life-cycle integration gathers papers that treats about the product technology, materials, and engineering.

4.3   Vertical Integration

In the vertical integration category we have researches that collects real-time data from plant-floor equipments and to impact in the factory energy consumption. They are five works that present technological alternatives for machines connection, and a framework to exploit this data. These are all non empirical works.

Tao et al. [46] is the only one that includes the emission reduction in the objectives of the system. He presents the manufacturing as the most important polluting business, and its reduction, together with the energy consumption reduction, can have an important impact to the environment.

Park and Jeong [33] addresses the energy problem in a indirect way, using the lean for green approach. The system is designed to reduce failures and non productive time in integrated circuits test lines, avoiding electrical energy waste.

The internet of things concept is used in two researches [42, 46] for sensor and automation systems communication. The cloud computing is proposed by Park and Jeong [33] to facilitate the system adoption. Lee et al. [26] proposes a modification on the Manufacturing Execution System, architecture for integration between plan-floor systems, and Enterprise Resource Systems, with sustainable objectives.

4.4 Horizontal Integration

The horizontal Integration in the most studied subject in the selected papers, with 16 publications. The systems models and objectives are more diverse than the vertical integration, and there is some empirical researches, with three surveys, that characterizes a more mature field.

The problem of closed loop supply chain management is addressed by five researches, with more general management system for collaboration [9] and flow optimization [10], reverse logistics management [15, 19] and product recovery for recycling and repair [19, 48]. Two complexities of the closed loop supply chain are treated by most of the authors. The first is the planning uncertainty due to irregular return of after use products from the field, with a non predictable material availability for manufacturing. The second is the recovery of the discarded products from customers, that are very geographically disperse and with no predictable scheduler.

Three of the publications are about traceability in the food supply chain [30, 32, 51]. The agricultural practices and supply chain transparence for the end user is important in this market for safety and for the social and environmental sustainability reasons. The agriculture is showed as an important soil and water polluter, and uses more child and slave work than other industries, so the conscious consumers are becoming an important market. Appelhanz et al. [3] presents a very similar traceability system, but for wood products, aiming to reach the premium market.

There are three works that uses quantitative methodologies to investigate the impact of ITC in the green supply chain. Dao et al. [11] finds the ITC as an important element to implement the sustainability culture in a supply chain, that is a necessary element of the collaboration to drive environmental impacts. Schniederjans and Hales [39] concludes that the cloud computing adoption has a positive effect on the supply chain sustainability. Green et al. [23] studies the green supply chain collaboration, and considers the ICT an necessary precursor of the green purchasing.

Wang et al. [50] and Marett et al. [29] research systems that can reduce the CO_2 emission monitoring and optimizing the road freight of a logistical system. Ahmad and Mehmood [2] presents an integrated concept of Supply Chain Management System and Smart Cities, with a system that can coordinate big and small industries to reduce the impact in the city environment.

In this group of article, the adopted technology for each system is less studied than the last, but we have two papers about cloud computing [2, 39] and the use of Radio Frequency Identification (RFiD) [30].

4.5 Life-Cycle Integrations

This group of papers deals mostly with product design, with great emphasis in meeting regulatory standards and enable collaboration.

Taghaboni-Dutta et al. [45] proposes a information system to record and exchange data about green products and parts to enable designers and manufacturers increase the use of green parts. Gong and Wang [17], Chen et al. [8, 9] design a system to deal with the complexity of design and manufacture products for customers in different regions, with different sustainability standards.

Guo et al. [20] studies a model to use the Enterprise Information System to green design assessment of electromechanical products. Eun et al. [13] proposes an information system for on-line Life Cycle Assessment.

4.6 Other ICT Applications

The seven papers selected for this group treats about very general or very specific use of ICT that cannot be classified in previous sections.

Quantitative analysis about supply chain and manufacturing sustainability and the presence of more general or informative ICT are presented in four papers. Pondeville et al. [34] studies the adoption of environmental management control systems, and concludes that it is mainly motivated by regulatory stakeholders. Ryoo and Koo [37] finds that the alignment of green practices and ICT have a positive effect in the coordination between green practices, manufacturing and marketing, with environmental performance improvement.

Benitez-Amado et al. [5] concludes that the capability of innovativeness mediates the effect of ICT adoption in green manufacturing. Gimenez et al. [16] propose the ICT as an ally to environmental practices to get better environmental performance.

Two papers studies about sustainability assessment. Muñoz et al. [31] proposes an ontology to develop systems for manufacturing and supply chain sustainability evaluation. Kusi-Sarpong et al. [24] design an algorithm to evaluate green supply chain practices in the mining industry.

Lee [27] is the only study about the adoption of environmental cost accounting systems in manufacturing, and finds a strong barrier to change existent accounting systems.

5 Discussion

The ICT for supply chain sustainability literature shows a very diverse group of areas, fields, and researchers, that proposes very similar applications than Smart

Manufacturing programs. Considering the operational categories of opportunities Table 1, proposed by Davis et al. [12], we can clearly find parallels for the five items:

- Smart machine line operations: The vertical integration researches deals with plant-floor integration and the exploitation of this data to smarter manufacturing operations decisions.
- In-production high-fidelity modeling: Still in a preliminary phase, models of the manufacturing are used by horizontal, life-cycle integration, and assessment works for planning and material management.
- Dynamic decisions, and Enterprise and supply chain decisions: The horizontal integration deals data from manufacturing and supply chain to planning and decisions.
- Design, planning, and model development: The life-cycle works addressed some of these challenges.

The advantages of the modern ICT technology capabilities are the main promise of the Smart Manufacturing [22, 40]. In the next sessions, these technologies are going to be discussed using the revised articles (excluding the additive manufacturing, since it is not an ICT technology).

5.1 Internet of Things and Sensors

The internet of things and sensors technology are connected in their application. The internet of things is the use of embedded sensors and communication technologies in the machinery, products, and vehicles, allowing data collection and coordination between "things" [22, 52].

This technology can already be found in the ICT literature, mainly for vertical integration, collecting energy consumption and material from the plant-floor equipments [42, 46]. Although the authors didn't identified as IoT, trucks real-time and autonomous monitoring are very similar applications [29, 50]. The possibility of measure and communicate physical quantities, from virtually any kind of equipment, can impact a lot of presented challenges.

The recovering in the closed loop supply chain can take advantage of this technology collecting data from the product about its using profile and disposing, being user independent for the reverse logistics [19, 48]. Sustainability reporting and assessment, highly dependent of the human process and adoption to obtain success [24, 27, 31], can be automated with data grabbed from the manufacturing and logistics processes.

5.2 Cloud Computing

The cloud computing allows the development very scalable, modular and inclusive systems, adequate for the coverage of Smart Manufacturing [12, 22]. It can accelerate the adoption of inter-company systems, since the infrastructure and geographic location are not important barriers.

Some studied papers uses this technology, as Park and Jeong [33], that implements a failure detection system using cloud computing, for easier adoption, Ahmad and Mehmood [2] that adopts cloud to scale a system that can coordinate smart cities and the industrial supply chain.

5.3 Big Data, Machine Learning, and Data Mining

With cloud computer infrastructures and sensor capturing all data from products and machines, it is needed algorithms that can deal with this big volume of data. The big data, machine learning, and data mining are the technologies responsible for recording, recover, analyze, and suggest actions based on data [22].

Machine learning and data mining are the core of the system intelligence, with prediction and adaptation capabilities [36]. They can support human decision makers with models based on process data, with better use of the human knowledge [1].

These technologies can address the problems studied in the literature review with an integrated manner. The same product data, gathered and communicated by an IoT system, can be analyzed for the supply chain optimization and new product development, just changing the context.

6 Conclusions

The Smart Manufacturing is a new research field, with few empirical examples and divergent models. The adherence of the ICT for sustainable supply chain literature with some of its proposals can generate new insights about the real potential of this model, a help to conduce future researches.

The application of ICT to improve the sustainability has shown a very positive effect in previous research. Dao et al. [11] and Schniederjans and Hales [39] are examples of papers that found the ICT as an important element do improve the collaboration in supply chain and keep the company engaged in achieving the sustainability goals.

This work intents to contribute to academic literature offering a map of previous works contextualized with the Smart Manufacturing concepts, that can be the basis of future empirical researches. For business readers, can be a guide to the principal

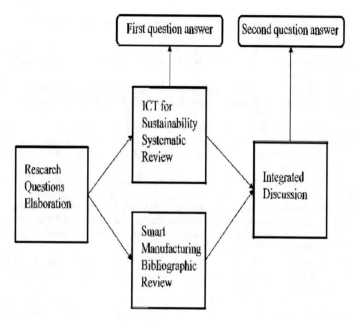

Fig. 1 Research process

concepts in Smart Manufacturing, and examples of real applications of ICT for sustainability.

This review has to be read considering its limitations. The papers selected to be studied are always a partial sample of the research in the field. Publications in non English language, not indexed and presented in scientific events wasn't considered.

References

1. Agarwal, N., & Brem, A. (2015). Strategic business transformation through technology convergence: Implications from General Electric's industrial internet initiative. *International Journal of Technology Management, 67*(2–4), 196–214. https://doi.org/10.1504/IJTM.2015.068224.
2. Ahmad, N., & Mehmood, R. (2015). Enterprise systems: Are we ready for future sustainable cities. *Supply Chain Management-an International Journal, 20*(3, SI), 264–283. https://doi.org/10.1108/SCM-11-2014-0370
3. Appelhanz, S., Osburg, V. S., Toporowski, W., & Schumann, M. (2016). Traceability system for capturing, processing and providing consumer-relevant information about wood products: System solution and its economic feasibility. *Journal of Cleaner Production, 110,* 132–148. https://doi.org/10.1016/j.jclepro.2015.02.034.
4. Archambault, É., Campbell, D., Gingras, Y., & Lariv ière, V. (2009). Comparing bibliometric statistics obtained from the web of science and Scopus. *Journal of the American Society for Information Science and Technology, 60*(7), 1320–1326. https://doi.org/10.1002/asi.21062.
5. Benitez-Amado, J., Perez-Arostegui, M., & Tamayo-Torres, J. (2010). Information technology-enabled innovativeness and green capabilities. *Journal of Computer Information Systems, 51*(2).

6. Buchholz, S. (2011). Factories of the future. In *Manufacturing chemist* (Vol. 82). https://doi.org/10.2777/29815.

7. Buda, A., Främling, K., Borgman, J., Madhikermi, M., & Kubler, S. (2015). Data supply chain in Industrial Internet. In *11th IEEE World Conference on Factory Communication Systems (WFCS 2015)*, (May).

8. Chen, M.-K., Tai, T.-W., & Hung, T.-Y. (2012). Component selection system for green supply chain. *Expert Systems with Applications, 39*(5), 5687–5701. https://doi.org/10.1016/j.eswa.2011.11.102.

9. Chen, Y. C., Chu, C. N., Sun, H. M., Chen, R. S., Chen, L. C., & Chen, C. C. (2015). Application of green collaboration operation on network industry. *International Journal of Precision Engineering and Manufacturing - Green Technology, 2*(1), 73–83. https://doi.org/10.1007/s40684-015-0010-2.

10. Chung, C.-J., & Wee, H.-M. (2010). Green-product-design value and information-technology investment on replenishment model with remanufacturing. *International Journal of Computer Integrated Manufacturing, 23*(5), 466–485. https://doi.org/10.1080/09511921003667714.

11. Dao, V., Langella, I., & Carbo, J. (2011). From green to sustainability: Information technology and an integrated sustainability framework. *Journal of Strategic Information Systems, 20*(1), 63–79. https://doi.org/10.1016/j.jsis.2011.01.002.

12. Davis, J., Edgar, T., Graybill, R., Korambath, P., Schott, B., Swink, D., et al. (2015). Smart manufacturing. *Annual Review of Chemical and Biomolecular Engineering, 6*, 141–160. https://doi.org/10.1146/annurev-chembioeng-061114-123255.

13. Eun, J.-H., Son, J.-H., Moon, J.-M., & Chung, J.-S. (2009). Integration of life cycle assessment in the environmental information system. *The International Journal of Life Cycle Assessment, 14*(4), 364–373. https://doi.org/10.1007/s11367-009-0076-6.

14. Fahimnia, B., Sarkis, J., & Davarzani, H. (2015). Green supply chain management: A review and bibliometric analysis. *International Journal of Production Economics, 162*, 101–114. https://doi.org/10.1016/j.ijpe.2015.01.003.

15. García-Rodríguez, F. J., Castilla-Gutiérrez, C., & Bustos-Flores, C. (2013). Implementation of reverse logistics as a sustainable tool for raw material purchasing in developing countries: The case of Venezuela. *International Journal of Production Economics, 141*(2), 582–592. https://doi.org/10.1016/j.ijpe.2012.09.015.

16. Gimenez, C., Sierra, V., Rodon, J., & Andres Rodriguez, J. (2015). The role of information technology in the environmental performance of the firm the interaction effect between information technology and environmental practices on environmental performance. *Academia-Revista Latinoamericana de Administracion, 28*(2), 273–291. https://doi.org/10.1108/ARLA-08-2014-0113.

17. Gong, D. C., & Wang, Y. T. (2011). UML presentation of a conceptual green design control system to react to environmental requirements. *International Journal of Advanced Manufacturing Technology, 52*(5–8), 463–476. https://doi.org/10.1007/s00170-010-2758-4.

18. Gorbach, G., Polsonetti, C., & Chatha, A. (2014). Planning for the industrial internet of things industrial internet of things (IoT) enables new business models. *ARC Advisory Group*.

19. Guide, V. D. R., Jayaraman, V., Srivastava, R., & Benton, W. C. (2000). Supply-chain management for recoverable manufacturing systems. *Interfaces, 30*(3), 125–142. https://doi.org/10.1287/inte.30.3.125.11656.

20. Guo, J., Zhou, M., Li, Z., & Xie, H. (2014). Green design assessment of electromechanical products based on group weighted-AHP. *Lecture Notes in Business Information Processing, 19*(8), 87–100. https://doi.org/10.1007/978-3-642-00670-8.

21. Henning, K., Wolfgang, W., & Johannes, H. (2013). Recommendations for implementing the strategic initiative INDUSTRIE 4.0. *Final Report of the Industrie 4.0 WG*, (April), 82.

22. Kang, H. S., Lee, J. Y., Choi, S., Kim, H., Park, J. H., Son, J. Y., et al. (2016). Smart manufacturing: Past research, present findings, and future directions. *International Journal of Precision Engineering and Manufacturing-Green Technology, 3*(1), 111–128. https://doi.org/10.1007/s40684-016-0015-5.

23. Green Jr, K. W., Zelbst, P. J., Meacham, J., Bhadauria, V. S., Zelbst, P. J., Bhadauria, V. S., et al. (2012). Green supply chain management practices: Impact on performance. *Supply Chain Management: An International Journal, 17*(3), 290–305. https://doi.org/10.1108/135985412 11227126.

24. Kusi-Sarpong, S., Bai, C., Sarkis, J., & Wang, X. (2015). Green supply chain practices evaluation in the mining industry using a joint rough sets and fuzzy TOPSIS methodology. *Resources Policy, 46,* 86–100. https://doi.org/10.1016/j.resourpol.2014.10.011.

25. Lee, C. K. M., & Lam, J. S. L. (2012). Managing reverse logistics to enhance sustainability of industrial marketing. *Industrial Marketing Management, 41*(4), 589–598. https://doi.org/10.1016/j.indmarman.2012.04.006.

26. Lee, H., Ryu, K., Son, Y. J., & Cho, Y. (2014). Capturing green information and mapping with MES functions for increasing manufacturing sustainability. *International Journal of Precision Engineering and Manufacturing, 15*(8), 1709–1716. https://doi.org/10.1007/s12541-014-0523-6.

27. Lee, K. H. (2011). Motivations, barriers, and incentives for adopting environmental management (cost) accounting and related guidelines: A study of the republic of Korea. *Corporate Social Responsibility and Environmental Management, 18*(1), 39–49. https://doi.org/10.1002/csr.239.

28. Longoni, A., & Cagliano, R. (2015). Environmental and social sustainability priorities: Their integration in operations strategies. *International Journal of Operations & Production Management, 35*(2), 216–245. https://doi.org/10.1108/IJOPM-04-2013-0182.

29. Marett, K., Otondo, R. F., & Taylor, G. S. (2013). Assessing the effects of benefits and institutional influences on the continued use of environmentally munificent bypass systems in long-haul trucking. *MIS Quarterly, 37*(4), 1301–1312.

30. Martínez-Sala, A. S., Egea-López, E., García-Sánchez, F., & García-Haro, J. (2009). Tracking of returnable packaging and transport units with active RFID in the grocery supply chain. *Computers in Industry, 60*(3), 161–171. https://doi.org/10.1016/j.compind.2008.12.003.

31. Muñoz, E., Capón-García, E., Laínez, J. M., Espuña, A., & Puigjaner, L. (2013). Considering environmental assessment in an ontological framework for enterprise sustainability. *Journal of Cleaner Production, 47,* 149–164. https://doi.org/10.1016/j.jclepro.2012.11.032.

32. Opara, L. U., & Mazaud, F. (2001). Food traceability from field to plate. *Outlook on Agriculture, 30*(4), 239–247. https://doi.org/10.5367/000000001101293724.

33. Park, J. H., & Jeong, H. Y. (2013). Cloud computing-based jam management for a manufacturing system in a Green IT environment. *Journal of Supercomputing, 69*(3), 1054–1067. https://doi.org/10.1007/s11227-013-1007-7.

34. Pondeville, S., Swaen, V., & De Rongé, Y. (2013). Environmental management control systems: The role of contextual and strategic factors. *Management Accounting Research, 24*(4), 317–332. https://doi.org/10.1016/j.mar.2013.06.007.

35. Prause, G. (2015). Sustainable business models and structures for industry 4.0. *5*(2). https://doi.org/10.9770/jssi.2015.5.2(3).

36. Qu, S., Jian, R., Chu, T., Wang, J., & Tan, T. (2015). Computational reasoning and learning for smart manufacturing under realistic conditions. In *Proceedings of 2014 IEEE International Conference on Behavioral, Economic, Socio-Cultural Computing, BESC 2014.* https://doi.org/10.1109/BESC.2014.7059529.

37. Ryoo, S. Y., & Koo, C. (2013). Green practices-IS alignment and environmental performance: The mediating effects of coordination. *Information Systems Frontiers, 15*(5), 799–814. https://doi.org/10.1007/s10796-013-9422-0.

38. Schmidt, N., Lüder, A., Rosendahl, R., Foehr, M., & Vollmar, J. (2015). Characterizing integration approaches: Identifying integration approach candidates for use in industrie 4.0. In *IEEE International Conference on Industrial Informatics* (pp. 527–532).

39. Schniederjans, D. G., & Hales, D. N. (2016). Cloud computing and its impact on economic and environmental performance: A transaction cost economics perspective. *Decision Support Systems, 86,* 73–82. https://doi.org/10.1016/j.dss.2016.03.009.

40. Sergi, B. S. (2015). Strategic factor analysis for industry 4.0. *Journal of Security and Susteinability Issues, 8*(2), 159–169.

41. Seuring, S., & Müller, M. (2008). From a literature review to a conceptual framework for sustainable supply chain management. *Journal of Cleaner Production, 16*(15), 1699–1710. https://doi.org/10.1016/j.jclepro.2008.04.020.

42. Shrouf, F., & Miragliotta, G. (2015). Energy management based on Internet of Things: Practices and framework for adoption in production management. *Journal of Cleaner Production, 100,* 235–246. https://doi.org/10.1016/j.jclepro.2015.03.055.

43. Stock, T., & Seliger, G. (2016). Opportunities of sustainable manufacturing in industry 4.0. *Procedia CIRP, 40*(Icc), 536–541. https://doi.org/10.1016/j.procir.2016.01.129

44. Subramanian, N., & Gunasekaran, A. (2015). Cleaner supply-chain management practices for twenty-first-century organizational competitiveness: Practice-performance framework and research propositions. *International Journal of Production Economics, 164,* 216–233. https://doi.org/10.1016/j.ijpe.2014.12.002.

45. Taghaboni-Dutta, F., Trappey, A. J. C., & Trappey, C. V. (2010). An XML based supply chain integration hub for green product lifecycle management. *Expert Systems with Applications, 37*(11), 7319–7328. https://doi.org/10.1016/j.eswa.2010.04.025.

46. Tao, F., Zuo, Y., Xu, L. D., Lv, L., & Zhang, L. (2014). Internet of things and BOM-based life cycle assessment of energy saving and emission-reduction of products. *IEEE Transactions on Industrial Informatics, 10*(2), 1252–1261. https://doi.org/10.1109/TII.2014.2306771.

47. Toro, C., Barandiaran, I., & Posada, J. (2015). A perspective on knowledge based and intelligent systems implementation in industrie 4.0. *Procedia Computer Science, 60,* 362–370. https://doi.org/10.1016/j.procs.2015.08.143.

48. Toyasaki, F., Wakolbinger, T., & Kettinger, W. J. (2013). The value of information systems for product recovery management. *International Journal of Production Research, 51*(January), 1214–1235. https://doi.org/10.1080/00207543.2012.695090.

49. Tranfield, D., Denyer, D., & Smart, P. (2003). Towards a methodology for developing evidence-informed management knowledge by means of systematic review *. *British Journal of Management, 14,* 207–222. https://doi.org/10.1111/1467-8551.00375.

50. Wang, Y., Sanchez Rodrigues, V., & Evans, L. (2015). The use of ICT in road freight transport for CO_2 reduction—An exploratory study of UK's grocery re tail industry. *The International Journal of Logistics Management, 26*(1), 2–29. https://doi.org/10.1108/IJLM-02-2013-0021.

51. Wognum, P. M., Bremmers, H., Trienekens, J. H., Van Der Vorst, J. G. A. J., & Bloemhof, J. M. (2011). Systems for sustainability and transparency of food supply chains—Current status and challenges. *Advanced Engineering Informatics, 25*(1), 65–76. https://doi.org/10.1016/j.aei.2010.06.001.

52. Yu, C., Xu, X., & Lu, Y. (2015). Computer-integrated manufacturing, cyber-physical systems and cloud manufacturing—Concepts and relationships. *Manufacturing Letters, 6,* 5–9. https://doi.org/10.1016/j.mfglet.2015.11.0.

Measuring and Evaluating Best Practices in Agile Testing Environment Using AHP

Abhishek Srivastava, Deepti Mehrotra, P. K. Kapur, and Anu G. Aggarwal

1 Introduction

Agile testing is a subset of evolutionary and iterative methods and they are based on opportunistic development and iterative enhancement processes. The Agile Manifesto clearly prioritizes 'individuals and the interactions among them over the tools and processes used, customer collaboration and intensive involvement over the contract negotiation, working software in the form of periodic deliverables over comprehensive documentation, and responding to changes according to the customer requirements over following a pre-determined plan. These agile principles intrinsically encourage the flexibility which common thing among these agile methods is that the implementation of software development is an empirical process in all these methods. Being from the same family of iterative and incremental approach, there are so many shared stuffs among these methods but still they do differ when it comes to their practices, processes and basic principles [1], the further insight into these agile

A. Srivastava (✉) · D. Mehrotra
Amity School of Engineering and Technology, Noida, India
e-mail: abhishek.sri13@gmail.com

D. Mehrotra
e-mail: mehdeepti@gmail.com

P. K. Kapur
Amity Center for Interdisciplinary Research, Amity University, Noida, Uttar Pradesh, India
e-mail: pkkapur1@gmail.com

A. G. Aggarwal
Department of Operational Research, Faculty of Mathematical Sciences, University of Delhi, Delhi, India
e-mail: anuagg17@gmail.com

© The Author(s), under exclusive license to Springer Nature Singapore Pte Ltd. 2021   475
P. K. Kapur et al. (eds.), *Advances in Interdisciplinary Research in Engineering and Business Management*, Asset Analytics,
https://doi.org/10.1007/978-981-16-0037-1_37

development methods is out of the scope of this paper. Based upon the variances in their processes and practices, few parameters is considered for this study which are argued later in the paper.

2 Literature Review

During this revolutionary phase of customization and evolving technology, competition is at its all-time high. Software companies are focussing on fulfilling all possible requirements of the client for, numerous methods are being reconnoitered with supplementary features. To address this, methods like quality function deployment, rapid application development, web surveys, etc. are being used to handpick the requirements of decision making [2, 3]. Several researchers have published their works discussing about these modeling techniques and several decision making processes has been discovered to overcome the shortfalls. Techniques like Fuzzy Logic [4], Analytic Network Process (ANP) [5], interpretive structural modeling(ISM) [1, 6], have been used to manifest their usability in diverse areas so as to make up for the current shortfalls in the framework. As per the available literature, the authors designed a method to effectively apply the agile software development but lesser work was done to improve the testing efficiency.

Hannola et al. [7] revealed a case study to adopt requirement engineering and discussed how it could meet the requirement of software to become agile and dynamic in its functionality. The author conducted the interviews, meeting, and discussions with the company managers and employees to collect the data required about the requirements and deeper understanding of the requirements of the software. This research indicated that the company must take steps like regular training programs and discussions to implement better practice of requirements engineering. This will ensure more involvement and engagement in requirement engineering process to achieve success in the software development process.

Similarly, Klendauer et al. [8] performed a qualitative study to penetrate more deeply into the requirement engineering through competency studies. He conducted interviews to develop his critical thinking towards sixteen major competencies with integration of situational and related factors along with diverse set of results required as per the circumstances. Through this research, the researcher developed a software based on the competency model using interpretive approach along with the critical incident technique.

Simultaneously Kassab [3] highlighted the knowledge gap about the required areas of improvements in requirements engineering by conducting surveys in the years 2003, 2008 and also same in 2013 to calibrate the developments in the practices of requirement engineering and gathered some crucial evidences through a comparative analysis of the collected data about the developing industry. This survey results majorly emphasized on the stakeholders' related requirements database and the qualitative aspect of projects handled, highlighting the need of more such qualitative research before moving ahead into development phase.

Latha and Suhanthi [9] also conducted in an empirical study with software experts to conclude that business perspective is the most significant of all factors deciding the requirements criteria for a software project. The researchers worked towards requirements engineering using AHP approach for developing product understanding first as a streamlined approach.

Researcher Kim et al. [10] focussed on building a framework for modelling architecture and fulfilling the domain requirements with better analysis of the domain requirements and architecture structure as a part of requirement engineering. Kim's model considered four general concepts i.e. domain requirement analysis, AHP, matrix technique and architecture styles for creating better dynamics among themselves for a better framework.

Whereas Cheong et al. [11] in his study developed a framework for the software development to be deployed on web platform. For this the Visual Studio.Net compiler was used. This is an Analytic Hierarchy Process (AHP) framework and is based on the fuzzy logic technique. This technique helps in complex decision making where multiple criteria are in consideration, like a semi-structured domain or an unstructured problem domain. The researcher ensured that the tool should be able to do consistency checks of the responses also so that data collected is validated at the same time. An elaborative process for final decision making has been taken up in this framework which has been divided into different steps.

Kamalrudin et al. [12] highlighted that the requirements templates must be used to deliver better quality and improve the requirements expression especially when the experts work on natural language requirement patterns to develop standard requirement [13]. The researcher elicited the 3C's approach for requirements specification, i.e. Consistency, Completeness and Correctness for evaluating and analyzing the viability of requirements templates through the end user study.

Pantoquilho et al. [14] delved deeper into the requirements patterns and regular problems faced during the software development process. The researchers identified problems in the existing approaches and upgraded template was developed based on the gap analysis. Researchers have developed several methods of upgrading the traditional templates used in requirements engineering to make them more focussed and useful.

In the same direction, the researcher Rajagopal et al. [15] decided to perform exhaustive research of the software requirements focussed on end user characteristics by performing a stepwise study and decided to overcome the barriers through measures like stakeholders trainings, setting priorities about software dynamics based on domain, developing and validating the sample of the model without overlooking the risks combined with cost effectiveness.

Lai et al. [16] studied the effectiveness of selection process of products through Analytic Hierarchy Process (AHP) in a group based decision making environment using multimedia processes. The researcher performed a study with AHP trained software experts exploring the multimedia authorization system (MAS) to select best out of its MAS products, based on both the essential requirements of the stakeholders.

3 Multi-criteria Decision Making

Multi criteria is used in resolving the conflicting criteria while we have decision making process in picture. To forecast better decisions it is recommended to structure complex problems and taking into consideration the multiple criteria explicitly. There are number of advances in this field of the multiple-criteria decision making discipline since early 1960s. The process of decision making is improving day after day, as the new methods are evolving and providing the substantial base for making them more and more reliable. The problem of decision making is all about choosing the best possible optimal solution among several conflicting alternatives. This process of finding the optimal solution not only depends merely upon the criteria itself but also influenced by the preferences of the decision maker. It came into existence by Brans et al. [17], and is used widely among several other available outranking methods. The Analytic hierarchy process (AHP) was coined by Saaty in 1980 [18]. The decision making process involves several conflicting criteria in one hand to choose among different alternatives available on the other hand.

3.1 *Analytical Hierarchical Process*

Analytical Hierarchical Process is an effective tool to help requirement engineering-related decision making when the decision makers have to facilitate the software developers with focussed information as per stakeholders' priorities. With the literature available, it is clear that AHP clarifies and provides all possible opportunities to the decision makers for multi-criteria based complex problems. Also, the decision makers could study the conflicting requirements in a hierarchical manner and finding alternates for sorting the problems in an organized manner [5].

AHP does a discrete analysis in hierarchical structural approach for giving opportunity of better decision making and acquiring uniformity in judging the requirements of attributes as per set priorities. Analytic Hierarchical Process is applicable in several areas of technology, management, health sector, financial decision and many others as it is highly organized and has clear approach i.e. select, prioritize, categorize, standardize and quality plaid.

There are some specified phases for application of Analytic Hierarchical Process.

Step 1: AHP has three phases: disintegration, relative decision making and setting priorities. Disintegration refers to break the objectives into lower level, with sub-criteria and alternatives listed.

During this step, the requirements are matched with criteria-wise attributes of software. This comparison in pairs based on stepwise analysis helps experts to draw broad conclusions on quality matrix and its varied determinants.

This is followed by identification and scaling of values attained by numerical valuation of user inclination towards attributes. Values assigned shall be 1–9 (Saaty, 1980; as cited in 2004 [18]).

Scale value are denoted as:

Equal (EQ) = 1.

Moderately Strong (MS) = 3.

Strong (S) = 5.

Very Strong (VS) = 7.

Extremely Strong (ES) = 9.

Intermediate values = 2, 4, 6, 8.

'EQ' refer to, two attributes with same importance.

'MS' means one attribute is moderately important than the other attribute.

'S' means an attribute is essential than others.

'VS' means an attribute is relatively stronger choice than other attributes.

'ES' suggests the attribute has extreme importance in comparison to others.

Step 2: Numerical values are incorporated in the judgement matrix (Xij) and initial matrix is normalized, followed by calculating the sum of row values. Below Equation has been used in this paper to calculate priority weight (w), where j is column and i is row.

$$W_i = \frac{\sum_{i=1}^{I} \left(\frac{a_{ij}}{\sum_{j=1}^{J} a_{ij}} \right)}{J} \tag{1}$$

where $1 \le i \le n$ and $1 \le j \le n$.

Step 3: Post normalization, the sum of rows is evaluated and find the AHP matrix of current matrix.

AHP matrix is also called priority weight or values.

Priority weight ($W1$, $W2$, Wj) for each factor is further calculated;

$$W_j = \frac{\sum_{j=1}^{n} X_{ij}}{n} \tag{2}$$

Step 4: The Consistency Check (CC) or consistency ratio (CR) is evaluated by finding the ratio of Consistency Index (CI) and Random Consistency Index (RCI).

Consistency Check (CC) gives the Consistency of responses on the compared pairs.

Consistency ratio (CR) or check (CC) is equal to ratio of CI and RCI.

i.e. CC or CR = CI/RCI

Consistency index (CI) is a measure of consistency, as degree of consistency, measured as,

$$CI = \frac{(\lambda_{max} - n)}{(n - 1)} \tag{3}$$

where, n is the number of attributes judged as decision makers.

λ_{max} is the highest Eigen value.

Eigen values are measured as the Sum of products of each element of Eigen vector and Sum of columns of reciprocal matrix.

RCI is calculated from multiple random matrices.

Consistency ratio (CR) should be <0.1 or below 10%.

If CR > 0.1, then judgement matrix is rejected and must be restructured.

As a result, Combination of priorities is obtained which helps in computing complex weight for each alternative set of priorities, in relation to attribute inclinations determined from the matrix. Final judgement values are used to find overall weights of the chosen requirement attributes. These qualitative judgements are transformed into measurable values.

Step 5: Utility theory is used to get the feasibility measure for the software, pertaining to the studied requirement attributes. The software experts list the essential and non-essential requirement attributes by means of interviews and discussions.

Utility theory helps to gauge the significance of specific attribute for the requirements in keeping with the software efficiency and stakeholders' inclination along with measuring the general effectiveness of utility obtained through all possible combination of attributes.

4 Proposed Approach and Experimental Setup

It has always been a critical assignment for a project analyst to select the utmost suitable agile testing process for a given project among several available agile methods, in the absence of any empirical approach.

To achieve the precise objective mentioned in the earlier section initially the AHP is cast-off to rank the different agile testing methods criteria for different methodologies and then to compensate the subjective behaviour of the decision maker. AHP is used to rank the diverse criteria for different methodology. Rank aggregation methods are used to aggregate the ranks produced from these methods and thus, at the final level the most appropriate agile method is selected according to the requirements of the given project.

The Analytic hierarchy process is commonly known as AHP; the role of Eigen vector is not only for objective evaluation but also takes care of the subjective nature of the human judgement. The Eigen value is further used for the verification of the evaluation consistency. As decision making involves different criteria according to a given problem, thus, agile manifesto and agile principles for selecting the criteria is considered. The criteria preferred to take care of every aspect of the project, like project analyst, team, customer, etc. The value given to each criterion lies in between 1 and 10, where 1 is used for least importance, and 10 indicates the highest importance. After deep analysis and study, the major six criteria which are perfectly in tune with the agile values, also mentioned in the Agile Manifest criteria, are: Reliability (R), Agile Testing Team (AT), Lead Time (LD), Customer Feedback (CF), Iterative Development (ID), Costs (C) is considered.

Each project differs from others in some respect, thus, every project has different requirements accordingly. Therefore, project analyst can make an addition to this list of criteria and can also remove some criteria according to the need of the project. Also for every project requirement we have computed the priority of different criteria's i.e. shown in Tables 1, 3, 5, 7 and 9. Qualitative judgement values were obtained by the expert opinion. The qualitative values were converted into quantitative values and the final priority weights for the six criteria were obtained realizing from Eq. (1).

Table 1 Comparision matrix for SCRUM methodology

Goal criteria	R	A	LD	C	ID	C
Reliability	1	0.2	3	0.333333	2	3
Agile testing team	5	1	7	4	3	4
Lead time	0.333333	0.142857	1	0.333333	2	0.333333
Customer feedback	3	0.25	3	1	5	3
Iterative development	0.5	0.333333	0.5	0.2	1	0.333333
Costs	0.333333	0.25	3	0.333333	3	1
Total	10.1667	2.1762	17.5000	6.2000	16.0000	11.6667

After extensive literature review, we have categorized five different testing methodology (Goals) that are usually used by enterprise these are SCRUM, LEAN, XP (Extreme Programming), DSDM (Dynamic systems development method) and FDD (feature driven development). The weights for various criteria were calculated using Analytic hierarchy process. For calculating ranks and weights, a crisp comparison matrix is filled from five industry experts, and a consolidated matrix is computed with the help of a weighted geometric mean of these participants, as shown in Table 2 for SCRUM methodology, Table 4 for LEAN methodology, Table 6 for CRYSTAL methodology, Table 8 for DSDM methodology and Table 10 for FDD methodology. Interestingly we have seen that every project has different priority while its execution.

It is quite clearly evident from the outcome that pairwise comparison among selected attributes with SCRUM process, the criteria Expandability plays significant role than other criteria. Interestingly it was observed and verified that significant measures being altered with every other agile testing process.

In the same manner the qualitative values were converted into quantitative values and the final priority weights for the six criteria were obtained using Eq. (1). And the elements of the normalized decision matrix were multiplied by the weights of the criteria determined in Table 3 and the weighted normalized decision matrix is obtained. The resultant weighted normalized decision matrix is illustrated in Table 4 for LEAN methodology and we found that reliability criteria plays important role in LEAN Methodology.

The resultant weighted normalized decision matrix is illustrated in Table 6. And here in case of CRYSTAL testing methodology Communication overhead is the most important criteria for the testing.

Equally, in case of XP testing methodology criteria Manageability plays an important role.

In case of DSDM again the study predicts that the criteria Lead time plays an important role.

Remarkably for FDD also the lead time overhead plays an imperative role as same with for DSDM.

Table 2 Weights and ranks of criteria with normalized values of selection criteria for SCRUM methodology

	R	A	LD	C	ID	C	P.E.V	P.E	CI	CR	Sum	Priority
Reliability	0.0984	0.0919	0.1714	0.0538	0.1250	0.2571	0.1329	6.8835	0.1767	0.133857	0.7976	13.29
Agile testing team	0.4918	0.4595	0.4000	0.6452	0.1875	0.3429	0.4211				2.5268	42.11
Lead time	0.0328	0.0656	0.0571	0.0538	0.1250	0.0286	0.0605				0.3629	6.05
Customer feedback	0.2951	0.1149	0.1714	0.1613	0.3125	0.2571	0.2187				1.3123	21.87
Iterative development	0.0492	0.1532	0.0286	0.0323	0.0625	0.0286	0.0590				0.3543	5.9
Costs	0.0328	0.1149	0.1714	0.0538	0.1875	0.0857	0.1077				0.6461	10.77
Total	1	1	1	1	1	1	1					

Table 3 Comparison matrix for LEAN methodology

	R	A	LD	CF	ID	C
Reliability	1	1	3	1	5	3
Agile testing team	0.5	1	1	0.2	1	0.333333
Lead time	0.333333	0.25	1	1	3	1
Customer feedback	0.333333	0.333333	3	1	1	2
Iterative development	0.2	0.333333	0.5	2	1	1
Costs	0.2	0.2	0.333333	3	3	1
SUM	2.566667	3.116667	8.833333	8.2	14	8.333333

Now with the above study practitioners can easily make the priority for the appropriate criteria. As project varies based on requirements insists project analyst either to add or subtract to the list of criteria. From computational point of view a crisp comparison matrix need to be obtained from industry experts, and a consolidated matrix is computed with the help of a weighted geometric mean of these participants, as mentioned in Tables 1, 3, 5, 7, 9 and 11.

The Consistency Ratio is calculated based upon this comparison matrix and if it is under 10% then the judgement is accepted, otherwise we have to modify the preferences. In our case, the consistency ratio has come out to be in range, which is quite a good approximation. The ranks and weights are calculated as shown in Tables 2, 4, 6, 8, 10 and 12. The higher value of priority indicator reflects higher preference for a particular criteria and vice-versa.

5 Conclusion

This work orients towards by implementing a multilevel approach for selecting best criteria in agile testing method selection according to the requirement of a particular project. As there was not much empirical work available that insists authors to apply AHP, thus providing the much awaited authenticity and reliability, which sometimes is questioned in case of the agile testing approach. There were very few studies that can quantitatively measure the criteria and select the most suitable criteria for a testing methodology in an organization. In this paper an approach of AHP is applied that can be opted by organization for different testing criteria (Reliability, Agile testing team, Lead time, Customer feedback, Iterative development, Costs) and the most suitable criteria for every testing methodology(Scrum, Lean, Crystal, XP, DSDM and FDD) was determined on the bases of ranking. A priority indicator is used to measure the criteria for ranking. The work would open a new horizon in this field of agile testing and will prove to be a pivotal topic for agile development method selection, and will help to generate better results in the future for this arena [19, 20].

Table 4 Weights and ranks of criteria with normalized values of selection criteria for LEAN methodology

Goal criteria for LEAN	R	A	LD	C	ID	C	P.E.V	P.E	CI	CR	Sum	Priority
Reliability	0.3896	0.3209	0.3396	0.1220	0.3571	0.3600	0.3149	6.5345	0.1069	0.080984	1.8892	31.49
Agile testing team	0.1948	0.3209	0.1132	0.0244	0.0714	0.0400	0.1274				0.7647	12.74
Lead time	0.1299	0.0802	0.1132	0.1220	0.2143	0.1200	0.1299				0.7795	12.99
Customer feedback	0.1299	0.1070	0.3396	0.1220	0.0714	0.2400	0.1683				1.0098	16.83
Iterative development	0.0779	0.1070	0.0566	0.2439	0.0714	0.1200	0.1128				0.6768	11.28
Costs	0.0779	0.0642	0.0377	0.3659	0.2143	0.1200	0.1467				0.8800	14.67
Total	1	1	1	1	1	1	1					

Table 5 Comparison matrix for CRYSTAL methodology

Goal criteria for CRYSTAL	R	A	LD	CF	ID	C
Reliability	1	0.333333	0.333333	0.333333	4	0.333333
Agile testing team	3	1	0.2	0.333333	3	4
Lead time	3	5	1	5	7	3
Customer feedback	3	3	0.2	1	5	3
Iterative development	0.25	0.333333	0.142857	0.2	1	2
Costs	3	0.25	0.333333	0.333333	0.5	1
SUM	13.25	9.916667	2.209524	7.2	20.5	13.33333

Table 6 Weights and ranks of criteria with normalized values of selection criteria for CRYSTAL

Goal criteria for CRYSTAL	R	A	LD	C	ID	C	P.E.V	P.E	CI	CR	Sum	Priority
Reliability	0.0755	0.0336	0.1509	0.0463	0.1951	0.0250	0.0877	7.4323	0.2865	0.217015	0.5264	8.77
Agile testing team	0.2264	0.1008	0.0905	0.0463	0.1463	0.3000	0.1517				0.9104	15.17
Lead time	0.2264	0.5042	0.4526	0.6944	0.3415	0.2250	0.4074				2.4441	40.74
Customer feedback	0.2264	0.3025	0.0905	0.1389	0.2439	0.2250	0.2045				1.2272	20.45
Iterative development	0.0189	0.0336	0.0647	0.0278	0.0488	0.1500	0.0573				0.3437	5.73
Costs	0.2264	0.0252	0.1509	0.0463	0.0244	0.0750	0.0914				0.5482	9.14
Total	1	1	1	1	1	1	1					

Table 7 Comparison matrix for XP methodology

Goal criteria for XP	R	A	LD	CF	ID	C
Reliability	1	1.333333	0.25	0.25	0.943396	1
Agile testing team	0.75	1	0.2	0.142857	5	5
Lead time	4	5	1	0.333333	0.943396	0.917431
Customer feedback	4	7	3	1	0.420168	0.666667
Iterative development	1.06	0.2	1.06	2.38	1	0.5
Costs	1	0.2	1.09	1.5	2	1
SUM	11.81	14.73333	6.6	5.60619	10.30696	9.084098

Table 8 Weights and ranks of criteria with normalized values of selection criteria for XP methodology

Goal criteria for XP	R	A	LD	C	ID	C	P.E.V	P.E	CI	CR	Sum	Priority
Reliability	0.0847	0.0905	0.0379	0.0446	0.0915	0.1101	0.0765	9.2627	0.6525	0.49435	0.4593	7.65
Agile testing team	0.0635	0.0679	0.0303	0.0255	0.4851	0.5504	0.2038				1.2227	20.38
Lead time	0.3387	0.3394	0.1515	0.0595	0.0915	0.1010	0.1803				1.0816	18.03
Customer feedback	0.3387	0.4751	0.4545	0.1784	0.0408	0.0734	0.2601				1.5609	26.01
Iterative development	0.0898	0.0136	0.1606	0.4245	0.0970	0.0550	0.1401				0.8405	14.01
Costs	0.0847	0.0136	0.1652	0.2676	0.1940	0.1101	0.1392				0.8351	13.92
Total	1	1	1	1	1	1	1					

Table 9 Comparison matrix for DSDM methodology

Goal criteria for DSDM	R	A	LD	CF	ID	C
Reliability	1	3	3	1.45	0.125	0.14
Agile testing team	0.2	1	2	0.333333	0.2	0.5
Lead time	8	4	1	5	7	0.8
Customer feedback	5	5	1	1	1	0.14
Iterative development	1	2	2	2	1	1.07
Costs	0.14	0.333333	0.2	1	0.142857	1
SUM	15.34	15.33333	9.2	10.78333	9.467857	3.65

Table 10 Weights and ranks of criteria with normalized values of selection criteria for DSDM methodology

Goal criteria for DSDM	R	A	LD	C	ID	C	P.E.V	P.E	CI	CR	Sum	Priority
Reliability	0.0652	0.1957	0.3261	0.1345	0.0132	0.0384	0.1288	10.3923	0.8785	0.665506	0.7730	12.88
Agile testing team	0.0130	0.0652	0.2174	0.0309	0.0211	0.1370	0.0808				0.4847	8.08
Lead time	0.5215	0.2609	0.1087	0.4637	0.7393	0.2192	0.3855				2.3133	38.55
Customer feedback	0.3259	0.3261	0.1087	0.0927	0.1056	0.0384	0.1662				0.9974	16.62
Iterative development	0.0652	0.1304	0.2174	0.1855	0.1056	0.2932	0.1662				0.9973	16.62
Costs	0.0091	0.0217	0.0217	0.0927	0.0151	0.2740	0.0724				0.4344	7.24
Total	1	1	1	1	1	1	1					

Table 11 Comparison matrix for FDD methodology

Goal Criteria for FDD	R	A	LD	CF	ID	C
Reliability	1	0.333333	0.333333	0.5	4	0.333333
Agile testing team	3	1	0.2	0.333333	3	4
Lead time	3	5	1	1	3	3
Customer feedback	2	3	0.2	1	5	3
Iterative development	0.25	0.333333	0.142857	0.2	1	2
Costs	3	0.25	0.333333	0.333333	0.5	1
SUM	12.25	9.916667	2.209524	3.366667	16.5	13.33333

Table 12 Weights and ranks of criteria with normalized values of selection criteria for DSDM methodology

Goal criteria for DSDM	R	A	LD	C	ID	C	P.E.V	P.E	CI	CR	Sum	Priority
Reliability	0.0816	0.0336	0.1509	0.1485	0.2424	0.0250	0.1137	7.0088	0.2018	0.152842	0.6820	11.37
Agile testing team	0.2449	0.1008	0.0905	0.0990	0.1818	0.3000	0.1695				1.0171	16.95
Lead time	0.2449	0.5042	0.4526	0.2970	0.1818	0.2250	0.3176				1.9055	31.76
Customer feedback	0.1633	0.3025	0.0905	0.2970	0.3030	0.2250	0.2302				1.3814	23.02
Iterative development	0.0204	0.0336	0.0647	0.0594	0.0606	0.1500	0.0648				0.3887	6.48
Costs	0.2449	0.0252	0.1509	0.0990	0.0303	0.0750	0.1042				0.6253	10.42
Total	1	1	1	1	1	1	1					

References

1. Garg, A., Shukla, B., & Kendall, G. (2015). Barriers to implementation of IT in educational institutions. *The International Journal of Information and Learning Technology, 2*(2), 94–108. https://doi.org/10.1108/IJILT-11-2014-0026
2. Paetsch, F., Eberlein, A., & Maurer, F. (2003). Requirements engineering and agile software development. In *Proceedings. Twelfth IEEE International Workshops on Conference: Enabling Technologies: Infrastructure for Collaborative Enterprises, WET ICE 2003*
3. Kassab, M. (2015). The changing landscape of requirements engineering practices over the past decade. In *IEEE Fifth International Workshop on Empirical Requirements Engineering (EmpiRE)*, Ottawa, ON (pp. 1-8). https://doi.org/10.1109/EmpiRE.2015.7431299.
4. Enabling Technologies: Infrastructure for Collaborative Enterprises (pp. 308–313) Linz, Austria (2003). https://doi.org/10.1109/ENABL.2003.1231428.
5. Alessio, I., & Ashraf, L. (2011). Review of the main developments in the analytic hierarchy process. *Journal of Expert Systems with Applications, 38*(11).
6. Srivastava, A., Kapur, P. K., & Mehrotra, D. (2017). Modelling fault detection with change-point in agile software development environment. *Infocom Technologies and Unmanned Systems (Trends and Future Directions) (ICTUS)*, 303–308. https://doi.org/10.1109/ICTUS.2017.828 6023
7. Hannola, L., Oinonen, P., & Nikula, U. (2011). Assessing and improving the front end activities of software development. *International Journal of Business Information Systems, 7*(1), 41–59. https://doi.org/10.1504/IJBIS.2011.037296.
8. Klendauer, R., Berkovich, M., Gelvin, R., Leimeister, J. M., & Krcmar, H. (2012). Towards a competency model for requirements analysts. *Information Systems Journal, 22*(6), 475–503. https://doi.org/10.1111/j.1365-2575.2011.00395.x
9. Latha, T. J., & Suhanthi, L. (2015). An empirical study on creating software product value in India – an analytic hierarchy process approach. *International Journal of Business Information Systems, 18*(1).
10. Kim, J. T., Park, S. Y., & Sugumaran, V. (2007). A framework for domain requirements analysis and modeling architectures in software product lines. *Journal of Systems and Software*
11. Cheong, C. W., Jie, L. H., Meng, M. C., & Lan, A. L. H. (2008). Design and development of decision making system using fuzzy analytic hierarchy process. *American Journal of Applied Sciences, 5*(7), 783–787. https://doi.org/10.3844/ajassp.2008.783.787.
12. Kamalrudin, M., Hosking, J., & Grundy, J. (2011). Improving requirements quality using essential use case interaction patterns. In *2011 33rd International Conference on Software Engineering (ICSE), Honolulu, HI* (pp. 531–540). https://doi.org/10.1145/1985793.1985866.
13. Toro, A. D., Jimenez, B. B., Cortes, A. R., & Toro Bonilla, M. (1999). A requirements elicitation approach based in templates and patterns. In *Proceedings of the Ibero-American Workshop on Requirements Engineering*.
14. Pantoquilho, M., Raminhos, R., et al. (2003). Analysis patterns specifications: Filling the gaps. In *The Second Nordic Conference on Pattern Languages of Programs, ViKingPLoP 2003*, Bergen. Norway.
15. Rajagopal, P., Lee, R., Ahlswede, T., Chia-Chu Chiang, C., & Karolak, D. (2005). A new approach for SoftwareR equirements elicitation. In *Proceedings of the 6th IEEE International Conference on Software Engineering, Artificial Intelligence, Networking and Parallel/Distributed Computing*. https://doi.org/10.1109/SNPD-SAWN.2005.5
16. Lai, V. S., Trueblood, R. P., & Wong, B. K. (1999). Software selection: a case study of the application of the analytical hierarchical process to the selection of a multimedia authoring system. *Information & Management, 36*(4), 221–232. ISSN 0378–7206, https://doi.org/10.1016/S0378-7206(99)00021-X.
17. Brans, J. P., Mareschal, B., & Vincke, P. (1984). PROMETHEE: A new family of outranking methods in multicriteria analysis. In J.P. Barns (Ed.), *Operational research, IFORS 84* (pp. 477–490). North Holland, Amsterdam.

18. Saaty, T. L. (2004). Decision making—The analytic hierarchy and network processes (AHP/ANP). *Journal of Systems Science and Systems Engineering, 13,* 1–35. https://doi.org/10.1007/s11518-006-0151.
19. Sharma, A., & Sharma, R. (2015). A systematic review of agile software development methodologies. In *Proceedings of the National Conference on Innovation and Development in Engineering and Management.*
20. Srivastava, A., Kapur, P. K., Mehrotra, D., & Majumdar, R. (2019). Modelling fault detection using SRGM in agile environment and ranking of models. *Journal of Cases on Information*

Printed in the United States
by Baker & Taylor Publisher Services